Nanocomposite Membrane Technology

Fundamentals and Applications

Nanocomposite Membrane Technology

Fundamentals and Applications

P.K. Tewari

CRC Press
Taylor & Francis Group
Boca Raton London New York

CRC Press is an imprint of the
Taylor & Francis Group, an **informa** business

CRC Press
Taylor & Francis Group
6000 Broken Sound Parkway NW, Suite 300
Boca Raton, FL 33487-2742

First issued in paperback 2017

ISBN-13: 978-1-4665-7682-7 (hbk)
ISBN-13: 978-1-138-74912-2 (pbk)

Visit the Taylor & Francis Web site at
http://www.taylorandfrancis.com

and the CRC Press Web site at
http://www.crcpress.com

Contents

Preface

This book is intended to serve as an introduction to the field of nanocomposite membranes as well as provide a comprehensive overview of its fundamental aspects and application areas. There are several books available on membrane science and technology. However, the subject of nanocomposite membrane technology is yet to be addressed and explored. Hence, there is a definite need for a book of this kind for students, academia, and industry.

Chapters 1–3 cover the basics of nanocomposite membrane technology, including introductory aspects, synthesis, and characterization. Chapters 4–6 deal with applications in the field of water treatment and gas separation as well as biomedical applications. Challenges in processing and health, environment, safety and societal aspects have been presented in Chapters 7–8.

Chapter 1 outlines the historical background of membranes as well as types of membranes and membrane technologies. It covers transport mechanism, membrane preparation techniques, and different configurations. It highlights the requirement of nanostructured materials and nanocomposites with polymeric/inorganic matrices.

Chapter 2 deals with raw materials, processing methods, and challenges involved. Nanostructured raw materials in the form of nanoparticles, nanotubes, nanofibers, and nanoplates as well as routes of their embedding in the host matrix have been discussed in this chapter. Processing methods such as phase inversion method, sol–gel method, and *in situ*/interfacial polymerization have been covered.

Chapter 3 presents the characterization of membrane morphology using instrumental techniques. It also deals with an evaluation of its performance.

Chapter 4 discusses the application of nanocomposite membranes to water treatment. Nanomaterials in membranes have potential to offer next-generation membranes, with better selectivity, higher throughput, and less fouling (more importantly, less biofouling). This chapter outlines the potential of nano-embedded membranes to decrease energy consumption in conventional water treatment operations.

Chapter 5 covers the potential of nanocomposite membranes in gas separation. The desirable characteristics of gas separation membranes are highlighted and different gas separation mechanisms are elaborated. Synthesis procedures for dense membranes (such as palladium/palladium–silver composites) and porous membranes (such as silica) are discussed. The role of nanocomposites in enhancing gas separation is covered. Permeation models for mixed matrix membranes are presented.

Nanotechnology has great potential to play a critical role in biomedical applications, particularly in drug delivery. Nanotechnology encompasses better understanding and treatment of living systems, revolutionary biotechnology processes, synthesis of new drugs and their targeted delivery, regenerative medicine, neuromorphic engineering, and biocompatible materials for sustainable environment. Chapter 6 highlights the current efforts and key research challenges in the development of these materials for use in potential biomedical applications, including drug delivery.

Nanocomposite materials have the potential to redefine the field of traditional composite materials in terms of both performance and applications. Polymer nanocomposites have tremendous market potential both as replacement for current composites and in the creation of new markets through their outstanding properties. Availability of nanomaterials, integration of nanomaterials into membrane systems, and societal implications, including health and environment risks posed by nanomaterials, are the key challenges in the development of integrated nano-based membrane systems. Development of the process-manufacturing technologies in terms of quantity and quality for commercialization is one of the biggest challenges. Although nanocomposites have a bright future, there are a few issues that warrant concern, for example, the mass commercialization of nano-based systems. Chapter 7 presents the challenges in processing of nanocomposite membranes and materials, including scalability issues.

The text is presented with clarity for the beginner, yet it is sufficiently comprehensive and thorough to be a valuable source of information for the researcher and experienced nanocomposite membranologist.

This book will be the guide of choice for the scientist and engineer, whether a student approaching the subject for the first time or a seasoned expert exploring the challenges and opportunities in the field of nanocomposite membrane technology. Both industry and academia will benefit from this book.

P.K. Tewari
Bhabha Atomic Research Centre

Author

Dr. P.K. Tewari is one of the renowned and distinguished scientists in the field of membrane technology as well as desalination and water purification. He has made outstanding contributions to many aspects of basic science and technology related to membrane and thermal processes. Dr. Tewari joined the Bhabha Atomic Research Centre (BARC) in 1977 and became the associate director of its Chemical Engineering Group in 2014. Currently, he is a Raja Ramanna Fellow in the BARC and consultant to the Office of Principal Scientific Adviser to the Government of India on water issues. He is a professor in the Homi Bhabha National Institute and guides PhD and postgraduate students.

Dr. Tewari received his PhD degree in chemical engineering from the Indian Institute of Technology Bombay, Mumbai, India, in 1987. He was chairman of the International Nuclear Desalination Advisory Group of the International Atomic Energy Agency (IAEA) from 2005 to 2008 and is currently chairman of Technical Working Group on Nuclear Desalination of the IAEA. Dr. Tewari is chairman of the Water Quality for Industrial Purposes Committee of the Bureau of Indian Standards. He is president of the Indian Desalination Association and member of the Board of Directors of the International Desalination Association.

Dr. Tewari has more than 200 research publications in peer-reviewed journals, proceedings, books, and encyclopedia. He is editor-in-chief of the *International Journal of Nuclear Hydrogen Production and Applications*. He is associate editor of the *International Journal of Nuclear Desalination* and member of the editorial board of the journal *Desalination and Water Treatment*. He is a recipient of several awards and felicitations as well as group achievement awards.

Author

1

Membrane Technology

1.1 Historical Background

Studies on membrane technology and its applications started in the eighteenth century.[1] The word *osmosis* was introduced by Abbé Nolet in 1748 to describe permeation of water through a diaphragm. During the nineteenth century, membranes were used only as laboratory tools to develop physical/chemical theories. For example, the measurements of solution osmotic pressure made with membranes by Traube and Pfeffer were used by van't Hoff in 1887 to develop the limit law, which explains the behavior of ideal dilute solutions. The work led to the van't Hoff equation. Around the same time, the concept of a perfectly selective semipermeable membrane was used by Maxwell and others in developing the kinetic theory of gases. Different types of diaphragm such as bladders of pigs, cattle, or fish and sausage casings made of animal gut were investigated by early membrane research workers. Later, nitrocellulose membranes were preferred due to reproducibility. Bechhold devised a technique in 1907 to prepare nitrocellulose membranes of graded pore size. Other workers such as Elford,[2] Zsigmondy and Bachmann,[3] and Ferry[4] improved Bechhold's technique to prepare nitrocellulose membranes. By the early 1930s, microporous membranes were commercially available. During the next 20 years, this microfiltration (MF) membrane technology was expanded to other polymers, such as cellulose acetate. Membranes found their first significant application in the testing of drinking water during the end of World War II when drinking water supplies in Germany and other parts of Europe were broken down, and filters were needed to test for water quality and safety. The research effort to develop these filters, sponsored by the U.S. Army, was later utilized by the Millipore Corporation (Billerica, MA).

Up until 1960, membranes were used in only a few laboratories and for specialized industrial applications. Membrane industry and membrane applications were almost nonexistent. In general, membranes suffered from four drawbacks limiting their widespread use as a separation process: (1) reliability issues, (2) slow transport, (3) selectivity issues, and (4) expensive. Drawbacks were overcome during the past 30–40 years.

Membrane-based separation processes are now used widely. The single most important work that transformed membrane separation from laboratory to industrial process was the development, in the early 1960s, of the Loeb–Sourirajan process for making defect-free, high-flux, anisotropic reverse osmosis (RO) membranes.[5] These membranes consist of an ultrathin, selective surface film on a much thicker but much more permeable microporous support, which provides the mechanical strength. The flux of the first Loeb–Sourirajan RO membrane was 10 times higher than that of any membrane then available and made RO a potentially practical method of desalting water. The work of Loeb and Sourirajan, and the financial support for research and development from the U.S. Department of Interior, Office of Saline Water (OSW), resulted in the commercialization of RO and was a major factor in the development of ultrafiltration (UF) and MF. The development of electrodialysis (ED) was also aided by OSW funding. Along with the development of these industrial applications of membranes was the independent development of membranes for medical separation processes, in particular, the artificial kidney. Kolf[6] demonstrated the first successful artificial kidney in the Netherlands in 1945. It took almost 20 years to refine the technology for use on a large scale. Since 1960s, the use of membranes in artificial organs has become a major life-saving procedure. The sales of these devices comfortably exceed the total industrial membrane separation market. Another important medical application of membranes is as membrane blood oxygenator and controlled drug delivery system. The membrane techniques developed by Alza, a company founded by Alex Zaffaroni, are widely used in the pharmaceutical industry to improve the efficiency and safety of drug delivery.

Lot of development in membrane technology took place during the period 1960–1980. Based on the Loeb–Sourirajan technique, other membrane preparation techniques, including interfacial polymerization and multilayer composite casting and coating, were developed for membrane giving high performance. Membranes with selective layers as thin as 0.1 micron (μm) or less are now being produced by several companies. Methods of packaging large membrane area and different configurations such as spiral wound, hollow fine fiber, capillary, and plate-and-frame modules were also developed. Improvement and advances have been made in enhancing the membrane stability. By 1980, MF, UF, RO, and ED were well established in industries.

The first major development of industrial membrane gas separation process was the Monsanto Prism membrane for hydrogen separation, introduced in 1980.[7] Dow started producing systems to separate nitrogen from air. Cynara and Separex were producing systems to separate carbon dioxide from natural gas. The first commercial pervaporation system was introduced for dehydration of alcohol by Gesselschaft Fur Trenntechnik (GFT), a German engineering company in the 1980s. A number of ethanol and isopropanol pervaporation-based dehydration plants have since been installed.

1.2 Membranes and Membrane Types

Membranes occupy an important place in industries and are used in a wide spectrum of applications. The ability to control the permeation rate of chemical species transporting through the membrane is utilized in a gainful manner. The objective in the membrane-based separation process is to allow one component of a mixture to permeate the membrane preferentially, while hindering permeation of the other component.

A semipermeable membrane separation system separates a feed stream into two streams known as the permeate and the concentrate. The permeate is the portion of the fluid that has passed through the semipermeable membrane, whereas the concentrate contains the constituents that have been rejected by the membrane as shown in Figure 1.1.

The membrane separation process enjoys numerous industrial applications with the following advantages:

- Appreciable energy savings
- Environmentally benign
- Clean technology with operational ease
- Operational simplicity
- High-quality products
- Greater flexibility in designing the systems

However, the following problems are also experienced:

- Concentration polarization
- Membrane fouling
- Membrane life
- Flux
- Selectivity

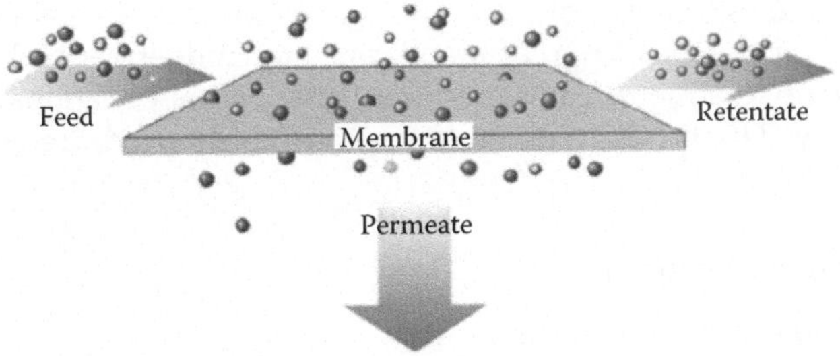

FIGURE 1.1
Schematic diagram of semipermeable membrane process.

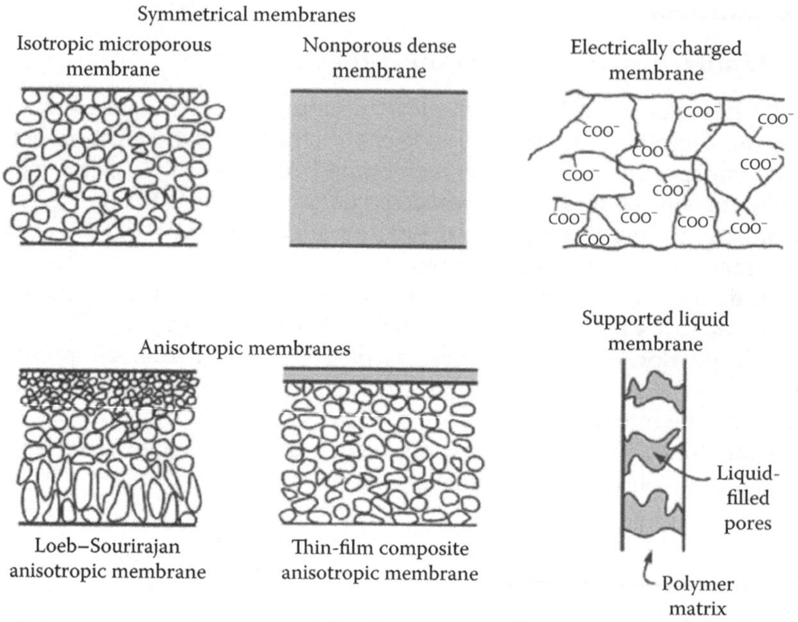

FIGURE 1.2
Schematic diagrams of the commonly used membranes. (From Baker, R.W.: Chapter 1: Overview of membrane science and technology. *Membrane Technology and Applications*. 2004. Copyright Wiley-VCH Verlag GmbH & Co. KGaA. Figures 1.1 and 1.6. Reproduced with permission.)

A membrane is a discrete thin interface that controls the permeation of chemical species in contact with it. This interface may be homogeneous, that is, uniform in composition and structure, or it may be chemically or physically heterogeneous, that is, containing holes or pores of finite dimensions or consisting of some form of layered structure. A conventional filter meets this definition of a membrane, but, by convention, the term filter is usually limited to structures that separate particulate suspensions larger than 10 μm.

The choice of a membrane depends on the specific application desired (i.e., particulate or dissolved solid removal), hardness reduction or ultra pure water production, removal of particular gas/chemical, and so on. The end use may also dictate selection of membranes for industries such as potable water, effluent treatment, gas separation, fuel cell applications, or control drug delivery. The types of membranes commonly used are shown schematically in Figure 1.2.

1.2.1 Isotropic Membranes

1.2.1.1 Microporous Membranes

A microporous membrane is similar to a conventional filter in structure and function. It has a rigid voided structure with randomly distributed interconnected pores. However, the pores differ from those in a conventional filter by being extremely small, on the order of 0.01–10 μm in diameter. Particles larger than the largest pore are completely rejected by the membrane. Particles smaller than the largest pores, but larger than the smallest pores, are partially rejected, which is governed by the pore size distribution of the membrane. Particles, which are smaller than the smallest pore, pass through the membrane. Thus, separation of solutes by microporous membranes is mainly a function of molecular size and pore size distribution. In general, only molecules that differ considerably in size can be separated effectively by microporous membranes, for example, in UF and MF.

1.2.1.2 Nonporous Membranes

Nonporous membrane consists of a dense film through which product is transported by diffusion under the driving force of a pressure, concentration, or electrical potential gradient. The separation of various components of a mixture depends on their relative transport rate through the membrane. The relative transport rate is determined by their diffusivity and solubility in the membrane material. Nonporous membrane can separate constituents of similar size if their concentration in the membrane material (i.e., their solubility) differs. Most of the membranes for gas separation, pervaporation, and RO use nonporous membrane.

1.2.1.3 Electrically Charged Membranes

Electrically charged membranes are normally microporous with the pore walls carrying fixed positively or negatively charged ions. However, these membranes can be nonporous or microporous. A membrane with fixed positively charged ions is referred as an anion-exchange membrane. Similarly, a membrane containing fixed negatively charged ions is called a cation-exchange membrane. Separation with charged membranes is achieved by exclusion of ions of the same charge as the fixed ions of the membrane structure, and to a much lesser extent by the pore size. The separation is affected by the charge and concentration of the ions in solution; for example, monovalent ions are excluded less effectively than bivalent ions. Selectivity decreases in solutions of high ionic strength. Electrically charged membranes are used for processing electrolyte solutions in ED.

1.2.2 Anisotropic Membranes

The development of membrane preparation methodology for producing anisotropic membrane structures was one of the major breakthroughs of membrane technology. Anisotropic membrane consists of an extremely thin surface layer supported on a thick and porous substructure. The transport rate of species through a membrane is inversely proportional to the membrane thickness. High transport rates are desirable for economic reasons; therefore, the membrane should be as thin as possible. The surface layer and its substructure can be made in a single operation or separately. Conventional film fabrication technology limits the manufacture of mechanically strong, defect-free films to about 20 μm thickness. In composite membranes, the layers are usually made from different polymers. The separation properties and permeation rates of the membrane are determined exclusively by the surface layer. The substructure works as a mechanical support. The advantages of the high fluxes provided by anisotropic membranes are so great that almost all commercial processes use such membranes.

1.2.3 Ceramic, Metal, and Liquid Membranes

In recent years, interest in membranes made from less conventional materials has increased. Ceramic membranes, a special class of microporous membranes, are being used in UF and MF applications where solvent resistance and thermal stability are desirable. Nonporous metal membranes, particularly palladium membranes, are being considered for the separation of hydrogen from gas mixtures, whereas supported liquid films are being developed for carrier-facilitated transport processes.

1.3 Membrane Technologies

An overview of membrane technologies is given in Table 1.1. MF, UF, RO, and ED are well-established technologies. The pressure-driven membrane separation processes—RO, UF, and MF—are illustrated in Figure 1.3. The relative size of different solutes removed by each class of membrane is given. RO, UF, MF, and conventional filtration are similar processes differing mainly in the average pore diameter of the membrane. The mode of separation in case of UF and MF is molecular sieving through increasingly fine pores. MF membranes filter colloidal particles and bacteria from

TABLE 1.1

Industrial Membrane Technologies

S. No.	Category	Process
1	Membrane separation technologies well established in the industries	MF, UF, RO, ED
2	Upcoming membrane separation technologies for the industries	Gas separation, pervaporation
3	Membrane separation technologies of interest for the industries	Carrier-facilitated transport membrane contactors, piezodialysis, etc.

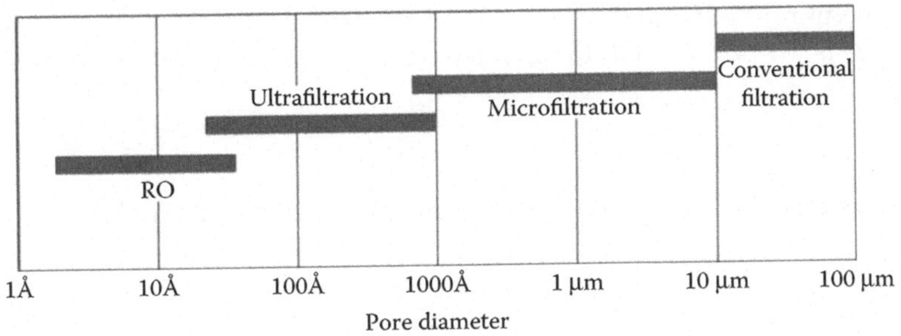

FIGURE 1.3
Pressure-driven membrane separation processes.

0.1 to 10 µm in diameter. UF membranes can be used to filter dissolved macromolecules, such as proteins from solutions. RO membranes are so dense that discrete pores do not exist. Transport occurs through the statistically distributed free volume areas. The pores of the membrane range from 3Å to 5Å in diameter, which is within the range of thermal motion of the polymer chains that form the membrane. The mechanism of transport through the RO membrane is governed by the solution–diffusion model. According to this model, solutes permeate the membrane by dissolving in the membrane material and diffusing down a concentration gradient. Separation occurs because of the difference in solubilities and mobilities of different solutes in the membrane.

Membrane can be considered as a series of cylindrical capillary pores of diameter d. The liquid flow rate (q) through a cylindrical capillary pore is given by Poiseuille's law as follows:

$$q = \frac{\pi d^4 \Delta p}{128\,\mu l} \tag{1.1}$$

where:

Δp is the pressure difference across the pore
μ is the liquid viscosity
l is the pore length

The flux J, or flow per unit membrane area, is the sum of all the flows through the individual pores and thus is given by

$$J = \frac{N\left(\pi d^4 \Delta p\right)}{128\mu l} \tag{1.2}$$

where N is the number of pores per unit area of membrane.

For membranes of equal pore area and porosity (ε), the number of pores per unit area is proportional to the inverse square of the pore diameter, that is,

$$N = \frac{4\varepsilon}{\pi d^2} \tag{1.3}$$

It follows that the flux, given by combining Equations 1.2 and 1.3, is

$$J = \frac{\Delta p \varepsilon d^2}{32 \mu l} \tag{1.4}$$

It may be noted that the flux (J) is proportional to the square of the pore diameter (d). This implies that flux per unit pressure difference ($J/\Delta p$) of MF membrane is higher than that of UF membranes, which in turn is higher than that of RO membranes. This plays an important role in selection of membrane process for particular application.

Table 1.1 shows gas separation with polymer membranes and pervaporation as upcoming membrane technologies. Gas separation with membranes has higher potential of application. Several companies worldwide use membrane based gas separation systems for a variety of applications. In gas separation, a gas mixture at an elevated pressure is transported across the surface of a membrane that is selectively permeable to one component of the feed mixture. Major current applications of gas separation membranes are the separation of hydrogen from nitrogen, argon, and methane in ammonia plants; the production of nitrogen from air; and the separation of carbon dioxide from methane in natural gas operations. Gas separation using a membrane process is an area of research interest with high potential of application.

In pervaporation, a liquid mixture is fed in the membrane system and the permeate in the form of vapor is removed. The driving force for the process is the low vapor pressure of the liquid mixture. The separation obtained is proportional to the rate of permeation of the particular component through the selective membrane. Pervaporation offers the possibility of separating closely boiling mixtures or azeotropes that are difficult to separate by distillation or other means. A schematic diagram of a simple pervaporation process using a condenser to generate the permeate vacuum is shown in Figure 1.4.

Currently, the main industrial application of pervaporation technology is for the dehydration of organic solvents, such as the dehydration of 90%–95% ethanol solutions, which is a difficult separation problem because of the ethanol–water azeotrope at 95% ethanol. Pervaporation membranes that selectively permeate water can produce more than 99.9% ethanol from these solutions. Pervaporation processes are also being developed for the removal of dissolved organics from water and for the separation of organic mixtures.

A number of other industrial membrane processes such as carrier-facilitated transport are under development, which often employs liquid membranes

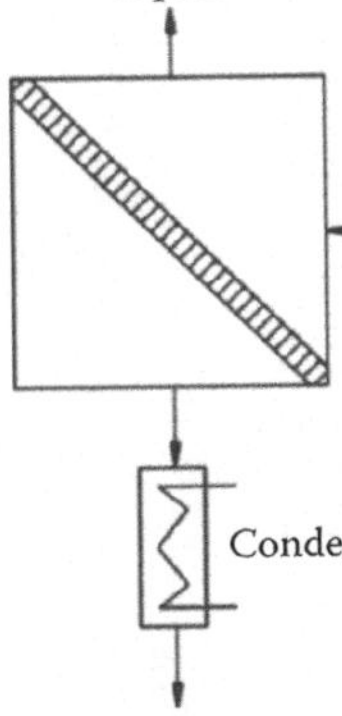

FIGURE 1.4
Schematic diagram of the basic pervaporation process.

containing a complexing or carrier agent. The carrier agent reacts with one component of a mixture on the feed side of the membrane and then diffuses across the membrane to release the permeant on the product side of the membrane. The reformed carrier agent then diffuses back to the feed side of the membrane. Thus, the carrier agent acts as a shuttle to selectively transport one component from the feed to the product side of the membrane.

Facilitated transport membranes can be used to separate gases. In this case, membrane transport is driven by a difference in the partial pressure across the membrane. Metal ions can also be selectively transported across a membrane, driven by a flow of hydrogen or hydroxyl ions in the other direction. This process is sometimes called coupled transport. Examples of carrier-facilitated transport processes for gas and ion transport are shown in Figure 1.5. The gas transport example shows the transport of oxygen across a membrane using hemoglobin as the carrier agent. The ion transport example shows the transport of copper ions across a membrane using a liquid ion-exchange reagent as the carrier agent.

As the carrier-facilitated transport process employs a reactive carrier species, very high membrane selectivities can be achieved. These selectivities are often far larger than the selectivities achieved by other membrane processes. This has maintained interest in facilitated transport for so many years. However, commercial deployment is yet to be deployed due to challenges faced with respect to (1) the physical instability of the liquid membrane and (2) the chemical instability of the carrier agent. In recent years, a number of potential solutions to this problem have been developed, which may yet make carrier-facilitated transport a viable process.

The membrane separation processes described earlier represent the bulk of the industrial membrane separation industry. Another process, dialysis,

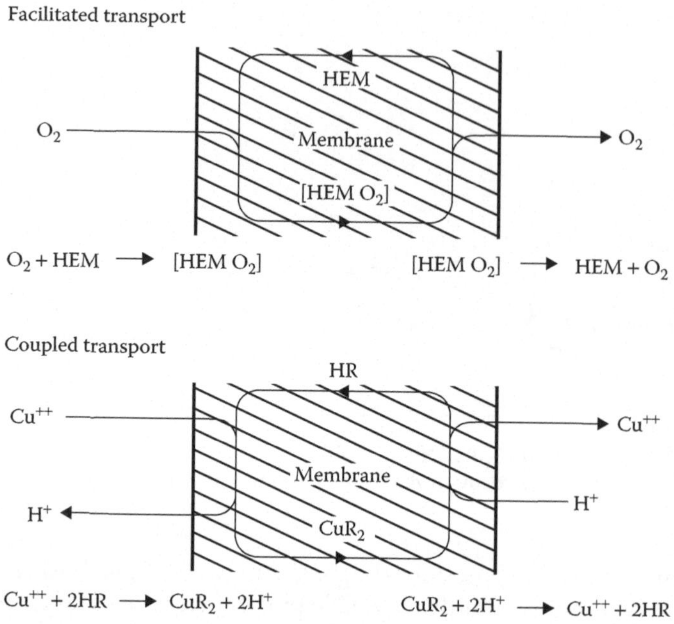

FIGURE 1.5
Schematic examples of carrier-facilitated transport of gas and ions. HEM, hemoglobin; HR, acid form of the reagent. (From Baker, R.W.: Chapter 1: Overview of membrane science and technology. *Membrane Technology and Applications*. 2004. Copyright Wiley-VCH Verlag GmbH & Co. KGaA. Figures 1.1 and 1.6. Reproduced with permission.)

is used on a large scale in the field of medical application to remove toxic metabolites from the blood in patients suffering from kidney failure. The first successful artificial kidney was based on cellophane (regenerated cellulose) dialysis membranes and was developed in 1945. Over the past 50 years, many advancements have been made. Currently, most artificial kidneys are based on hollow-fiber membranes formed into modules having a membrane area of about a square meter (1 m^2). Schematic diagram of a hollow-fiber artificial kidney dialyzer used to remove urea and other toxic elements from the blood is shown in Figure 1.6.

Blood is circulated through the center of the fiber, whereas isotonicsaline, the dialysate, is pumped counter currently around the outside of the fibers. Urea, creatinine, and other low-molecular-weight metabolites in the blood diffuse across the fiber wall and are removed with the saline solution. The process is quite slow, usually requiring several hours to remove the required amount of the metabolite from the patient, and repeats 1 or 2 times per week. In terms of the membrane area used and the dollar value of the membrane produced, artificial kidneys are the single largest application of membranes. Following the success of the artificial kidney, similar devices were developed to remove carbon dioxide and deliver oxygen to the blood. These so-called

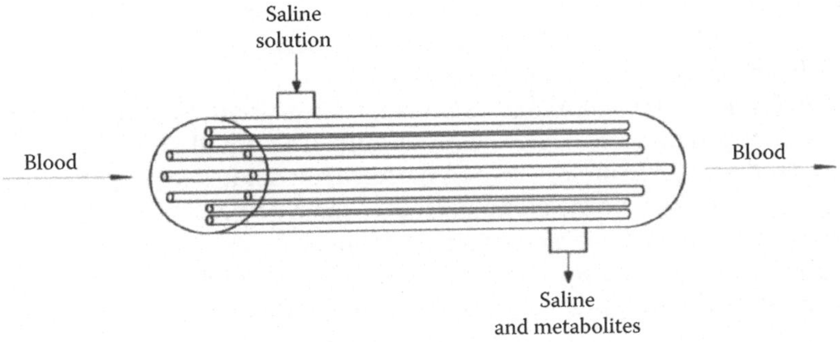

FIGURE 1.6
Schematic diagram of a hollow-fiber artificial kidney dialyzer.

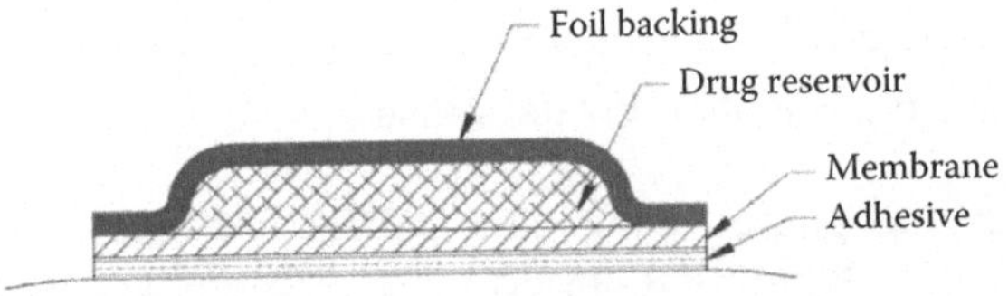

FIGURE 1.7
Schematic of transdermal patch in which the rate of delivery of drug to the body is controlled by a polymer membrane.

artificial lungs are used in surgical procedures during which the patient's lungs cannot function.

Another major medical use of membranes is in controlled drug delivery. Controlled drug delivery can be achieved by a wide range of techniques, most of which involve membranes. A simple example is illustrated in Figure 1.7. Systems, which operate using this principle, are used to moderate the delivery of drugs such as nitroglycerine (for angina), nicotine (for smoking cessation), and estradiol (for hormone replacement therapy) through the skin.

1.3.1 Reverse Osmosis

Conventional water filtration can remove suspended materials larger than 1 μm size, whereas the RO or hyperfiltration has the capability to remove the dissolved solids, bacteria, viruses, and other germs contained in the raw water. RO is essentially a pressure-driven membrane process for separating dissolved solutes. It is generally used for desalination of seawater or brackish water for its conversion into potable water. It operates at an ambient temperature and involves no phase change. It is relatively a low energy-consuming process compared to conventional distillation.

RO membranes are made of polymeric material such as cellulose acetate and aromatic polyamide. They are generally of two types: (1) asymmetric or skinned membranes and (2) thin-film composite (TFC) membranes. The support material is normally polysulfone based, whereas the thin film is made from various types of polyamines.

RO membranes have the smallest pore structure, with pore diameter ranging from approximately 3Å–5Å (0.3–0.5 nm). The small size of pores does not allow the organic molecules and monovalent solutes to pass through the semipermeable membrane along with the water. In practice, more than 99% of inorganic salts and charged organics are rejected by the membrane due to extremely small size of pores and charge repulsion at the membrane surface.

RO process finds extensive applications in the following areas:

- Potable water from sea or brackish water
- Ultrapure water for food processing and electronic industries
- Pharmaceutical grade water
- Water for chemical and pulp and paper industries
- Wastewater treatment and water reuse

The technology has gained wider acceptance as a viable water treatment option for different separation applications. Affordable cost and ability to remove organic contaminants and more than 99% of inorganic salts with minimal chemical requirements make RO an attractive technology for many industrial applications.

1.3.2 Nanofiltration

Nanofiltration (NF) can operate at lower pressures compared to RO. It is good for applications where high organic removal and moderate inorganic removal are desired. It is capable of concentrating sugars, bivalent salts, proteins, dyes, and other constituents that have a molecular weight greater than 1000 Da.

Membranes used for NF are made of cellulosic acetate and aromatic polyamide having characteristics of giving salt rejection from 95% for bivalent salts to 40% for monovalent salts and an approximate 300 molecular-weight cutoff (MWCO) for organics. An advantage of NF over RO is that NF can typically operate at higher recoveries, thereby conserving the total water usage due to a lower concentrate stream flow rate. NF is not effective for separation of monovalent ions and small molecular-weight organics.

1.3.3 Ultrafiltration

UF membranes are capable of retaining species in the range of 300–500,000 Da of molecular weight, with pore sizes ranging from 20 to 1000 Å. These are mostly described by their nominal MWCO, which means the smallest molecular-weight species for which the membranes have more than 90% rejection. The driving force for transport across the membrane is a pressure differential operating at about 2–10 bar pressure. It is commonly used for clarification and fractionation of solutes.

The permeate rate is controlled by the rate of transport through the polarization layer rather than by membrane properties. Hence, UF throughput depends on physical properties of the membrane, such as permeability, thickness, and process and system variables such as feed consumption, feed concentration, system pressure, velocity, and temperature.

The important characteristics for membrane materials are porosity, morphology, surface properties, mechanical strength, and chemical resistance. Polymeric materials, such as polysulfone, polypropylene, nylon-6, polytetrafluoroethylene (PTFE), polyvinyl chloride (PVC), and acrylic copolymer, are used successfully for making UF membranes. Inorganic materials such as ceramics, carbon-based membranes, and zirconia are also used for making UF membranes.

UF has great potential for a wide range of applications as separation technology in the field of bioprocessing, pharmaceutical, oil and paints, food and dairy industries, protein harvesting as well as several other areas such as for concentration of biological macromolecules and macromolecular separation from energy-saving point of view.

1.3.4 Microfiltration

This is by far the most widely used membrane process with total sales greater than the combined sales of all other membrane processes. MF is a process of separating material of colloidal size. The pore size of the MF membrane varies from 0.1 to 10 μm.

MF is based on the sieving mechanism with distinct pore sizes for retaining larger size particles than the pore diameter. The MF membranes are made from natural or synthetic polymers such as cellulose nitrate or acetate, polyvinylidene difluoride (PVDF), polyamides, polysulfone, polycarbonate, polypropylene, and PTFE. The inorganic materials such as metal oxides (alumina), glass, and zirconia-coated carbon are also used for manufacturing the MF membranes.

The properties of membrane materials such as mechanical strength, temperature resistance, chemical compatibility, hydrophobility, hydrophilicity, permeability, and permselectivity play an important role in the selection process with respect to particular application.

MF has a wide array of applications in food, beverages, and chemical industries. It has high potential to grow in the field of biotechnology, water treatment, and several other industries.

1.3.5 Electrodialysis

ED is an electromembrane process in which the ions are transported through a membrane from one solution to another under the influence of an electrical potential. It can be utilized to perform several general types of separations such as separation and concentration of salts, the separation and concentration of monovalent ions from multiple charged components, or the separation of ionic compounds from uncharged molecules. ED membranes are usually made of cross-linked polystyrene that has been sulfonated. Anion membranes can be of cross-linked polystyrene containing quaternary ammonia groups. Usually, ED membranes are made of flat sheets containing about 30%–50% water by applying the cation- and anion-selective polymers to a fabric material.

The ED system consists of cation and anion membranes, which are placed in an electric field. The cation-selective membrane permits only the cations and the anion-selective membrane permits only the anions. The transport of ions across the membranes results in ion depletion in cells and ion concentration in alternate ones.

ED is widely used for producing potable water from brackish water, electroplating rinse recovery, desalting cheese whey, producing ultrapure water, and so on. The ED industry has experienced a steady growth rate. To ensure further growth beyond desalination and salt production, new areas of application are being explored in the areas of chemicals and pharmaceuticals, food, industrial and municipal effluent treatment, and so on. This can be achieved through slight modifications in the conventional process as well as extensive R&D work in the field.

1.3.6 Gas Separation

The membrane-based gas separation technology is one of the most significant unit operations. It competes well with technology alternatives such as adsorption and cryogenic distillation in niche application areas. The membrane-based gas separation process enjoys certain advantages, such as compactness and light in weight, modular design permitting easy scale-up or operation at partial capacity, and low-energy requirement. Membranes made of polymers and copolymers in flat and hollow-fiber configuration have been used for gas separation.

Membrane-based gas separation is based on the concept that different gases pass through the membrane at significantly different rates. The rate of permeation is proportional to the pressure differential across the membrane and inversely proportional to the membrane thickness. The rate of permeation is also proportional to the diffusivity of gas through the membrane.

Gas separation is affected by three key performance attributes of membranes: selectivity for the gases, membrane flux or permeability, and the life of the membrane. The membrane-based gas separation is used for hydrogen recovery, natural gas processing, air separation, nitrogen production, helium recovery, and so on.

1.3.7 Pervaporation

Pervaporation is a membrane-based process for separating miscible liquids. The process is based on the absorption of one of the components of the liquid by the membrane, diffusion of this component across the membrane, and evaporation as permeate vapor into the partial vacuum applied to the other side of the membrane. Transport through the membrane takes place by maintaining vapor pressure gradient across the membrane.

It offers effective separation and significant energy saving in applications which are difficult to separate by conventional techniques such as azeotropic mixtures or mixtures of close boiling components. It is used for separation of ethanol–water mixture, solvent recovery, separation of heat-sensitive products, or enrichment of organic pollutants.

Table 1.2 gives the characteristics of membranes used in different membrane separation processes.

TABLE 1.2
Characteristics of Membranes Used in Different Membrane Separation Processes

Process	Membrane Type and Pore Size	Membrane Material	Diving Force	Applications
MF	Symmetric microporous, 0.1–10 μm	Cellulose nitrate or acetate, PVDF, PTFE, metal oxides, etc.	Hydrostatic pressure difference at ~10–500 kPa	Separation of suspended solids, bacteria
UF	Asymmetric microporous, 2–100 nm	Polysulfone, polypropylene, Nylon 6, PTFE, PVC, acrylic copolymer	Hydrostatic pressure difference at ~0.1–0.5 Mpa	Separation of macromolecules; virus from solution
NF	Asymmetric skin-type, 0.5–2.0 nm	Cellulosic acetate, aromatic polyamide	Hydrostatic pressure difference at ~0.5–2 Mpa	Separation of bivalent ions and macromolecules from solution
RO	Asymmetric skin-type, 0.3–0.5 nm	Cellulosic acetate, aromatic polyamide	Hydrostatic pressure difference at ~2–10 Mpa	Separation of monovalent salts and microsolutes from solutions
ED	Cation- and anion-exchange membranes	Sulfonated cross-linked polystyrene	Electrical potential gradient	Desalting of ionic solutions
Gas separation	Asymmetric homogeneous polymer	Polymers and copolymers	Hydrostatic pressure and concentration gradients	Separation of gas mixtures
Pervaporation	Asymmetric homogenous polymer (a nonporous membrane)	Polyacrylonitrile, polymers	Vapor pressure gradient	Separation of azeotropic mixtures

1.4 Other Membrane Processes

1.4.1 Dialysis

Dialysis was the first membrane process to be used on an industrial scale.[8] It was also used in the laboratory in the 1950s and 1960s, mainly to purify biological solutions or to fractionate macromolecules. A schematic diagram of the laboratory dialyzer used by Craig is shown in Figure 1.8, which was used for separation of low-molecular-weight impurities from biological solutions.[9,10] The feed solution was circulated through the inside of the membrane tube. Solvent solution was circulated on the outside. Boundary layer formation was overcome by rotating the outer shell with a small motor.

Until UF membranes became available in the late 1960s, this device was the only way to separate many large-volume biological solutions. The major application of dialysis is its use as the artificial kidney. More than 100 million of these devices are used annually for this application. Dialysis process has certain limitations because it is based on principle of diffusion which is slow, to achieve a separation. UF or ED is at times due to the use of more selective membrane giving better and faster separation.

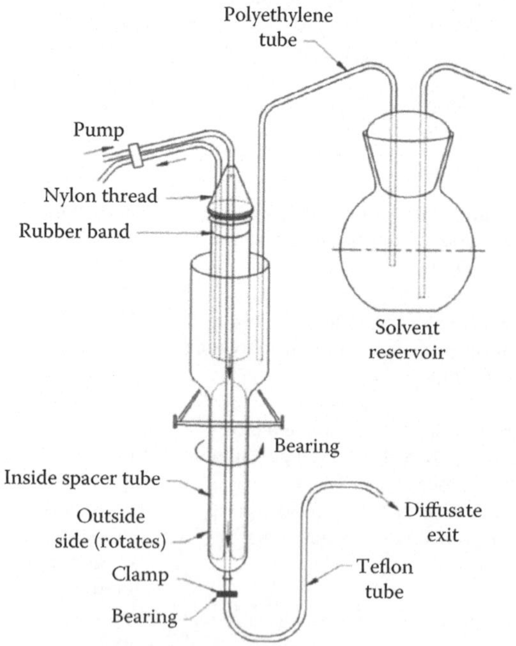

FIGURE 1.8
Schematic drawing of laboratory dialyzer developed by Craig to separate low-molecular-weight impurities from biological solutions.

1.4.1.1 Donnan Dialysis and Diffusion Dialysis

Donnan dialysis was first described as a separation technique in 1967 by Wallace,[11,12] who was interested in concentrating small amounts of radioactive metal ions. He used cation-exchange membranes to treat a large volume of nearly neutral feed solution containing small amount of metal salts such as uranyl nitrate $UO_2(NO_3)_2$. A small volume of 2 M nitric acid was used as the receiving solution. Because the membrane contained fixed negative charges, negative ions from the surrounding solutions were essentially excluded from the membrane, and only hydrogen ions (H^+) and uranyl ions (UO_2^{++}) could permeate the membrane.

Like coupled transport, Donnan dialysis can concentrate metal ions manyfold. The process is usually driven by an appropriate pH gradient. Because the membranes are normal cation- or anion-exchange membranes, the stability problem that plagues the liquid membranes used in coupled transport is overcome. However, coupled transport uses carriers selective for one particular ion, excluding others. This property allows coupled transport membrane to selectively transport one particular ion across the membrane, both concentrating and separating the target ion from similar ions in the feed solution. Donnan dialysis membranes are essentially nonselective—all ions of the same charge in the feed solution are transported to the product solution at about the same rate. It can be made more selective if a complexing agent specific to one of the ions being transported across the membrane is added to the feed solution.

A related dialysis process, diffusion dialysis, is used to recover acid from spent metal pickling agents such as sulfuric acid, hydrochloric acid, or nitric–hydrofluoric acid.[13] These pickling chemicals remove scale from metal parts and get contaminated with iron, chromium, copper, nickel, zinc, and other heavy metals. Acid recovery by ED is possible, but diffusion dialysis—a completely passive process—is often preferred because of its simplicity. The process utilizes the difference in permeability of hydrogen ions and multivalent metal ions through anion-exchange membranes.

1.4.1.2 Charge Mosaic Membranes

Donnan dialysis is a type of ion-exchange process. Ions of the same charge are redistributed across the membrane, but no net flow of salt from one side of the membrane to the other occurs. This is because ion-exchange membranes are quite impermeable to salts. Although counterions to the fixed charge groups in the membrane can easily permeate through the membrane, ions with the same charge as the fixed charge groups do not permeate. Sollner[14] proposed that, if ion-exchange membranes consisting of separated small domains of anionic and cationic membranes could be made, they would be permeable to anions and cations. These membranes are called charge mosaic membranes. The concept is illustrated in Figure 1.9. Cations permeate the cationic membrane domain, whereas anions permeate the anionic domain.

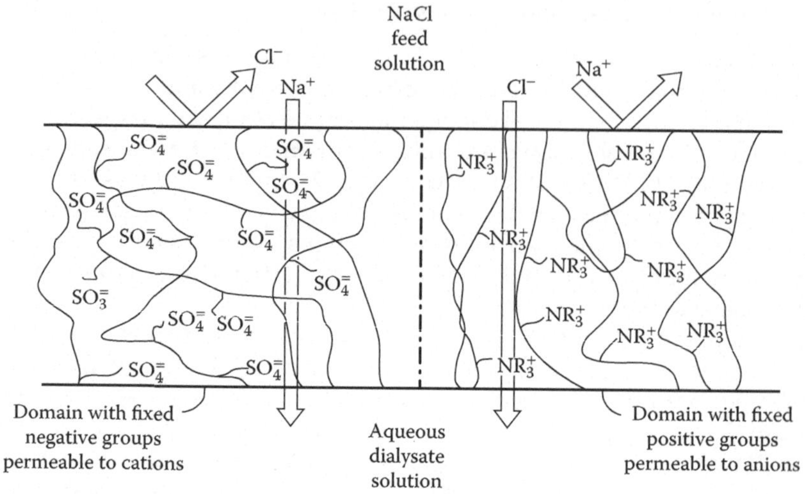

FIGURE 1.9
Charge mosaic membranes, consisting of finely dispersed domains containing fixed negatively and fixed positively charged groups. (From Baker, R.W.: Chapter 13: Other membrane processes. *Membrane Technology and Applications*. 2004. Copyright Wiley-VCH Verlag GmbH & Co. Figures 13.5, 13.9, and 13.13. Reproduced with permission.)

Charge mosaic membranes can preferentially permeate salts from water. This is because the principle of electroneutrality requires that the counterion concentration inside the ion-exchange regions be at least as great as the fixed charge density. Because the fixed charge density of ion-exchange membranes is typically greater than 1 M, dilute counterions present in the feed solution are concentrated ten- to hundredfold in the membrane phase. The large concentration gradient that forms in the membrane leads to high ion permeabilities. For charge mosaic membranes to work most efficiently, the cationic and anionic domains in the membrane must be close together to minimize charge separation effects.[15–17] The first charge mosaic membranes were made by distributing very small ion-exchange beads in an impermeable support matrix of silicone rubber.[18,19] A second approach, used by Platt and Schindler,[20] was to use the mutual incompatibility of most polymers that occurs when a solution containing a mixture of two different polymers is evaporated.

1.4.1.3 Membrane Contactors

In membrane contactors, the membrane functions as an interface between two phases but does not control the rate of passage of permeants across the membrane. Membrane contactors are typically shell-and-tube devices containing microporous capillary hollow-fiber membranes.[21–23] The membrane pores are sufficiently small that capillary forces prevent direct mixing of the

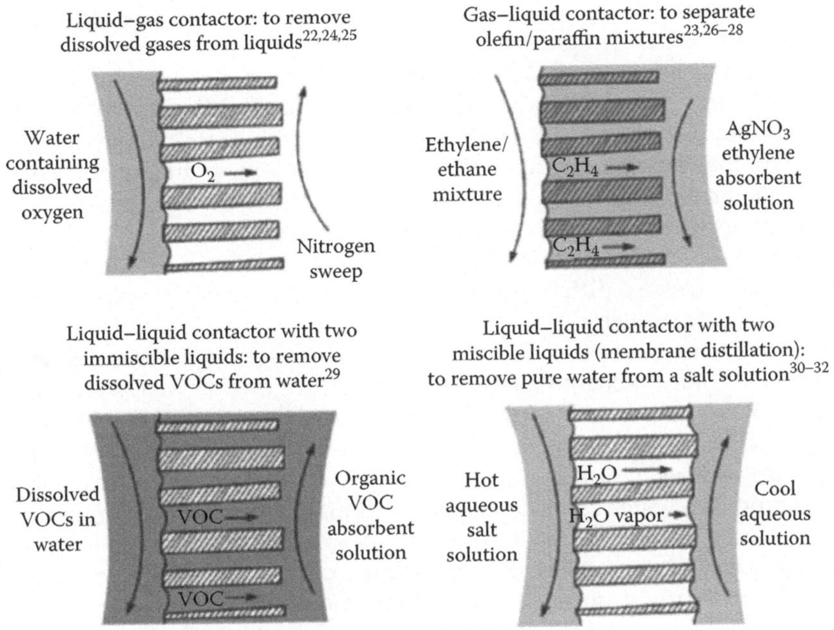

FIGURE 1.10
Membrane contactors and their applications. VOC, volatile organic compound. (From Baker, R.W.: Chapter 13: Other membrane processes. *Membrane Technology and Applications*. 2004. Copyright Wiley-VCH Verlag GmbH & Co. Figures 13.5, 13.9, and 13.13. Reproduced with permission.)

phases on either side of the membrane. The membrane contactor[24–29] separates a liquid and a gas phase in case of a liquid/gas contactor. Membrane contactors can also be used to separate two immiscible liquids (liquid/liquid contactors) or two miscible liquids (usually called membrane distillation). Contactors can also be used to selectively absorb one component from a gas mixture into a liquid (gas/liquid contactors). The various types of membrane contactors that have been used are illustrated in Figure 1.10. Contactors have a number of advantages compared to simple liquid/gas absorber/strippers or liquid/liquid extractors, particularly high surface area per volume.

Membrane contactor provides 10 times higher contactor area than equivalent-sized towers. This makes membrane contactor compact and lightweight. It is based on the physical separation of the counter-flowing phases by the membrane. The membrane area between the two phases is independent of their relative flow rates, so large flow ratio differences can be used without producing channeling or flooding or poor phase contact, and maximum advantage can be taken of the ability of counterflow to separate and concentrate the components crossing the membrane. Small volume of high-cost extractants can be used to treat large volumes of low-value feed. Separation of the two phases also eliminates entrainment of one phase into the other, as well as foaming. Finally, unlike traditional contactors, fluids of equal density can be used for the two phases.

The main disadvantages of contactors are related to the nature of the membrane interface. The membrane acts as an additional barrier to transport between the two phases, which introduces a resistance to the rate of separation. The membranes can foul during the operation, reducing the permeation rate, or develop leaks, thus allowing direct mixing of the two phases. The polymeric membranes are thin for reducing the resistance to flow and enhance the permeation rate. However, this introduces limitation to withstand large pressure differences across the membrane or exposure to harsh solvents and chemicals.

An important example of liquid/liquid membrane contactor is membrane distillation, shown schematically in Figure 1.11. In this process, a warm feed solution flows on one side of the membrane and pure distillate on the other side. Because the solutions are at different temperatures, their vapor pressures are different; as a result, water vapor flows across the hydrophobic membrane. The water vapor flux is proportional to the vapor pressure difference between the warm feed and the cold permeate. Because of the exponential rise in vapor pressure with temperature, the flux increases dramatically as the temperature difference across the membrane is increased.

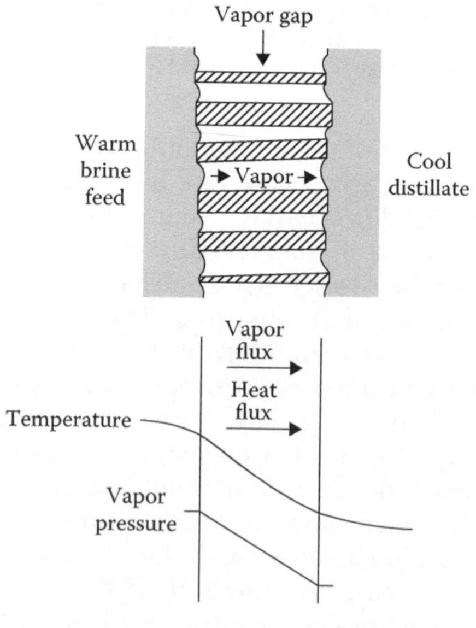

FIGURE 1.11
Schematic diagram of the membrane distillation process showing temperature and water vapor pressure gradients that drive the process. (From Baker, R.W.: Chapter 13: Other membrane processes. *Membrane Technology and Applications*. 2004. Copyright Wiley-VCH Verlag GmbH & Co. Figures 13.5, 13.9, and 13.13. Reproduced with permission.)

Membrane distillation[30] offers a number of advantages over alternative pressure-driven membrane processes such as RO. Because the process is driven by temperature gradient, low-grade waste heat can be used. Unlike RO, expensive high-pressure pump is not required. Membrane distillation can be used for concentrated feed where RO cannot be used. This is an advantage over RO. In RO, osmotic pressure increases with concentration which offers a practical limit on the concentration of a salt in the feed solution to be processed. However, the membrane distillation process is still the subject of academic interest.[31,32] There is need to direct the research efforts toward cost reduction strategies through technological innovations with respect to membrane modules for membrane distillation system.

1.4.1.4 Membrane Reactors

A membrane reactor integrates both membrane-assisted reaction and separation. It is also termed as membrane-based reactive separation. It is compact and plays an important role in enhancement of selectivity/yield in case of equilibrium-limited reaction. The conventional membrane reactor system is shown in Figure 1.12. Figure 1.13 shows a schematic diagram of an integrated membrane reactor system.

There are different configurations of membrane reactor[33] in order to combine the membrane separation module and the reactor into a single unit as shown in Figure 1.14. Six basic types of configurations are indicated in Table 1.3.

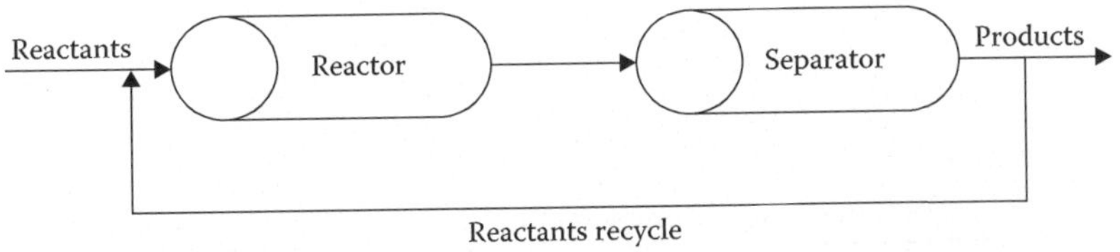

FIGURE 1.12
Conventional membrane reactor system.

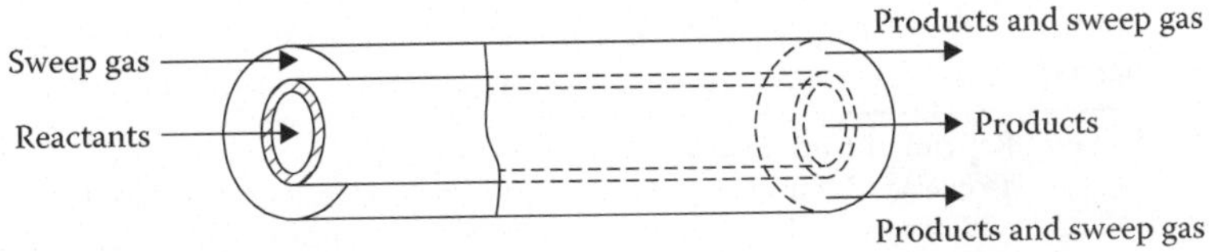

FIGURE 1.13
Integrated membrane reactor system.

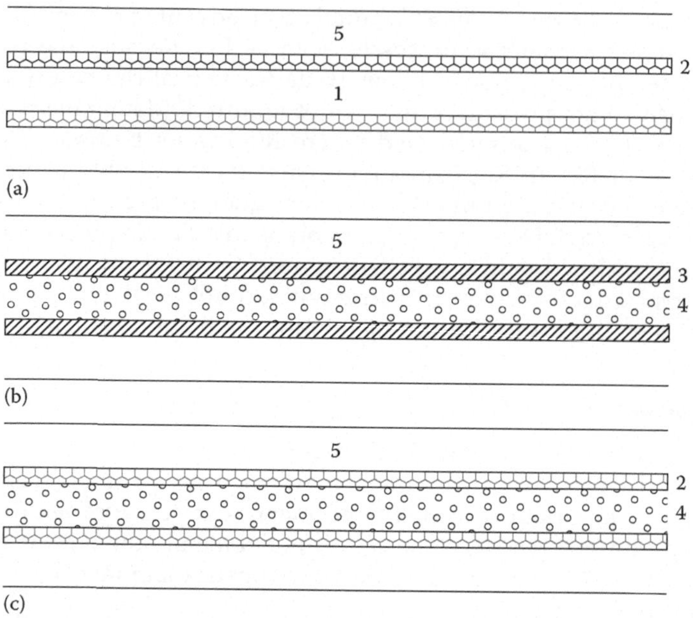

FIGURE 1.14
Different membrane reactor configurations: (a) CMR, CNMR; (b) PBMR, FBMR; (c) PBCMR, FBCMR. 1, tube side; 2, catalytic membrane; 3, inert membrane; 4, catalyst bed; 5, shell side. (Reprinted from *Fundamentals of Inorganic Membrane Science and Technology*, Membrane Science and Technology Series 4, Sanchez, J. and Tsotsis, T.T., Current developments and future research in catalytic membrane reactors, Burggraaf, A.J. and Cot, L. (eds.), 529–568, Copyright 1996, Figure 11.1 and Table 11.1, with permission from Elsevier.)

TABLE 1.3
Classification of Membrane Reactors

Acronym	Description
CMR	Catalytic membrane reactor
CNMR	Catalytic nonpermselective membrane reactor
PBMR	Packed-bed membrane reactor
PBCMR	Packed-bed catalytic membrane reactor
FBMR	Fluidized-bed membrane reactor
FBCMR	Fluidized-bed catalytic membrane reactor

Source: Reprinted from *Fundamentals of Inorganic Membrane Science and Technology*, Membrane Science and Technology Series 4, Sanchez, J. and Tsotsis, T.T., Current developments and future research in catalytic membrane reactors, Burggraaf, A.J. and Cot, L. (eds.), 529–568, Copyright 1996, Figure 11.1 and Table 11.1, with permission from Elsevier.

In packed-bed membrane reactor (PBMR) configuration, packed bed of catalyst is kept in contact with the membrane. The membrane itself is not catalytic. In the catalytic membrane reactor (CMR) configuration, intrinsically catalytic membrane or a membrane that has been made catalytic through activation is used. If the membrane used in PBMR configuration is catalytically active, then the configuration is termed as packed-bed CMR (PBCMR). When the packed bed is replaced by a fluidized bed, it is called the fluidized-bed membrane reactor (FBMR) configuration. The fluidized-bed CMR (FBCMR) uses both a catalytic bed and a permselective membrane. In the catalytic nonpermselective membrane reactor (CNMR) configuration, the membrane is not typically permselective and it is only used to provide a reactive surface.

Membrane reactors can be classified as staged membrane reactors, membrane reactors with multiple feed ports, and multimembrane reactors.[34,35] Research studies are pursued on minimizing reactant loss,[36,37] the use of sweep gas under concurrent or countercurrent operation,[38] and the use of a vacuum on the permeate side.

Membrane reactors are also classified[39–41] as reactive membrane extractors when their function is to remove one or more products. Such action could result in increasing the equilibrium yield. Membranes could also serve as a distributor for one of the reactants. Such membrane reactors find application for consecutive and parallel reactions for a better yield of the intermediate products and they do have a potential to avoid the thermal runaway phenomena, typically associated with highly exothermic reactions. The membrane in a membrane reactor improves the contact between different reactive phases where the membrane acts as a medium for providing the intimate contact between reactants, which are fed separately in either side of the membrane. Multiphase reactions involving a catalyst and liquid and gaseous reactants can also be studied in reactive membrane contactors. The primary advantage of the use of the membrane reactor is to decrease the mass transfer limitations, frequently encountered with such reactions in slurry or trickle-bed reactors.

Membrane reactors are being considered for many processes, and some of them are already being used on an industrial scale, such as cell culture and fermentation processes,[42–45] light hydrocarbon gas-phase catalytic reactions,[46–52] and chiral drug separation.[53,54]

1.4.1.5 Biomimetic Membranes

Biomimetic membranes incorporate biological elements, borrow concepts, and ideas, and get inspiration from natural biological systems. Such membranes utilize the proven performance evolved by nature over billions of years for improving transport efficiency and functioning. Biological membranes rely on the host of different mechanisms to evolve efficient separations as given in Table 1.4 providing inspiration for synthetic membrane

TABLE 1.4

Biological Membrane Separation Paradigm

Membrane Type	Dimensions or Mechanism	Location (in Biological Systems)	Relevant Membrane Applications
Surface layer membranes	2–8 nm of pore size	External membrane surface	UF membrane
Lipid bilayers	Nonporous, diffusion	Enveloping the cell and cell compartments	RO and forward osmosis membranes
Ionosphere-based membranes	Nonporous, diffusion by carrier	Hydrophobic region of lipid bilayers	Electrode sensors, liquid membranes
Membrane protein-facilitated lipid bilayers	0.3–1.5 nm of pore size	Transmembrane	Biomimetic desalination membranes, artificial channels
Biological antifouling surface	Surface physiochemical interactions, antifouling chemical signals, surface topography	Membrane surfaces (blood cell membrane, plant and animal skin)	Surface antifouling coating for separations

Source: Reprinted from *J. Membr. Sci.*, 454, Shen, Y.-X., Saboe, P.O., Sines, I.T., Erbakan, M., and Kumar, M., Biomimetic membranes: A review, 359–381, Copyright 2014, with permission from Elsevier.

design. As shown in Figure 1.15, gram-negative bacterial organisms perform size-graded membrane filtration, which is comparable to similar strategies used in membrane filtration. Structures on the outer membrane surface provide coarse filtration (surface layer proteins) and fouling resistance. The second level of filtration is through large outer membrane channels for macromolecular separations and specialized inner (or plasma membrane) membrane proteins for specific transport of solutes and water using pumps, channels, and transporters. The simple membrane bilayers allow for passive diffusion of water, gases, and specific solutes by solution diffusion as well as by carrier-mediated diffusion of ionophores in the hydrophobic membrane interior. These cells also employ antifouling strategies to prevent unwanted protein deposition on their surface and attachment by other microorganisms.[55]

Research and development in biomimetic membranes has reached an interesting phase. Several technologies are being developed that could advance the state of the art in membrane separation. However, there are many challenges to overcome for biomimetic membranes, such as (1) the fundamental understanding of the interaction between functional molecules and matrix materials, (2) the engineering approach for the synthesis of biomimetic membranes, and (3) the scale-up issues.

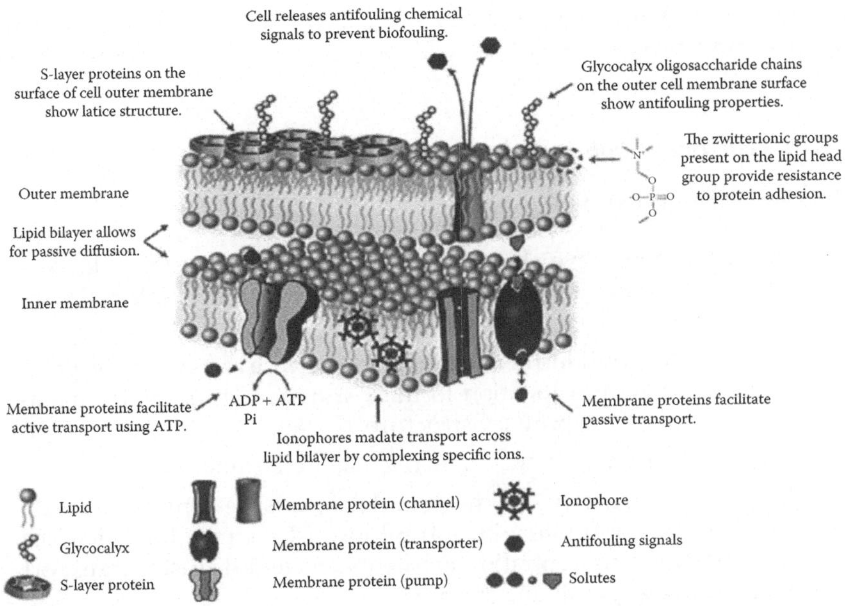

FIGURE 1.15
Biological membrane separation and antifouling strategies for an example of a gram-negative bacterial organism. (Reprinted from *J. Membr. Sci.*, 454, Shen, Y.-X., Saboe, P.O., Sines, I.T., Erbakan, M., and Kumar, M., Biomimetic membranes: A review, 359–381, Copyright 2014, with permission from Elsevier.)

1.5 Membrane Transport Theories

Theories of membrane transport phenomena for both liquids and gaseous separations have been described.

1.5.1 Membrane Transport Involving Liquid Systems

In membrane filtration, the separation process is accomplished by using a differential driving potential across a membrane having selective permeability. For example, the differential driving potential used to transport solvent across UF membrane is the hydrostatic pressure. UF is commonly used to separate suspended solids, colloids, and macromolecules from water. Whenever the solvent of a mixture flows through the membrane, retained species are locally concentrated at the membrane surface, thereby resisting the flow. This localized concentration of solute normally results in precipitation of a solute gel over the membrane. Hence, UF throughput depends on physical properties of the membrane, such as permeability, thickness, and process and system variables such as feed concentration, system pressure, velocity, and temperature. Two models with different approach are described below.

1.5.1.1 Gel Polarization Model

The basic assumptions of this model are as follows:

1. UF membranes have skin that offers minimum resistance to flow and the asymmetry of the pore virtually eliminates internal pore fouling.
2. Concentration buildup at the membrane surface rises up to the point of incipient gel precipitation forming a dynamic secondary membrane on top of the primary structure.
3. The secondary membrane offers the major resistance to flow.
4. The gel layer grows in thickness until the pressure-activated convective transport of solute with solvent toward the membrane surface just equals the concentration gradient-activated diffusive transport away from the membrane surface.
5. Beyond a certain threshold pressure, increase in pressure does not improve the flux because the gel layer grows thicker to offer more resistance to the increased driving force. This is called critical flux.

$$\text{Water flux } (J_w) = \text{TMP}/(R_c + R_m)$$

where:
TMP is the transmembrane pressure
R_c is the resistance of the deposited cake
R_m is the hydraulic resistance of the membrane

6. Eventually, the concentration at the membrane surface will be high enough to form a gel.
7. In the steady state, the convective transport to the membrane must equal the back-diffusive transport away from the membrane.

$$J = -D\left(\frac{dC}{dx}\right) \tag{1.5}$$

where:
J is the solvent flux through the membrane
C is the concentration of solutes or colloids retained in membrane
D is the solute diffusivity
x is the distance from the membrane surface

Upon integration,

$$J = \left(\frac{D}{\delta}\right) \cdot \ln\left(\frac{C_g}{C_b}\right) = k \cdot \ln\left(\frac{C_g}{C_b}\right) \tag{1.6}$$

where:

k is the mass transfer coefficient

δ is the boundary layer thickness considering D as constant

C_g and C_b represent the maximum solute concentration in the gel layer and the concentration of solutes in the bulk of feed, respectively

8. Lower solute concentration (C_b) will have higher threshold pressure because much higher flux is required to transport enough solute to the membrane to begin to form a gel.

1.5.1.2 Resistance Model

The mechanism of separation in UF involves not only the size exclusion but also the adsorption and surface charge characteristics of membranes. In the absence of a solute, the water flux through a microporous membrane is defined by Darcy's law, which states that pure solvent flux is directly proportional to the applied pressure differential (ΔP) and inversely proportional to pure solvent viscosity (μ_w).

$$J_w = \frac{\Delta P_{appl}}{(R_m \mu_w)} \tag{1.7}$$

where R_m is the membrane hydraulic resistance which is a function of pore size, tortuosity, membrane thickness, and porosity.

If the feed solution contains solutes which are retained at the membrane interface, the water flux in UF is generally lower than pure water flux. A number of phenomena have been suggested to account for this flux reduction, such as resistance due to gel layer formation, resistance due to concentration polarization, and resistance due to an absorption layer and pore plugging. For a macromolecular solute of high molecular weight at low concentration, the osmotic pressure effect can be neglected. The effect of the gel layer can be represented as

$$J_{uf} = \frac{\Delta P_{appl}}{\mu_w (R_m + R_p)} \tag{1.8}$$

where R_p is the resistance due to gel polarization.

The time-dependent case of Equation 1.8 can be represented as

$$J_{uf}(t) = \frac{\Delta P_{appl}}{\mu_w [R_m + R_p(t)]} \tag{1.9}$$

After testing, if the membrane is thoroughly washed with appropriate washing solution and the pure water flux (J_w) is determined at the same ΔP_{appl}, it may be found to be less than J_w but still greater than J_{uf}. The difference between J_w and J_{uf} may be accounted for by the irreversible fouling due to adsorption of solute on the membrane and this loss in flux can be visualized as additional resistance to the flux (R_a). Hence,

$$J_w = \frac{\Delta P_{appl}}{\mu_w (R_m + R_a)} \tag{1.10}$$

Incorporating R_a, Equation 1.9 can be written as

$$J_{uf}(t) = \frac{\Delta P_{appl}}{\mu_w \left[R_m + R_a + R_p(t) \right]} \tag{1.11}$$

It is noted that J_{uf} reaches an almost constant final flux $J_{uf}(F)$ and the time corresponding to this $J_{uf}(F)$ is $t(F)$. At this stage, $R_p(t)$ becomes constant $R_p(F)$.

$$J_{uf}(F) = \frac{\cdot \, \Delta P_{appl}}{\mu_w \left[R_m + R_a + R_p(F) \right]} \tag{1.12}$$

There also exists a concentration polarization resulting from the relative rate of solute transport to the membrane surface by convection and the back-diffusive solute flux. Although both concentration polarization and fouling reduce the membrane flux, they have opposing effects on the observed percent rejection. Another way to distinguish the two phenomena is through their time dependence. Concentration polarization is dependent on operating parameters such as pressure, temperature, feed concentration, and velocity but is not a function of time. Fouling is partially dependent on these variables, particularly feed concentration, but is also a function of time.

The change of flux with time due to different kinds of resistances is given in Figure 1.16 for a typical UF membrane.[56] It shows asymptotic behavior after a particular duration of time.

The mass transfer coefficient, k, can be calculated from the following equation:

$$k = \frac{\mathrm{PR}}{3600 M_B S \left[\frac{1 + m(1-f) M_A}{1000} \right] c(1 - X_3) \left[\ln \left(\frac{X_2 - X_3}{X_1 - X_3} \right) \right]} \tag{1.13}$$

where:

PR refers to the product rate

f refers to the solute separation with reference to the chosen reference solute

M_A and M_B refer to the molecular weights of solute and water, respectively

S refers to the membrane area

m is the solute molarity

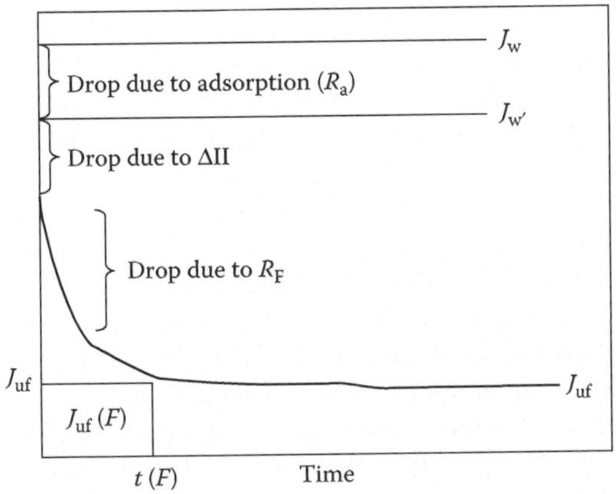

FIGURE 1.16
Change of flux with time for UF membrane showing asymptotic behavior after a particular duration of time.

c is the molar density of feed
X_1, X_2, and X_3 refer to the mole fractions of solute at bulk, membrane–solution interface, and membrane-permeated product, respectively

1.5.2 Membrane Transport Involving Gaseous Systems

There are two main membrane permeation mechanisms: (1) through the bulk of the material (dense membranes) and (2) through the pores (porous membranes). Dense membranes usually have high selectivity and low membrane flux. This principle also applies to small-pore membranes. Larger pores increase the flux but decrease the selectivity.

1.5.2.1 Dense Membrane Separation Mechanism

The solution diffusion mechanism is the commonly used physical model to describe gas transport through dense membranes. A gas molecule is adsorbed on one side of the membrane, dissolves in the membrane material, diffuses through the membrane, and desorbs on the other side of the membrane. If diffusion through the membrane takes place in the form of ions and electrons (i.e., proton-exchange transport) or as atoms (e.g., for hydrogen transport through dense metal), the molecule needs to split up after adsorption and recombine after diffusing through the membrane.

1.5.2.2 Porous Membrane Separation Mechanisms

Four types of transport mechanisms[57] exist with respect to separation in porous membranes: Knudsen diffusion, surface diffusion, capillary condensation, and molecular sieving. In some cases, molecules can move through the membrane by more than one mechanism. Knudsen diffusion gives relatively low separation selectivity compared to surface diffusion and capillary condensation. Shape-selective separation or molecular sieving can yield high selectivities. The separation factor for these mechanisms depends on pore size distribution, temperature, pressure, and interactions between gases being separated and the membrane surfaces.

Knudsen (or free-molecule) diffusion takes place for the cases with large Knudsen numbers. The Knudsen number K_n is defined as the ratio of the mean free path of the gas molecules (average distance between collisions) and a representative physical length scale (e.g., the pore radius).

If the pore radius is used as representative physical length scale, the mean free path lengths are substantially higher than pore radii at Knudsen numbers larger than 10. The result is that mainly the lighter molecules permeate through the pores. Selectivity, however, is limited and can be calculated with the square root of the ratio of the molar masses of the gasses involved. The smaller the Knudsen number, the larger the pores become (relative to the mean free path of the gas molecules), the lower selectivity becomes. For Knudsen numbers <1, the dominant transport mechanism is viscous flow, which is nonselective.

Surface diffusion may occur in parallel with Knudsen diffusion. Gas molecules are adsorbed on the pore walls of the membrane and migrate along the surface. Surface diffusion increases the permeability of the components adsorbing strongly to the membrane pores. At the same time, the effective pore diameter is reduced. Consequently, transport of nonadsorbing components is reduced and selectivity is increased. This positive contribution of surface diffusion works for certain temperature ranges and pore diameters.

Capillary condensation occurs if a condensed phase partially fills the membrane pores. If the pores are completely filled with condensed phase, only the species soluble in the condensed phase can permeate through the membrane. Fluxes and selectivities are generally high for capillary condensation. The appearance of capillary condensation, however, depends on gas composition, pore size, and uniformity of pore sizes.

If pore sizes become sufficiently small (3–5 Å), the mechanism of molecular sieving is applicable in separating molecules that differ in kinetic diameter. The pore size becomes so small that only the smaller gas molecules can permeate through the membrane.

1.5.3 Transport Mechanism in ED Membrane

The transport mechanism in the ED membrane is shown in Figure 1.17 showing concentration and potential gradients in a well-stirred ED cell. In this example, chloride ions easily permeate the anionic membranes containing fixed positive groups and are stopped by the cationic membranes

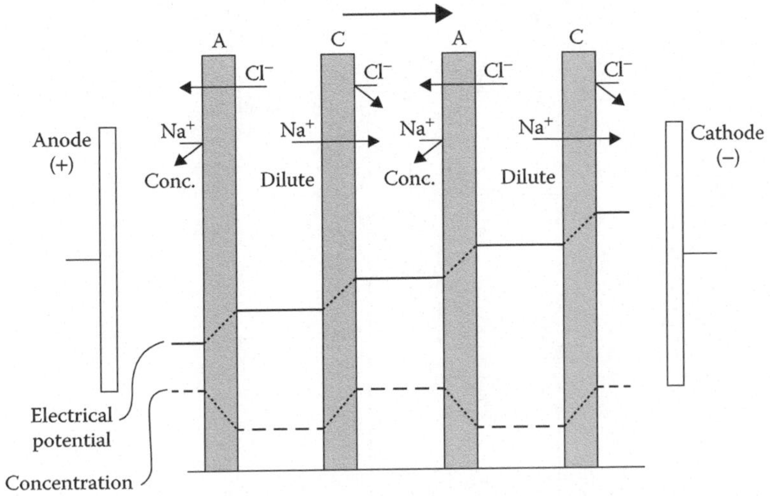

FIGURE 1.17
Schematic of the concentration and potential gradients in a well-stirred ED cell. (From Baker, R.W.: Chapter 10: Ion Exchange membrane processes-electrodialysis. *Membrane Technology and Applications*. 2004. Copyright Wiley-VCH Verlag GmbH & Co. Figures 10.7 and 10.8. Reproduced with permission.)

containing fixed negative groups. Similarly, sodium ions permeate the cationic membranes but are stopped by the anionic membranes. The overall result is increased salt concentration in alternating compartments, whereas the other compartments are simultaneously depleted of salt. The voltage potential drop caused by the electrical resistance takes place entirely across the ion-exchange membrane. In a well-stirred cell, the flux of ions across the membranes, and hence the productivity of the ED system can be increased without limit by increasing the current across the stack. In practice, however, the resistance of the membrane is often small in proportion to the resistance of the water-filled compartments, particularly in the dilute compartment where the concentration of ions carrying the current is low. In this compartment, the formation of ion-depleted regions next to the membrane places an additional limit on the current, and hence the flux of ions through the membranes. Ion transport through this ion-depleted aqueous boundary layer generally controls ED system performance.

The formation of concentration gradients caused by the flow of ions through a cationic membrane is shown in Figure 1.18.

It shows the concentration gradient of univalent sodium ion next to a cationic membrane. Exactly equivalent gradient of anion, such as chloride ion, forms adjacent to the anionic membranes in the stack. The ion gradient

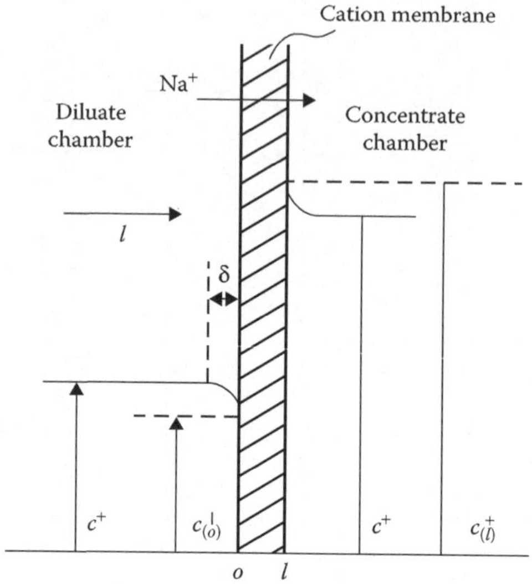

FIGURE 1.18
Schematic of the concentration gradients adjacent to a single cationic membrane in an ED stack. (From Baker, R.W.: Chapter 10: Ion exchange membrane processes-electrodialysis. *Membrane Technology and Applications*. 2004. Copyright Wiley-VCH Verlag GmbH & Co. Figures 10.7 and 10.8. Reproduced with permission.)

formed on the left dilute side of the membrane can be described by Fick's law. Thus, the rate of diffusion of cations to the surface is given by

$$J^+ = \frac{D^+\left(c^+ - c^+_{(o)}\right)}{\delta} \tag{1.14}$$

where:

D^+ is the diffusion coefficient of the cation in water
c^+ is the bulk concentration of the cation in the solution
$c^+_{(o)}$ is the concentration of the cation in the solution adjacent to the membrane surface (o)
δ is film thickness

The rate at which the cations approach the membrane by electrolyte transport is (t^+I/F). The total flux of sodium ions to the membrane surface (J^+) is the sum of these two terms.

$$J^+ = \frac{D^+\left(c^+ - c_{(o)}{}^+\right)}{\delta} + \left(\frac{t^+ I}{F}\right) \tag{1.15}$$

Transport through the membrane is also the sum of two terms (1) due to the voltage difference and (2) due to the diffusion caused by the difference in ion concentrations on each side of the membrane. Thus, the ion flux through the membrane can be written as

$$J^{+} = \left(\frac{t^{+}_{(m)}I}{F}\right) + \frac{P^{+}\left(c^{+}_{(o)} - c^{+}_{(1)}\right)}{1} \tag{1.16}$$

where P^{+} is the permeability of the sodium ions in a membrane of thickness l.

The quantity $P^{+}(c^{+}_{(o)} - c^{+}_{(1)})/1$ is much smaller than transport due to the voltage gradient, so Equations 1.15 and 1.16 can be combined and simplified to

$$\frac{D^{+}\left(c^{+} - c^{+}_{(o)}\right)}{\delta} + \left(\frac{t^{+}I}{F}\right) = \left(\frac{t^{+}_{(m)}I}{F}\right) \tag{1.17}$$

For a selective cationic ion-exchange membrane for which $t^{+}_{(m)} = 1$, Equation 1.17 can be further simplified to

$$I = \left(\frac{FD^{+}}{1-t^{+}}\right)\frac{\left(c^{+} - c^{+}_{(o)}\right)}{\delta} \tag{1.18}$$

This important equation has a limiting value when the concentration of the ion at the membrane surface is zero $\left(c^{+}_{(o)} = 0\right)$. At this point, the current reaches its maximum value. The limiting current is given by the equation:

$$I_{\text{lim}} = \frac{FD^{+}c^{+}}{\delta\left(1-t^{+}\right)} \tag{1.19}$$

The limiting current, I_{lim}, is the maximum current that can be employed in an ED process. If the potential required to produce this current is exceeded, the extra current will be carried by other processes, first by transport of anions through the cationic membrane and, at higher potentials, by hydrogen and hydroxyl ions formed by dissociation of water. Both of these undesirable processes consume power without producing any separation. This decreases the current efficiency of the process.

The limiting current density for an ED system operated at the same feed solution flow rate is a function of the feed solution salt concentration, as shown in Equation 1.19. As the salt concentration in the solution increases, more ions are available to transport current in the boundary layer, so the limiting current density also increases. For this reason, large ED systems with several ED stacks in series will operate with different current densities in each stack, reflecting the change in the feed water concentration as salt is removed.

1.6 Membrane Preparation Technique

Synthetic membranes can be classified in five basic groups: (1) microporous media, (2) homogeneous solid films, (3) asymmetric structures, (4) electrically charged barriers, and (5) liquid films with selective carriers. This classification, however, is rather arbitrary and there are many structures which would fit more than one of the abovementioned classes, for example, a membrane may be microporous, asymmetric in structure, and carry electrical charges.

1.6.1 Neutral Microporous Membranes

The neutral microporous films represent a very simple form of a membrane, which closely resembles the conventional fiber filter as far as the mode of separation and the mass transport are concerned. These membranes consist of a solid matrix with defined holes or pores. Separation of various chemical components is achieved strictly by a sieving mechanism with the pore diameters and the particle sizes being the determining parameters. Microporous membranes can be made from various materials, such as ceramics, graphite, metal or metal oxides, and various polymers. Their structure may be symmetric, that is, the pore diameters do not vary over the membrane cross section, or they can be asymmetrically structured, that is, the pore diameters increase from one side of the membrane to the other by a factor of 10–1000. The properties and areas of application of various microporous filters are summarized in Table 1.5.

1.6.1.1 Sintered Membranes

Sintered membranes are normally made from ceramic materials, glass, graphite, and metal powders such as stainless steel and tungsten.[58] The particle size of the powder is the main parameter determining the pore sizes

TABLE 1.5

Microporous Membranes, Their Preparation, and Applications

Membrane Type	Membrane Material	Pore Size (μm)	Manufacturing Process	Application
Microporous membrane	Ceramic, metal, or polymer powder	0.1–20	Pressing and sintering of powder	MF
	Homogeneous polymer sheets (PE, PTFE)	0.5–10	Stretching of extruded polymer sheets	MF, burn dressings, artificial blood vessels
	Homogeneous polymer sheets (polycarbonate)	0.02–10	Track-etching	MF
	Polymer solution (cellulose acetate)	0.01–5	Phase inversion	MF, UF, sterilization

of the final membrane. The membrane can be made in the form of discs, candles, or fine bore tubes. Sintered membranes are used for the filtration of colloidal solutions and suspensions. This type of membrane is suitable for gas separation. It is also used for the separation of radioactive isotopes such as uranium.

1.6.1.2 Stretched Membranes

Another method for preparing microporous membranes is the stretching of a homogeneous polymer film of partial crystallinity. This technique is employed with films of polyethylene (PE) or PTFE, which are extruded from a polymer powder and then stretched perpendicular to the direction of extrusion.[59,60] This leads to relatively uniform pores. These membranes, which have a very high porosity, up to 90%, and a fairly regular pore size, are widely used for MF of acid and caustic solutions, and organic solvents. They have to a large extent replaced the sintered materials used earlier in this application. Stretched membranes can be produced as flat sheets as well as tubes and capillaries. This membrane type has a high permeability for gases and vapors due to high porosity, and, because of the hydrophobic nature of the basic polymer, it is impermeable to aqueous solutions. This membrane has also been used for a novel process, generally referred to as membrane distillation for removing ethanol[61] from fermentation broths and for desalination of saline water and in medical applications.

1.6.1.3 Capillary Pore Membranes

Microporous membranes with very uniform, nearly perfectly round cylindrical pores are obtained by a process generally referred to as track-etching.[62] The membranes are made in a two-step process. During the first step, a homogeneous thick polymer film is exposed to charged particles in an irradiator. As the particles pass through the film, they leave sensitized tracks where the chemical bonds in the polymer backbone are broken. In the second step, the irradiated film is placed in an etching bath. In this bath, the damaged material along the tracks is preferentially etched forming uniform cylindrical pores.

The pore density of a track-etched membrane is determined by the residence time in the irradiator, whereas the pore diameter is controlled by the residence time in the etching bath. The minimum pore diameter of these membranes is ~0.01 μm. The maximum pore size that can be achieved in track-etched membranes is determined by the etching procedure. With exposure time in the etching medium, the pore size increases and the thickness of the film is correspondingly reduced. Capillary pore membranes are prepared mainly from polycarbonate and polyester films. The advantage of these polymers is that they are commercially available in very uniform thickness (10–15 μm), which is the maximum penetration depth of collimated particles obtained from irradiator of energy 0.8–1 MeV. Particles with higher energy, up to 10 MeV, may be obtained in an accelerator. They are used to irradiate thick polymer films up to 50 μm thickness or inorganic materials such as mica.[63] Because of their narrow pore size distribution and low tendency to plug, capillary pore membranes made from polycarbonate and polyester have found application on a large scale in analytical chemistry and microbiological laboratories, and in medical diagnostic procedures.[64] The membranes are also used in standard clinical tests for red blood cell deformability studies.

1.6.1.4 Phase Inversion Membranes

Most of the microporous membranes are made by phase inversion technique.[65] In phase inversion process, a polymer is dissolved in an appropriate solvent and spread as film of about 20–200 μm thickness. Water is added causing separation of the homogeneous polymer solution into a solid polymer phase and a liquid solvent phase. The precipitated polymer forms a porous structure containing a network of more or less uniform pores. This type of membrane can be made from almost any polymer which is soluble in an appropriate solvent and can be precipitated in a nonsolvent.[66] By varying the polymer, the polymer concentration, the precipitation medium, and the precipitation temperature, microporous phase inversion membranes can be made of different pore sizes (from less than 0.1 to more than 20 μm) with varying chemical, thermal, and mechanical properties. These membranes were originally prepared from cellulosic polymers by precipitation at room temperature in an atmosphere of ~100% relative humidity.[67] Lately, symmetric microporous membranes are also prepared from Nylon 66, Nomex, polysulfone, and PVDF by precipitation of a cast polymer solution in aqueous liquid.[68] The symmetric, microporous polymer membranes made by phase inversion are widely used for separations on a laboratory and industrial scale.[69] Typical applications range from the clarification of turbid solutions to the removal of bacteria or enzymes, the detection of pathological components, and the detoxification of blood in an artificial kidney.

1.6.1.5 Asymmetric Microporous Membranes

An asymmetric membrane consists of a thin (0.1–1 μm) selective skin layer on a highly porous (100–200 μm thick) substructure. The thin skin represents the membrane. Its separation characteristics are determined by the nature of the polymer and the pore size, whereas the mass transport rate is determined by the membrane thickness, because the mass transport rate is inversely proportional to the thickness of the actual barrier layer. The porous sublayer serves as a support for the thin skin and has little effect on separation characteristics or the mass transfer rate of the membrane. It provides mechanical strength. Asymmetric membranes are used primarily in pressure-driven membrane processes, such as RO, UF, or gas separation due to high mass transfer rate and good mechanical stability.[70] In addition to high filtration rates, asymmetric membranes are fouling resistant. Asymmetric membranes are surface filters and retain all rejected materials on the surface, which are removed by the shear force applied by the feed solution moving parallel to the membrane surface.

Two techniques are used to prepare asymmetric membranes. The first technique utilizes the phase inversion process. The second technique is based on getting composite structure by depositing an extremely thin polymer film on a microporous substructure.

The development of the asymmetric phase inversion membrane was a major breakthrough in the development of UF and RO. These membranes were made from cellulose acetate and yielded fluxes 10–100 times higher than symmetric structures with comparable separation characteristics. Asymmetric phase inversion membranes can be prepared from cellulose acetate and many other polymers by the following general preparation procedure[71]:

- A polymer is dissolved in an appropriate solvent to form a solution containing 10–30 wt.% polymer.
- The solution is cast into a film.
- The film is quenched in a nonsolvent typically water or an aqueous solution.

During the quenching process, the homogeneous polymer solution separates into two phases: a polymer-rich solid phase, which forms the membrane structure, and a solvent-rich liquid phase, which forms the liquid-filled membrane pores. Generally, the pores at the film surface, where precipitation occurs first and most rapidly, are much smaller than those in the interior or the bottom side of the film, which leads to the asymmetric membrane structure. There are different variations to this general preparation procedure; for example, Loeb and Sourirajan used an evaporation step to increase the polymer concentration in the surface of the cast polymer solution and an annealing step during which the precipitated polymer film is exposed for a certain time period to hot water of 70°C–80°C.[72] The detailed formation mechanism of microporous symmetric or asymmetric membranes is given in literatures.[73,74] Both quantitative and qualitative description and correlation of the various preparation parameters with membrane structures and properties are also described in literatures.[75–81]

1.6.2 Homogeneous Membranes

The mass transport in homogeneous membranes occurs primarily by diffusion; thus, permeability is rather low. Homogeneous membranes should, therefore, be as thin as possible. The separation of various components in a solution is directly related to their transport rates within the membrane phase, which is determined mainly by their diffusivity and concentration in the membrane matrix.[82–86]

1.6.2.1 Homogeneous Polymer Membranes

Polymer chemistry is capable of tailoring polymers of mechanical and thermal stability as well as chemical compatibility. In general, mass transfer is greater in amorphous polymers than in highly cross-linked polymers.[87] Thus, crystallization and orientation are avoided as much as possible when high permeability and transmembrane flux are desired. However, physical properties and, in particular, the mechanical strength of the polymer as well as its selectivity are adversely affected, and the final product will represent a compromise between necessary strength, selectivity, and mass transfer rates.[88] The two basic membrane configurations are flat sheets and hollow fibers.[89,90] Flat sheets can be prepared by casting from solution, by extruding from a polymer melt, or by blow and press molding. Hollow fibers are generally made by extrusion with central gas injection.[91] Because of their high selectivity for different chemical components, homogeneous membranes are used in various applications, which in general involve the separation of different low-molecular-weight components with identical or nearly identical molecular dimensions. The important applications of homogeneous polymer membranes are in gas separation, pervaporation, and RO.

1.6.2.2 Homogeneous Metal and Glass Membranes

The permeability of hydrogen in palladium, palladium alloys, and several other metals such as platinum, silver, iron, and nickel is several orders of magnitude higher than that of any other gases. The permeability of hydrogen in palladium alloy membranes is highly temperature dependent. The separation is, therefore, carried out at elevated temperature (400°C).[92] The palladium or palladium alloy membrane used for the separation and purification of hydrogen is produced by electroless deposition.[93] Because of their high selectivity, these membranes are used for production of high-purity hydrogen (>99.99% H_2). The same is true for the use of homogeneous silica glass membranes, which appears promising as selective barrier especially for the separation of helium. Like metal membranes, glass membranes are operated at elevated temperature.

1.6.2.3 Liquid Membranes

Liquid membranes in combination with the facilitated transport is gaining increasing significance. It utilizes selective *carriers* transporting certain components such as metal ions selectively and at a relatively high rate across the liquid membrane interphase.[94,95] It is relatively easy to form a thin fluid film. It is difficult, however, to maintain and control this film and its properties during the separation process. In order to avoid a breakup of the film, some type of reinforcement is necessary to support such a weak membrane structure. Two different techniques are used today for the preparation of liquid membranes. In the first case, the selective liquid barrier material is stabilized as a thin film by a surfactant in an emulsion-type mixture.[96,97]. In the second technique for making liquid membranes, a microporous polymer structure is filled with the liquid membrane phase.[98,99] In this configuration, the microporous structure provides the mechanical strength and the liquid-filled pores provide the selective separation barrier. Both types of membranes are used on a pilot plant stage for the selective removal of heavy metal ions or certain organic solvents from industrial waste streams. They have also been used rather effectively for the separation of oxygen and nitrogen.[100]

1.6.2.4 Ion-Exchange Membranes

Ion-exchange membranes consist of highly swollen gels carrying fixed positive or negative charges. The properties and preparation procedures of ion-exchange membranes are closely related to those of ion-exchange resins.[101] As with resins, there are many possible types with different polymer matrices and different functional groups to confer ion-exchange properties on the product.

There are two types of ion-exchange membranes: (1) cation-exchange membrane that contains negatively charged groups fixed to the polymer matrix and (2) anion-exchange membrane that contains positively charged groups fixed to the polymer matrix. In a cation-exchange membrane, the fixed anions are in electrical equilibrium with mobile cations in the interstices of the polymer, as indicated in Figure 1.19. It shows schematically the matrix of a cation-exchange membrane with fixed anions and mobile cations, which are referred to as counterions. By contrast, the mobile anions, called co-ions, are more or less completely excluded from the polymer matrix because of their electrical charge which is identical to that of the fixed ions. Due to the exclusion of the co-ions, a cation-exchange membrane permits transfer of cations only. Anion-exchange membranes carry positive charges fixed on the polymer matrix permeating anions only. The most required properties from ion-exchange membranes are as follows:

- High permselectivity, that is, an ion-exchange membrane should be highly permeable for counterions, but should be impermeable to co-ions
- Low electrical resistance, that is, the permeability of an ion-exchange membrane for the counterions under the driving force of an electrical potential gradient should be as high as possible
- Good mechanical and form stability: the membrane should be mechanically strong and should have a low degree of swelling or shrinking in transition from dilute to concentrated ionic solutions

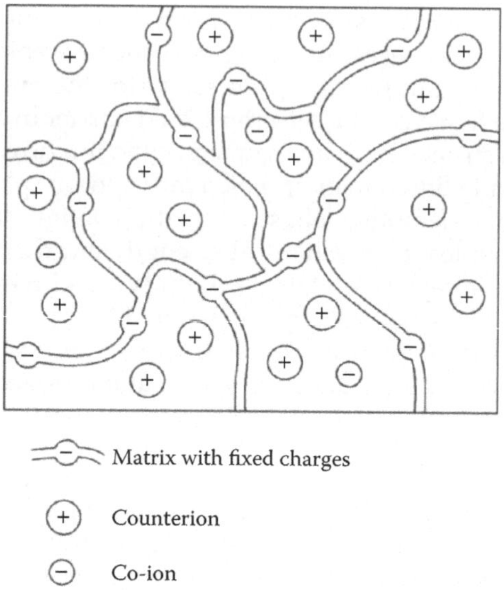

FIGURE 1.19
Schematic diagram of the structure of a cation-exchange membrane showing the polymer matrix with the negative fixed charges, the positive counterions, and the negative co-ions.

- High chemical stability: the membranes should be stable over a wide pH range from 1 to 14 and in the presence of oxidizing agents

It is often difficult to optimize the properties of ion-exchange membranes because the parameters determining the different properties often act contrary. For instance, a high degree of cross-linking improves the mechanical strength of the membrane but also increases its electrical resistance. A high concentration of fixed ionic charges in the membrane matrix leads to a low electric resistance but, in general, causes a high degree of swelling combined with poor mechanical stability. The properties of ion-exchange membranes are determined by two parameters, that is, the basic polymer matrix and the type and concentration of the fixed ionic moiety. The basic polymer matrix determines to a large extent the mechanical, chemical, and thermal stability of the membrane. Very often the matrix of an ion-exchange membrane consists of hydrophobic polymers such as polystyrene, PE, or polysulfone.[102] Although these basic polymers are insoluble in water and show a low degree of swelling, they may become water soluble by the introduction of the ionic moieties. Therefore, the polymer matrix of ion-exchange membranes is very often cross-linked. The degree of cross-linking then determines to a large extent the degree of swelling, and the chemical and thermal stability, but also has a large effect on the electric resistance and the permselectivity of the membrane.[103,104]

The type and the concentration of the fixed ionic charges determine the permselectivity and the electrical resistance of the membrane, but they also have a significant effect on the mechanical properties of the membrane. The degree of swelling, especially, is affected by the fixed charge concentration. The following moieties are used as fixed charges in cation-exchange membranes[105–109]:

$$-SO_3^-, -COO^-, -PO_3^{2-}, -AsO_3^{2-}$$

In anion-exchange membranes, fixed charges may be

$$-NH_3^+, >NH_2^+, \geqslant N^+, >S^+$$

These differently charged groups have a significant effect on the ion-exchange behavior of the membrane.

Most commercial ion-exchange membranes can be divided, according to their structure and preparation procedure, into two major categories: homogeneous and heterogeneous membranes. In general, heterogeneous ion-exchange membranes have several disadvantages, the most important of which are relatively high electrical resistance and poor mechanical strength.

Homogeneous ion-exchange membranes have significantly better properties in this respect, because the fixed ion charges are distributed homogeneously over the entire matrix. The methods of making homogeneous ion-exchange membranes can be summarized by three different basic procedures:

1. Polymerization or polycondensation of monomers[110]
2. Introduction of anionic or cationic moieties into a preformed film by techniques such as imbibing styrene into polymer films, polymerizing the imbibed monomer, and then sulfonating the styrene[111]
3. Introduction of anionic or cationic moieties into a polymer chain such as polysulfone, followed by dissolving the polymer and casting it into a film[112]

Heterogeneous ion-exchange membranes consist of fine colloidal ion-exchange particles embedded in an inert binder such as PE, phenolic resins, or PVC.[113] The membranes can be prepared by calendaring ion-exchange particles into an inert plastic film. Another procedure is dry molding of inert film-forming polymers and ion-exchange particles, and then milling the mold stock. Also, ion-exchange particles can be dispersed in a solution containing a film-forming binder, and the solvent evaporated to give an ion-exchange membrane. Similarly, ion-exchange particles can be dispersed in a partially polymerized binder polymer, and the polymerization subsequently completed. Heterogeneous membranes with usefully low electrical resistances contain more than 65% by weight of the cross-linked ion-exchange particles. Because these ion-exchange particles swell when immersed in water, it has been difficult to achieve adequate mechanical strength combined with low electrical resistance. Most heterogeneous membranes that possess adequate mechanical strength generally show poor electrochemical properties.

1.6.3 Composite Membranes

In processes such as RO, gas separation, and pervaporation, the actual mass separation is achieved by a solution-diffusion mechanism in a homogeneous polymer layer. Because the diffusion process in a homogeneous polymer matrix is relatively slow, these membranes should be as thin as possible. Therefore, an asymmetric membrane structure is mandatory for these processes. Unfortunately, many polymers with satisfactory selectivity and permeability for the various components in gas mixtures or liquid solutions are not well suited for the preparation of asymmetric membranes by the phase inversion process. This has led to the development of the so-called composite membranes. A composite membrane is shown schematically in Figure 1.20. It is composed of a 20–100 nm thin dense polymer barrier layer formed over an ~100 μm-thick microporous film. Composite membranes differ from asymmetric membranes by the mode of preparation which consists of two steps: (1) casting of the microporous support and (2) deposition of the barrier layer on the surface of this microporous support layer. This preparation mode leads to significant advantages of the composite membrane compared to the integral asymmetric membrane. In an integral asymmetric membrane, the selective barrier layer and the microporous support always consist of the same polymer. In a composite membrane, different polymers in general are used for the microporous support and the selective barrier layer. This means that polymers which show the desired selectivity for a certain separation problem, but have poor mechanical strength or poor film-forming properties, and which are therefore not suited for preparation into integral asymmetric membranes,

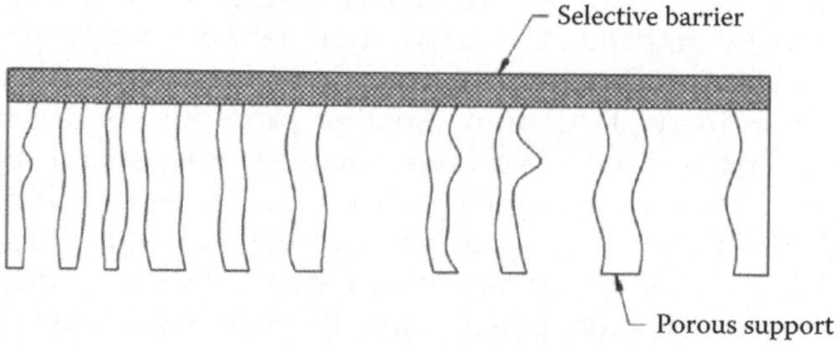

FIGURE 1.20
Schematic diagram of an asymmetric composite membrane showing the microporous support structure and the selective skin layer.

may well be utilized as the selective barrier in composite membranes. This enhances the choice of available materials for the preparation of semipermeable membranes.

1.6.3.1 Preparation Procedure

The preparation of composite membrane requires a suitable microporous support and the barrier layer and lamination to the surface of the support film. The performance of a composite membrane depends on the proper ties of the selective barrier layer as well as the properties of the microporous support film.

The preparation of composite structure involve the following[114–118]:

- Casting of the barrier layer, for example, on the surface of a water bath followed by lamination to the microporous support film
- Dip coating of the microporous support film in a polymer, a reactive monomer, or prepolymer solution followed by drying or curing with heat or radiation
- Gas-phase deposition of the barrier layer of the microporous support film
- Interfacial polymerization of reactive monomers on the surface of the microporous support film

Casting an ultrathin film of cellulose acetate on a water surface and transferring the film on a microporous support was one of the earliest techniques used for preparing composite RO membranes for water desalination. Dip coating a microporous support membrane in polymer or prepolymer solution was also first developed for the preparation of RO membranes. Here, a microporous membrane prepared from mixed cellulose esters was first coated by a protective layer of polyacrylic acid to prevent the solvent of the casting solution of the barrier layer, which consisted, for example, of cellulose triacetate in chloroform, from dissolving the support membrane. This technique was later improved by using a microporous sublayer, which had better overall mechanical and thermal stability and which was insoluble in the solvent of the barrier layer, such as an *open* polysulfone UF membrane. Dip coating is applied mainly for the preparation of composite membranes to be used in gas separation and pervaporation.[119] Gas-phase deposition of the barrier layer on a dry microporous support membrane by plasma polymerization was also successfully used for the preparation of RO membranes.[120] Large-scale industrial production utilizing plasma polymerization for the preparation of composite membrane seems to be difficult.

Important technique for preparing composite membranes is the interfacial polymerization of reactive monomers on the surface of a microporous support film. The first membrane produced on a large scale with excellent RO desalination properties was developed in the early 1970s in the North Star Research Institute under the code name NS100[121] involving two steps. In the first step, polyethyleneimine reacts rapidly at the interphase with the toluene diisocyanate to form a polyamide surface skin whereas amine groups below this surface remain unreacted. In the second step known as heat treatment step, internal cross-linking of polyethyleneimine takes place. Thus, the final membrane has three distinct layers of increasing porosity: (1) the dense polyamide surface skin that acts as the actual selective barrier, (2) a thin cross-linked polyethyleneimine layer that extends into the pore of the support film, and (3) the actual polysulfone support membrane.

1.6.3.2 Sol–Gel Route

The sol–gel route is another method for preparation of functional oxide layers. Two sol–gel routes are usually followed[122]: One is based on colloid chemistry in aqueous media and the other has to do with the chemistry of metal organic precursors in organic solvent, both being able to produce porous materials. These two routes can be used to prepare supported ceramic membranes. The porous structure is influenced by the steps involved in the process. The general method for membrane preparation through sol–gel route is shown in Figure 1.21.[123]

1.6.4 Hollow-Fiber Membranes

Hollow-fiber membranes have high membrane surface areas per unit volume. The modules are compact. The diameter of hollow fibers varies over a wide range, from 50 to 3000 μm. Fibers can be made with a uniformly dense structure, but preferably are formed as a microporous structure having a dense selective layer on either the outside or the inside surface. The dense surface layer can be either integral with the fiber or a separate layer coated onto the porous support fiber. Many fibers are packed and potted into tube making a membrane module. Module with a surface area of even a few square meters requires kilometers of fibers. Because a module should not contain broken or defective fibers, hollow-fiber production requires stringent quality control and high reproducibility.

Two methods are used to prepare hollow fibers: solution spinning and melt spinning.[124,125] The common process is solution spinning or wet spinning, in which a 20–30 wt.% polymer solution is extruded and precipitated into a nonsolvent, generally water. Fibers made by solution spinning have the anisotropic structure of membranes. This technique is generally used to make relatively large, porous hemodialysis and UF fibers. In the alternative technique of melt spinning, a hot polymer melt is extruded from an appropriate die and is then cooled and solidified in air prior to immersion in a quench tank. Melt-spun fibers are usually denser and have lower fluxes than

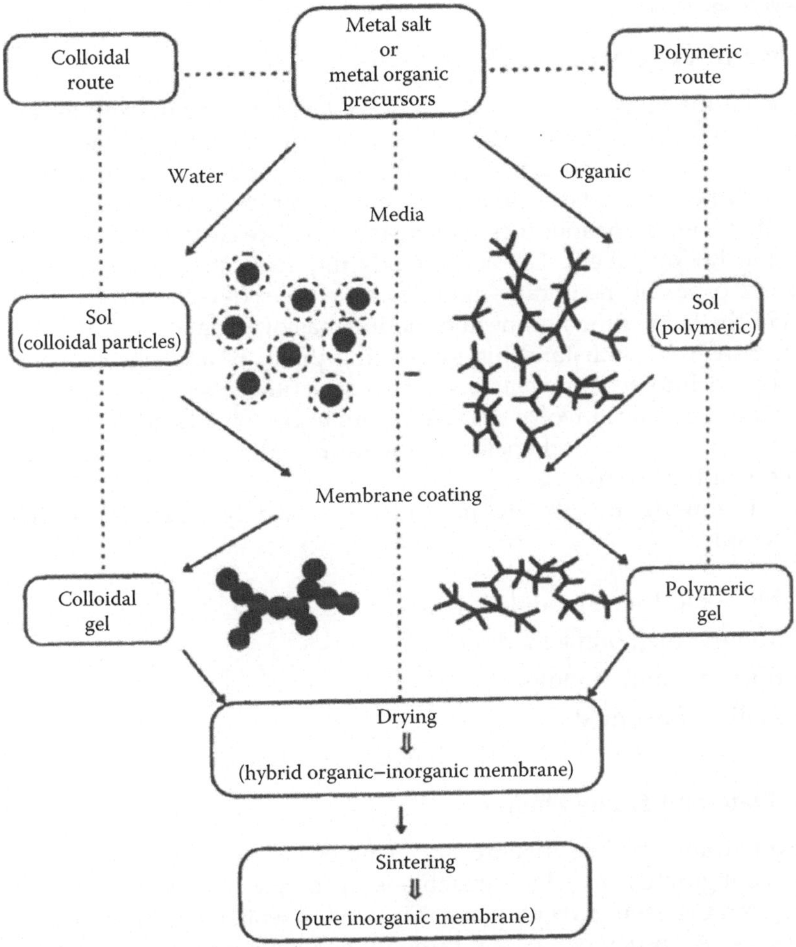

FIGURE 1.21
Schematic of two sol–gel routes used in inorganic membrane preparation. (Reprinted from *Fundamentals of Inorganic Membrane Science and Technology*, Membrane Science and Technology Series 4, Sanchez, J. and Tsotsis, T.T., Current developments and future research in catalytic membrane reactors, Burggraaf, A.J. and Cot, L. (eds.), 227–258, Copyright 1996, Figure 7.1, with permission from Elsevier.)

solution-spun fibers, but, because the fibers can be stretched after they leave the die, very fine fibers can be made. Melt-spun fibers can also be produced at high speeds. The technique is usually used to make hollow fine fibers for high-pressure RO and gas separation applications, and is also used with polymers such as poly(trimethylpentene), which are not soluble in convenient solvents and are difficult to form by wet spinning.

1.7 Membrane Modules

Industrial membrane plants often require large membrane surface area to perform the separation required on a useful scale. Before a membrane separation can be performed industrially, therefore, methods of economically and efficiently packaging large areas of membrane are required. These packages are called membrane modules. The membranes are cast as flat sheets, tubes, and fine hollow fibers. For accommodating such shapes and structures, different types of membrane modules have been developed. The development in the field of membrane and module has overcome several challenges making them to withstand high operating pressure and physical compaction. The techno-economic factors for the selection, design, and operation of membrane modules involve several factors to be considered such as cost of supporting materials and enclosure (pressure vessel), energy consumption, and ease of replacement.

The following membrane modules are widely used for industrial applications:

- Plate-and-frame module
- Tubular membrane module
- Spiral wound module
- Hollow-fiber module

1.7.1 Plate-and-Frame Module

Plate-and-frame modules (Figure 1.22[126]) are one of the earliest types of membrane configuration in which membrane, feed spacers, and product spacers are layered between two end plates. The feed is sent across the surface of the membrane. A portion of it passes through the membrane, enters the permeate channel, and makes its way to a central permeate collection manyfold. Plate-and-frame modules are used in ED and pervaporation systems. A modified version of plate-and-frame module known as disc and tube module configuration has become a popular wastewater application for highly fouling feed streams.

1.7.2 Tubular Membrane Module

Tubular modules are generally limited to MF and UF applications, for which the benefit of resistance to membrane fouling due to good fluid hydrodynamics outweighs their high cost. Typically, the tubes consist of a porous paper or fiberglass support with the membrane on the inside of the tubes. In a typical tubular membrane system, a large number of tubes are arranged in series. The permeate stream from each tube is collected in the permeate collection header. A tubular system is shown in Figure 1.23.

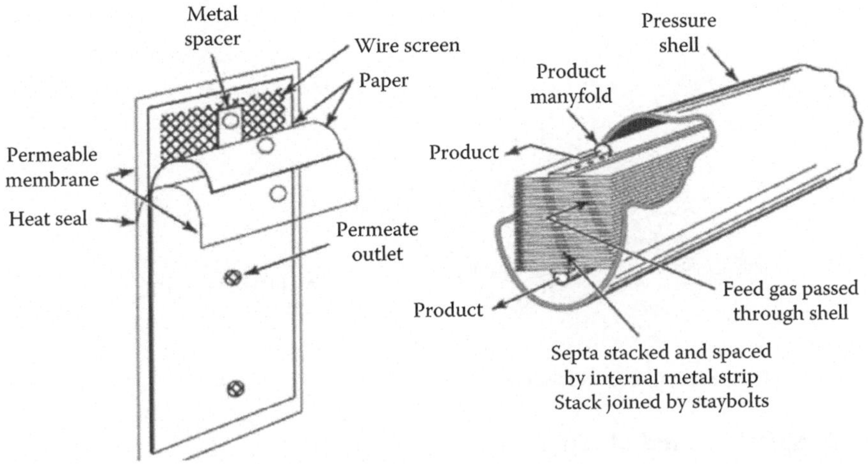

FIGURE 1.22
Early plate-and-frame design for the separation of helium from natural gas. (From Baker, R.W.: Chapter 3: Membranes and modules. *Membrane Technology and Applications*. 2004. Copyright Wiley-VCH Verlag GmbH & Co. Figures 3.38, 3.41, and 3.46. Reproduced with permission.)

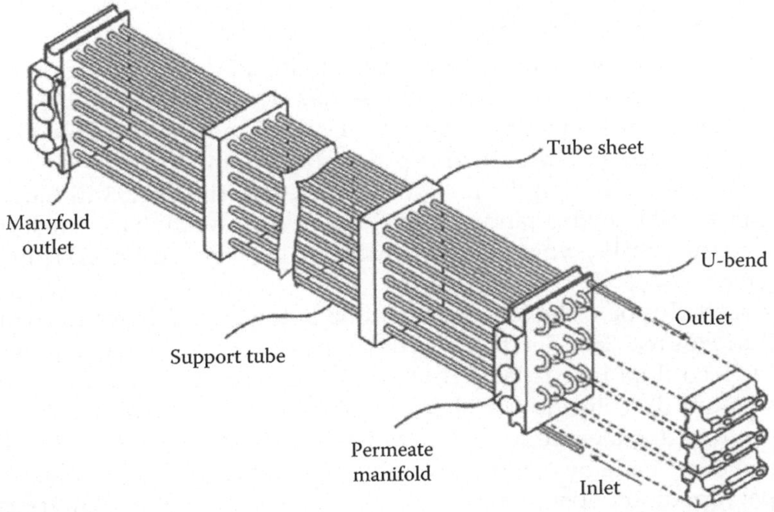

FIGURE 1.23
Tubular UF system in which tubes are connected in series. (From Baker, R.W.: Chapter 3: Membranes and modules. *Membrane Technology and Applications*. 2004. Copyright Wiley-VCH Verlag GmbH & Co. Figures 3.38, 3.41, and 3.46. Reproduced with permission.)

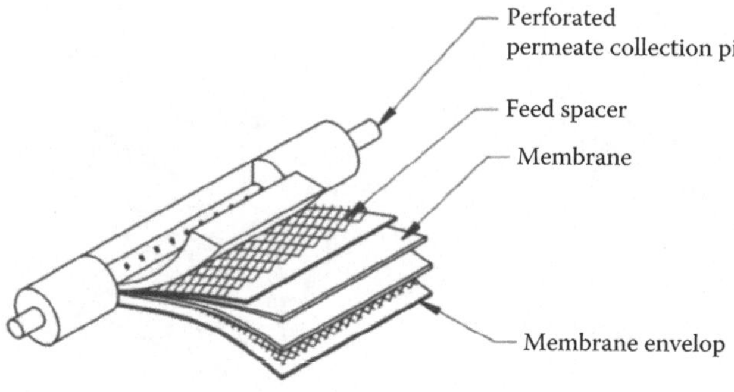

FIGURE 1.24
Exploded view of spiral wound module.

1.7.3 Spiral Wound Module

In the spiral wound module, the support fabric, feed spacer, and permeate carrier encase the membrane, providing structural integrity, as shown in Figure 1.24. Feed solution passes across the membrane surface. A portion passes through the membrane and enters the membrane envelop where it spirals inward to the central perforated collection pipe. The feed enters the module. The permeate stream and concentrate (reject) stream come out of the module. Spiral wound modules are commonly used by desalination industries for brackish and seawater desalination.

1.7.4 Hollow-Fiber Module

Hollow-fiber membrane modules can be classified into two categories based on feed arrangement. The first is the shell-side feed design illustrated in Figure 1.25a. In such a module, a loop or a closed bundle of fibers is contained in a pressure vessel. The system is pressurized from the shell side and the permeate passes through the hollow fiber. Because the fiber wall must support considerable hydrostatic pressure, the fibers usually have small diameters and thick walls, typically 50 μm internal diameter and 100–200 μm outer diameter.

The second type of hollow-fiber module is the bore-side feed type illustrated in Figure 1.25b. The fibers in this type of unit are open at both ends, and the feed fluid is circulated through the bore of the fibers. To minimize the pressure drop inside the fibers, the diameters are usually larger than those of the fine fibers used in the shell-side feed system. The hollow fibers are generally made by solution spinning. The modules are popular for UF and pervaporation operations. They are used for low- to medium-pressure gas applications. Feed pressures are usually limited to below 10 bar in this type of module. Capillary fiber, which is a modified version of hollow fiber, appears promising for several applications where concentration polarization and fouling are faced in hollow-fiber modules.

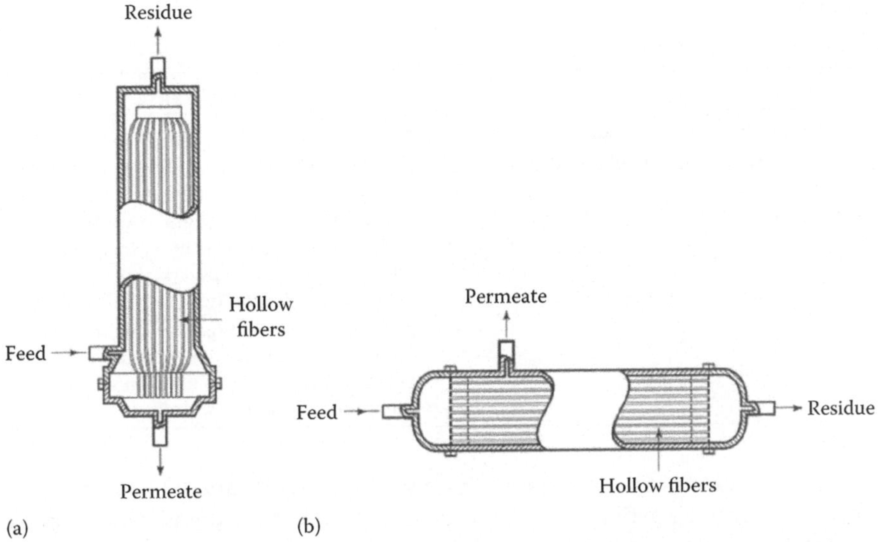

FIGURE 1.25
Two types of hollow-fiber modules: (a) Shell-side feed, and (b) bore-side feed. (From Baker, R.W.: Chapter 3: Membranes and modules. *Membrane Technology and Applications*. 2004. Copyright Wiley-VCH Verlag GmbH & Co. Figures 3.38, 3.41, and 3.46. Reproduced with permission.)

1.8 Selection of Membrane Module

The selection of membrane module depends on a number of parameters. The principal module design parameters are summarized in Table 1.6.

Major factors determining module selection are concentration polarization control and resistance to fouling. Concentration polarization control is a particularly important issue in liquid separation such as RO and UF. In gas separation applications, concentration polarization is more easily controlled but is still a problem with high-flux, highly selective membranes.

TABLE 1.6
Parameters for Membrane Module Design

Parameter	Hollow Fine Fibers	Capillary Fibers	Spiral Wound	Plate and Frame	Tubular
Manufacturing cost (US$/m^2)	Low	Low	Moderate	High	High
Concentration polarization/ fouling control	Poor	Fair	Moderate	Good	Very good
Permeate-side pressure drop	High	Moderate	Moderate	Low	Low
Limited to specific types of membrane material	Yes	Yes	No	No	No

TABLE 1.7

Commonly Used Module Configurations

Application	Module Type
Seawater and brackish water desalination	Spiral wound and disc tube modules
Wastewater treatment as well as water recycle and reuse	Tubular, capillary, and spiral wound modules
Gas separation	Hollow fibers for high-volume applications with low-flux, low-selectivity membranes in which concentration polarization is easily controlled (nitrogen from air)
	Spiral wound when fluxes are higher, feed gases more contaminated and concentration polarization a problem (natural gas separations, vapor permeation)

Another factor is the ease with which various membrane materials can be fabricated into a particular module design. Almost all membranes can be formed into plate-and-frame, spiral wound, and tubular modules, but many membrane materials cannot be fabricated into hollow fine fibers or capillary fibers. The suitability of the module design for high-pressure operation and the relative magnitude of pressure drops on the feed and permeate sides of the membrane are important parameters in selection of membrane modules. The types of modules generally used in some of the major membrane processes are listed in Table 1.7.

1.9 Concentration Polarization and Fouling

Concentration polarization and membrane fouling are the major challenges being faced in membrane separation processes.[80,81]

1.9.1 Concentration Polarization

In membrane systems, a feed consisting of gas or liquid mixture passes through the feed side of the membrane, and permeate enriched in one of the components of the mixture is withdrawn from the other side of the membrane. Because the components in feed mixture permeate at different rates, concentration gradient is formed on both sides of the membrane. The phenomenon is called concentration polarization.

The layer of solution immediately adjacent to the membrane surface becomes depleted in the permeating solute on the feed side of the membrane and enriched in this component on the permeate side. Equivalent concentration gradients are also formed for the other components of the feed solution. This concentration polarization reduces the concentration difference across the membrane, thereby decreasing its flux and the membrane selectivity. Concentration polarization can significantly affect the performance of the membrane process.

Two approaches have been used to describe the effect of concentration polarization. One approach has its origins from heat transfer analogy. In this approach, the resistance to permeation across the membrane and the resistance in the fluid layers adjacent to the membrane are treated as resistances in series.

The second approach to concentration polarization is to model the phenomenon by assuming that a thin layer of unmixed fluid, thickness, exists between the membrane surface and the well-mixed bulk solution. The concentration gradients, which control concentration polarization, are formed in this layer. This boundary layer film model simplifies the fluid hydrodynamics occurring in membrane modules and contains one adjustable parameter, the boundary layer thickness. This simple model can explain most of the experimental data.

UF and pervaporation processes, used for the removal of organic solutes from water, are both seriously affected by concentration polarization. In UF, the low diffusion coefficient of macromolecules produces a concentration of retained solutes 70 times the bulk solution volume at the membrane surface. At these high concentrations, macromolecules precipitate, forming a gel layer at the membrane surface and reducing flux. In coupled transport and solvent dehydration by pervaporation, concentration polarization effects are generally modest and controllable.

The cross-, co- and counterflow schemes are illustrated in Figure 1.26, together with the concentration gradient across the membrane.

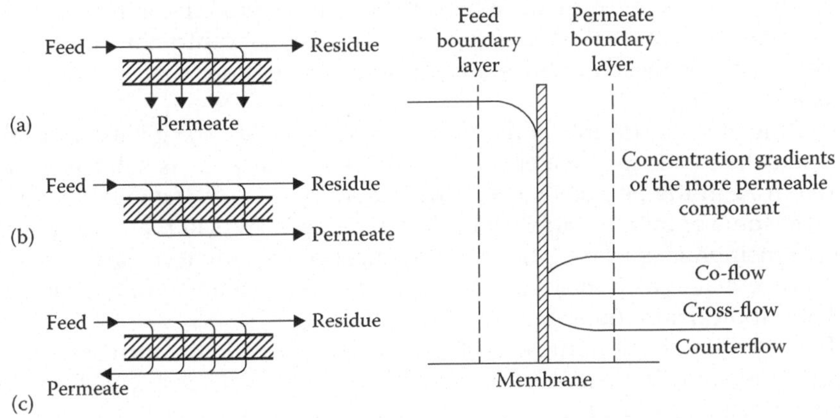

FIGURE 1.26
(a) Cross-, (b) co-, and (c) counterflow schemes in a membrane module and the changes in the concentration gradients that occur across a median section of the membrane. (From Porter, M.C. and Consultant Pleasanton (Eds.). 1990. *Handbook of Industrial Membrane Technology*, California Noyes Publications, Park Ridge, NJ. Figure 1.28; Baker, R.W.: Chapter 4: Concentration Polarisation. *Membrane Technology and Applications*. 2004. Copyright Wiley-VCH Verlag GmbH & Co. Figure 4.17. Reproduced with permission.)

It implies that the performance of membrane system can be improved by operating a module in an appropriate flow mode. Generally, counterflow appears to be promising and beneficial. However, in case of UF and RO processes, the selective side of the membrane faces the feed solution, and a microporous support layer faces the permeate. Concentration gradient easily builds up in this boundary layer, outweighing the benefit of counterflow. Counterflow module designs are limited to gas separation and pervaporation processes. In these processes, the permeate stream is a gas, and permeate-side concentration gradients are more easily controlled because diffusion coefficients in gases are high. The benefit obtained from counterflow depends on a case-to-case basis; however, it can often be substantial, particularly in gas separation and pervaporation processes.

1.9.2 Membrane Fouling

Membrane fouling is the main cause of flux decline and loss of product quality. The initial cause of membrane fouling is concentration polarization. The cause and prevention of fouling depend on the feed characteristics and appropriate control procedures. In general, sources of fouling can be divided into four principal categories: scaling, silting, bacteria, and organic material. More than one type of fouling may occur in a membrane system. Pretreatment of the feed water may be carried out to prevent fouling. Regular cleaning may also be required to handle any reversible fouling that still occurs. Fouling by particulates (silt), bacteria, and organics such as oil is generally controlled by a suitable pretreatment procedure. This type of fouling greatly affects the first few modules in the plant. Whereas the last few modules in the plant are greatly affected by scaling due to the increasing concentration of the feed stream because these modules are exposed to the more concentrated feed stream.

Fouling due to scaling and salting are caused by precipitation of dissolved salts in the feed water on the membrane surface. As salt-free water is removed in the permeate, the concentration of ions in the feed increases until at some point the solubility limit is exceeded. The salt then precipitates on the membrane surface. The salts that most commonly form scale are calcium carbonate, calcium sulfate, silica complexes, barium sulfate, strontium sulfate, and calcium fluoride.

Typical sources of fouling are silt due to organic colloids, corrosion products, algae, and fine particulate matter. The silt density index (SDI) of the feed water gives fair idea of probability of fouling due to silt. The SDI, an empirical measurement, is the time required to filter a fixed volume of water through a standard 0.45 μm pore size MF membrane.[127] Suspended material in the feed water that plugs the microfilter increases the sample filtration time, giving a higher SDI.

Biological fouling takes place due to deposition of biological matter on the membrane surface. The susceptibility of membranes to biological fouling is a strong function of the membrane composition. Cellulose acetate membranes are an ideal nutrient for bacteria and can be completely destroyed by a few weeks of uncontrolled bacterial attack. Polyamide-based hollow fibers are also somewhat susceptible to bacterial attack, but TFC membranes are generally quite resistant. Periodic treatment of such membranes with a bactericide usually controls biological fouling.

Organic fouling is the attachment of materials such as oil or grease onto the membrane surface. Such type of fouling is common in membrane processes used in industries, particularly effluent streams.

1.10 Materials for Different Membrane Processes

1.10.1 Filtration Membranes

1.10.1.1 UF Membranes

A number of materials[80,81] are used for making UF membrane, such as polyacrylonitrile, PVC–polyacrylonitrile copolymers, polysulfone, polyether sulfone, polyvinylidene fluoride, aromatic polyamides, and cellulose acetate. In general, hydrophilic membranes are more fouling resistant than hydrophobic materials. For this reason, water-soluble polymers such as polyvinyl pyrrolidone or polyvinyl methyl ether are often added to the membrane casting solutions used for hydrophobic polymers such as polysulfone or polyvinylidene fluoride. During the membrane precipitation step, most of the water-soluble polymer is leached from the membrane, but enough remains to make the membrane surface hydrophilic. The charge on the membrane surface is important. Many colloidal materials have a slight negative charge from carboxyl, sulfonic, or other acid groups. If the membrane surface also has a slight negative charge, adhesion of the colloidal gel layer to the membrane is reduced, which helps to maintain a high flux and inhibit membrane fouling. The effect of a slight positive charge on the membrane is the opposite. Charge and hydrophilic character can be the result of the chemical structure of the membrane material or can be applied to a preformed membrane surface by chemical grafting or surface treatment. The appropriate treatment depends on the application and the feed characteristics.

1.10.1.2 RO Membranes

RO membranes can be grouped into three main categories:

- Seawater and brackish water desalination membranes operated with 0.5–5 wt.% (5,000–50,000 ppm) concentration salt solution at a pressure of 15–90 bar

- Low-pressure NF membranes operated with 200–5000 ppm concentration salt solution at a pressure of 7–15 bar
- Hyperfiltration membranes used to separate solutes from organic solvent solutions

Cellulose acetate is the material discovered first for high-performance RO membrane. The flux and salt rejection of cellulose acetate membrane has now been taken over by interfacial composite membranes.[128–131] However, cellulose acetate membrane still maintains a market because of resistance to degradation by chlorine and other oxidants, which is a problem with interfacial composite membranes. Cellulose acetate membrane can tolerate up to 1 ppm chlorine content. This means that chlorination can be used to sterilize the feed water, which is a major advantage with feed streams having significant bacterial loading. The water and salt permeability of cellulose acetate membranes is extremely sensitive to the degree of acetylation of the polymer used to make the membrane.[132–134]

Subsequently, polyamide membranes have been developed by several groups. Aromatic polyamide membranes giving high salt rejection and flux have been successfully developed by Toray,[135] Chemstrad (Monsanto),[136] and Permasep (Du Pont)[137] in hollow-fiber configuration. Aliphatic polyamide membranes give low salt rejection and flux. Cadotte and his coworkers observed that high-flux, high-rejection RO membranes can be made by interfacial polymerization.[138–140] Interfacial composite membranes have significantly higher salt rejection and flux than cellulose acetate membranes. The first membranes made by Cadotte had salt rejection greater than 99% and flux of 18 gal/ft^2 day (gfd) at a pressure of 1500 psi for 3.5% concentration of sodium chloride solution (synthetic seawater). Current interfacial composite membranes are significantly better. Typical membranes, tested with 3.5% concentration sodium chloride solution, have a salt rejection of 99.5% and a water flux of 30 gfd at 800 psi, which is a significant improvement over cellulose acetate membranes.

1.10.1.3 NF Membranes

Typically, NF membranes have sodium chloride rejections between 20% and 80% and MWCOs of 200–1000 Da for dissolved organic solutes. These properties are intermediate between RO membranes with a salt rejection of more than 90% and MWCO of less than 50 and UF membranes with a salt rejection of less than 5%. Although some NF membranes are based on cellulose acetate, most of them are based on interfacial composite membranes. The preparation procedure used to make these membranes can result in acid groups attached to the polymeric backbone. The rejection of salts by NF membranes depends on molecular size and Donnan exclusion effects caused by the acid groups attached to the polymer backbone.

The neutral NF membrane rejects the salts in proportion to molecular size, so the extent of rejection is in this order: $Na_2SO_4 > CaCl_2 > NaCl$. The anionic NF membrane[141,142] has positive groups attached to the polymer backbone. These positive charges repel positive cations, particularly divalent cations such as Ca^{2+}, while attracting negative anions, particularly divalent anions such as SO_4^{2-}, that is, salt rejection for $CaCl_2 > NaCl > Na_2SO_4$. The cationic NF membrane[143,144] has negative groups attached to the polymer backbone. These negative charges repel negative anions, such as SO_4^{2-}, while attracting positive cations, particularly divalent cations such as Ca^{2+} implying the extent of salt rejection for $Na_2SO_4 > NaCl > CaCl_2$.

1.10.2 Gas Separation Membranes

Hydrogen permeates through metallic membranes such as tantalum, niobium, vanadium, nickel, iron, copper, cobalt, and platinum.[145] The metal membrane is operated at high temperatures (>300°C) to obtain useful permeation rates and prevent embrittlement and cracking of the metal by sorbed hydrogen. A breakthrough in metal permeation studies occurred in the 1960s, noting that palladium/silver alloy membranes show no hydrogen embrittlement even when used to permeate hydrogen at room temperature.[146] Attempts have been made to reduce the cost of palladium metal membranes by preparing composite membranes. In these membranes, a thin selective palladium layer is deposited onto a microporous ceramic, polymer, or base metal layer.[147,148] The palladium layer is applied by electrolysis coating, vacuum sputtering, or chemical vapor deposition.

Early gas separation membranes[149] were adapted from the cellulose acetate membranes produced for RO by the Loeb–Sourirajan phase separation process. These membranes are produced by precipitation in water; the water must be removed before the membranes can be used to separate gases. However, the capillary forces generated as the liquid evaporates cause collapse of the finely microporous substrate of the cellulose acetate membrane, destroying its usefulness. This problem has been overcome by a solvent exchange process in which the water is first exchanged for an alcohol, then for hexane. The surface tension forces generated as liquid hexane is evaporated are much reduced, and a dry membrane is produced. From 1978 to 1980, Henis and Tripodi,[150] then at Monsanto, devised an ingenious solution to the membrane defect problem; the Monsanto group made Loeb–Sourirajan hollow-fiber membranes (principally from polysulfone), then coated the membranes with a thin layer of silicone rubber. The coating plugged membrane defects in the polysulfone membrane and eliminated convective flow through these defects. The silicone rubber layer also protected the membrane during handling.

Ube Industries, Ltd., (Tokyo, Japan) introduced a polyimide membrane. These polyimide materials are good candidates for gas separation membranes owing to their high selectivities, high insolubilities, and high strengths and glass transition temperatures.[151–153] Signal's UOP Fluid Systems Division Spiragas (a trademark of Signal Co.) membrane is prepared from an ultrathin silicone material on a porous polysulfone flat sheet. A spiral wound module is prepared from this membrane. Silicones, typically, have very high permeabilities and low selectivities.[154]

During the past 15–20 years, ceramic- and zeolite-based membranes have begun to be used for a few commercial separations. These membranes are all multilayer composite structures formed by coating a thin selective ceramic or zeolite layer onto a microporous ceramic support. Ceramic membranes are prepared by the sol–gel technique. Zeolite membranes are prepared by direct crystallization, in which the thin zeolite layer is crystallized at high pressure and temperature directly onto the microporous support.[155]

1.10.3 Pervaporation Membranes

Most pervaporation membranes are composites formed by solution-coating the selective layer onto a microporous support. Microporous polyacrylonitrile coated with a 5–20 μm layer of cross-linked polyvinyl alcohol is the most commonly used commercial material.[156] Chitosan[157] and polyelectrolyte membranes such as Nafion[158,159] have equivalent properties. Membranes comprising silicone rubber coated onto polyimides, polyacrylonitriles, or other microporous support membranes are widely used.[160,161] Polyvinyl alcohol and cellulose acetate[162] have been used to separate alcohol from ether. Polyurethane–polyimide copolymers have been used for aromatic/aliphatic separations.[163] Chemically cross-linked polyvinyl alcohol, formed as a composite membrane by solution casting onto a polyacrylonitrile microporous support, was developed and has been used for a long time. This membrane has a water/alcohol selectivity of more than 200[164] and can achieve good separation of water from ethanol or isopropanol solutions. However, the membrane gets swollen and even dissolved by hot acid or base solutions such as hot acetic acid or hot aniline. Membranes stable to such feed solutions can be prepared by plasma polymerization.[165] Silicone rubber is easy to fabricate, is mechanically and chemically strong, and has good separation factors for many common organic compounds. Polyamide–polyether block copolymers have been used successfully with polar organics such as phenol and aniline.[166–168] The separation factors obtained with these organics are greater than 100, far higher than the separation factors obtained with silicone rubber.

1.10.4 Ion-Exchange Membranes

Ion-exchange membranes fall into two broad categories: homogeneous and heterogeneous. In homogeneous membranes, the charged groups are uniformly distributed through the membrane matrix. These membranes swell relatively uniformly when exposed to water, the extent of swelling being controlled by their cross-linking density. In heterogeneous membranes, the ion-exchange groups are contained in small domains distributed throughout an inert support matrix, which provides mechanical strength. Heterogeneous membranes can be made, for example, by dispersing finely ground ion-exchange particles in a polymer support matrix.

A number of early homogeneous membranes were made by simple condensation reactions of suitable monomers, such as phenol–formaldehyde condensation reaction:

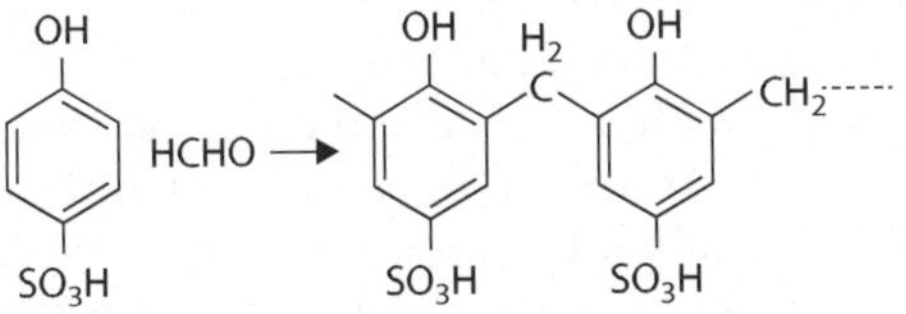

The mechanical stability and ion-exchange capacity of these condensation resins were modest. Another approach is followed to prepare a suitable cross-linked base membrane, which can then be converted to a charged form in a subsequent reaction. In a typical preparation procedure, a 60:40 mixture of styrene and divinyl benzene is cast onto a fabric web, sandwiched between two plates, and heated in an oven to form the membrane matrix. The membrane is then sulfonated with 98% sulfuric acid or a concentrated sulfur trioxide solution. The degree of swelling in the final membrane is controlled by varying the divinyl benzene concentration in the initial mix to control cross-linking density. The degree of sulfonation can also be varied.

Anion-exchange membranes can be made from the same cross-linked polystyrene membrane base by posttreatment with monochloromethyl ether and aluminum chloride to introduce chloromethyl groups into the benzene ring, followed by formation of quaternary amines with trimethyl amine.

The perfluorocarbon-type[169,170] ion-exchange polymer is made by DuPont under the trade name Nafion. The base polymer is made by polymerization of a sulfinol fluoride vinyl ether with tetrafluoroethylene. The copolymer formed is extruded as a film of about 120 μm thick, after which the sulfinol fluoride groups are hydrolyzed to form sulfonic acid groups:

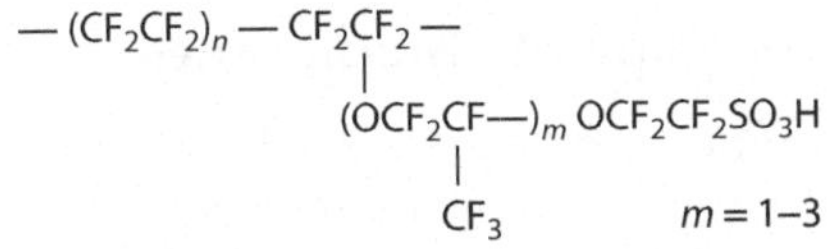

Asahi Chemical (Tokyo, Japan)[171] and Tokuyama Soda (Tokyo, Japan)[172] have developed similar chemistries in which the $–CF_2SO_2F$ groups are replaced by carboxylic acid groups. In these perfluoropolymers, the backbone is extremely hydrophobic, whereas the charged acid groups are strongly polar. Because the polymers are not cross-linked, some phase separation into different domains takes place. The hydrophobic perfluoropolymer domains provide a nonswelling matrix, ensuring the integrity of the membrane.

These perfluorocarbon membranes are inert to concentrated sodium hydroxide solutions and have been widely used in membrane electrochemical cells in the chloralkali industry.

The simplest form of heterogeneous membrane has finely powdered cation- or anion-exchange particles uniformly dispersed in polypropylene. A film of the material is then extruded to form the membrane. The mechanical properties of these membranes are often poor because of swelling of the relatively ion-exchange particles' large diameter (10–20 μm). A much finer heterogeneous dispersion of ion-exchange particles, and consequently a more stable membrane, can be made with a PVC plastisol. A plastisol of approximately equal parts of PVC, styrene monomer, and cross-linking agent in a dioctyl phthalate plasticizing solvent is prepared. The mixture is then cast and polymerized as a film. The PVC and polystyrene polymers form an interconnected domain structure. The styrene groups are then sulfonated by treatment with concentrated sulfuric acid or sulfur trioxide to form a very finely dispersed but heterogeneous structure of sulfonated polystyrene in a PVC matrix, which provides toughness and strength.

1.10.5 Biological Membranes

An important area of ongoing research is the development of functional synthetic membranes that mimic the function of the biological membrane.[173] Because of the heterogeneity of cell membranes, their specific functions are difficult to study directly. However, from lipids it is possible to construct model systems that can mimic biological membranes. Biocatalytic membranes, energy, and information-transducing membranes have already been produced on a laboratory scale[174] and are used as biosensors. Basic research in membrane technology combined with progress in molecular engineering should help to improve the properties of functional synthetic membranes.

1.11 Need for Nanocomposite Membranes

The majority of membranes used commercially are made of polymeric material. However, interest in membranes made from less conventional material is increasing. Ceramic membranes, a special class of microporous membranes, are used in UF and MF applications where solvent resistance and thermal stability are required. Dense metal membranes, particularly palladium membranes, appear promising for the separation of hydrogen from gas mixtures. Supported liquid membranes are developed for carrier-facilitated transport processes. The heart of a membrane process is the membrane itself. To fully utilize the growing opportunities in the field of liquid/gas separation as well as applications in the field of energy and medicine, strong interest exists in the identification of new membrane materials which can comply with current requirements and future potential. Durability, mechanical integrity, operating conditions, membrane flux, and separation efficiency are important parameters that are considered while selecting membrane material for a given separation process. Selectivity, permeation rate, concentration polarization, compaction, and antibiofouling characteristics are important properties of the membrane during its use in operation. For pure polymeric/inorganic materials, a general trade-off exists between permeability and selectivity with an upper limit. When materials with separation properties near this limit are modified based on the traditional structure–property relation, the resultant polymer has permeability and selectivity tracking along this line instead of exceeding it. However, the inorganic materials have properties lying far beyond the upper bound limit for the organic polymers. Though tremendous improvements have been achieved in tailoring polymer structure to enhance separation properties during the past two decades, further progress exceeding the trade-off line seems to present a severe challenge in the near future. Similarly, the immediate application of inorganic membranes is still hindered due to the technological challenges to make defect-free membranes and the high cost for the membrane production. A new approach is needed to provide an alternate and cost-effective membrane with separation properties well above the upper bound limit between permeability and selectivity. The answer comes from mixed matrix membranes. The mixed matrix membrane essentially calls for the adoption and usage of composite materials for achieving desirable separation. While considering the development of composite system, an unprecedented opportunity is being provided by nanostructured materials with the fact that the building blocks in this dimension make it possible to design and create unique materials and devices with significant improvements in the physical/chemical/physicochemical properties and flexibility.

Nanocomposites are a solid structure with nanometer-scale distances between the different phases that constitute the structure and typically consist of inorganic matrix embedded in organic phase or organic matrix embedded in inorganic phase. Nancomposite materials have the ability to demonstrate unique mechanical, chemical, electrical, optical, and catalytic properties. Though the idea of causing improvement and enhancing the properties of a material by making multiphase composites is not recent, the application of nanocomposite system to membrane science and technology is relatively new and evolving. The idea of utilizing the benefits of a nanocomposite as a membrane material is targeted to develop an ideal membrane with improved flux, reasonable selectivity, and other desirable characteristics, which varies on a case-to-case basis. The nanocomposite membranes offer preferential permeation for selective transport while acting as a barrier for undesired transport. The interfacial quality between the nanoparticles and the organic polymer plays an important role which can be improved by chemical modifications of the host polymer matrix and the inorganic nanofillers. Both glass- and rubber-based polymers have been utilized as the organic matrix in nanocomposites. Also, a wide variety of nanoparticles have been tested as inorganic nanofillers such as metal oxides (e.g., TiO_2, Al_2O_3, SiO_2, MgO, AgO, Fe_3O_4), pure metals (e.g., nanosilver), zeolites (e.g., ZSM-5, silicalate-1, zeolite 4A), nanosized macromer polyhedral oligomeric silsesquioxanes (POSSs), carbon nanoparticles (e.g., carbon nanotubes, C60 fullerenes), and mineral clays. These nanoscale fillers are typically prepared *ex situ* and then introduced to the casting mixture, but also in some cases it is possible to generate them *in situ* from precursors. Nanocomposite membranes have great potential in most of the membrane processes, including some novel membrane-based applications.

Nanocomposite can provide numerous advantages over conventional membranes, which are mentioned as follows:

- Increase in permeability over native polymer membranes
- Reduced pressure requirements
- Compact system
- Lower energy requirement
- Reduced membrane surface area
- Smaller footprint for industrial applications
- Enhanced mechanical stability
- Enhanced thermal stability

The importance of nanostructured materials and in turn the nanocomposites is evident from their potential benefits in liquid separation, gas separation, energy sector (fuel cell applications), and biomedical field.

References

1. Baker, R.W. 2004. *Membrane Technology and Applications*, 2nd Edition, Membrane Technology and Research Inc., Menlo Park, CA.
2. Elford, W.J. 1937. Principles governing the preparation of membranes having graded porosities. The properties of "gradocol" membranes as ultrafilters. *Trans. Faraday Soc.* 33: 1094.
3. Zsigmondy, R. and Bachmann, R. 1918. Über Neue filter. *Z. Anorg. Chem.* 103: 119.
4. Ferry, J.D. 1936. Ultrafilter membranes and ultrafiltration. *Chem. Rev.* 18: 373.
5. Loeb, S. and Sourirajan, S. 1963. Sea water demineralization by means of an osmotic membrane, *Saline Water Conversion—II*, R.F. Gould (ed.), Advances in Chemistry Series Number 38, American Chemical Society, Washington, DC, 117–132.
6. Kolf, W.J. and Berk, H.T. 1944. The artificial kidney: A dialyzer with great area. *Acta Med. Scand.* 117: 121–134.

7. Henis, J.M.S. and Tripodi, M.K. 1980. A novel approach to gas separation using composite hollow fiber membranes. *Sep. Sci. Technol.* 15: 1059–1068.
8. Cerini, L. 1929. Apparatus for the purification of impure solutions of caustic soda and the like on osmotic principals, US Patent 1,719,754 (July 1929) and US Patent 1,815,761.
9. Chen, H.-C., O'Neal, C.H., and Craig, L.C. 1971. Rapid dialysis for aminoacylation assay of tRNA. *Anal. Chem.* 43: 1017–1020.
10. Craig, L.C. and Stewart, K. 1964. Thin film counter-current dialysis. *Science* 144: 1093.
11. Wallace, R.M. 1967. Concentration and separation of ions by Donnan membrane equilibrium. *Ind. Eng. Chem. Process Des. Dev.* 6: 423–431.
12. Wallace, R.M. 1969. Concentration of ions using selective membranes, US Patent 3,454,490.
13. Oh, S.J., Moon, S.-H., and Davis, T. 2000. Effects of metal ions on diffusion dialysis of inorganic acids. *J. Membr. Sci.* 169: 95–105.
14. Sollner, K. and Mosaik, U. 1932. Membranen. *Biochem Z.* 244: 370.
15. Leitz, F.B. 1976. Piezodialysis, in *Membrane Separation Processes*, P. Meares (ed.), Elsevier Santi Publishing, Amsterdam, the Netherlands, 261–294.
16. Gardner, C.R., Weinstein, J.N., and Caplan, S.R. 1973. Transport properties of charge-mosaic membranes III. Piezodialysis. *Desalination* 12: 19–33.
17. Merten, U. 1966. Desalination by pressure osmosis. *Desalination* 1: 297–310.
18. Weinstein, J.N. and Caplan, S.R. 1970. Charge-mosaic membranes: Dialytic separation of electrolytes from nonelectrolytes and amino acids. *Science* 169: 296–298.
19. Caplan, S.R. 1971. Transport in natural and synthetic membranes in membrane processes, *Industry and Biomedicine*, M. Bier (ed.), Plenum Press, New York, 1–22.
20. Platt, K.L. and Schindler, A. 1971. Ionic membranes for water desalination. *Makromol. Chem.* 19: 135–155.
21. D'Elia, N.A., Dahuron, L., and Cussler, E.L. 1986. Liquid–liquid extraction with microporous hollow fibers. *J. Membr. Sci.* 29: 309–319.
22. Falk-Pedersen, O. and Dannstrom, H. 2001. Method for removing carbon dioxide from gases, US Patent 6,228,145.
23. Qi, Z. and Cussler, E.L. 1985. Microporous hollow fiber for gas absorption. *J. Membr. Sci.* 23: 321–332.
24. Wiesler, F. and Sodaro, R. 1996. Deaeration: Degasification of water using novel membrane technology. *Ultrapure Water* 35: 53–56.
25. Page, J.K.R. and Kalhod, D.G. 1996. Control of dissolved gases in liquids, US Patent 5,565,149 (October 1996); Honda, K. and Yamashita, M. Method for deaerating liquid products, US Patent 5,522,917.
26. Jansen, A.E., Feron, P.H.M., Hanemaaijer, J.H., and Huisjes, P. 2002. Apparatus and method for performing membrane gas/liquid absorption at elevated pressure, US Patent 6,355,092.
27. Tsou, D.T., Blachman, M.W., and Davis, J.G. 1994. Silver-facilitated olefin/paraffin separation in a liquid membrane contactor system. *Ind. Eng. Chem. Res.* 33: 3209–3216.
28. Valus, R.J., Eshraghi, R., Velikoff, A.E., and. Davis, J.C. October 1991. High pressure facilitated membranes for selective separation and process for the use thereof, US Patent 5,057,641.

29. Prasad, R. and Sirkar, K.K. 1992. Membrane based solvent extraction, *Membrane Handbook*, W.S.W. Ho and K.K. Sirkar (eds.), Van Nostrand-Reinhold, New York, 727–763.
30. Schneider, K., Hölz, W., Wollbeck, R., and Ripperger, S. 1988. Membranes and modules for transmembrane distillation. *J. Membr. Sci.* 39: 25–42.
31. Schofield, R.W., Fane, A.G., and Fell, C.J.D. 1990. Gas and vapor transport through microporous membranes. *J. Membr. Sci.* 53: 173.
32. Lawson, K.W. and Lloyd, D.R. 1997. Membrane distillation. *J. Membr. Sci.* 124: 25.
33. Sanchez, J. and Tsotsis, T.T. 1996. Current developments and future research in catalytic membrane reactors, *Fundamentals of Inorganic Membrane Science and Technology*, A.J. Burggraaf and L. Cot (eds.), Membrane Science and Technology Series 4, Elsevier Science, Amsterdam, the Nethelands, 532.
34. Tecik, M.N., Paunovic, R.N., and Ciric, G.M. 1994. Applicability of two membranes for improving performance of a membrane reactor. *J. Membr. Sci.* 96: 213.
35. Omorjan, R.P., Paunovic, R.N., and Tekic, M.N. 1998. Non-isothermal two-membrane reactors for reversible gas phase reactions. *J. Membr. Sci.* 138: 57.
36. Wu, J.S. and Liu, P.K.T. 1992. Mathematical analysis on catalytic dehydrogenation of ethylbenzene using ceramic membranes. *Ind. Eng. Chem. Res.* 31: 322.
37. Tiscareno-Lechuga, F., Hill Jr., C.G., and Anderson, M.A. 1996. Effect of dilution in the experimental dehydrogenation of cyclohexane in hybrid membrane reactors. *J. Membr. Sci.* 118: 85.
38. Itoh, N. 1995. Limiting conversions of dehydrogenation in palladium membrane reactors. *Catal. Today* 25: 351.
39. Dixon, A. 1999. Innovations in catalytic inorganic membrane reactors, *Catalysis*, James J. Spivey (ed.), Vol. 14, The Royal Society of Chemistry, Cambridge.
40. Sirkar, K.K., Shanbhag, P.V., and Kovvali, S. 1999. Membrane in a reactor: A functional perspective. *Ind. Eng. Chem. Res.* 38: 3715.
41. Julbe, A., Farrusseng, D., and Guizard, C. 2001. Porous ceramic membranes for catalytic reactors—Overview and new ideas. *J. Membr. Sci.* 181: 3.
42. Cheryan, M. and Mehaia, M.A. 1986. Membrane bioreactors, *Membrane Separations in Biotechnology*, W.C. McGregor (ed.), Marcel Dekker, New York, 255–302.
43. Cheryan, M. and Mehaia, M.A. 1985. Membrane bioreactors for high-performance fermentations, *Reverse Osmosis and Ultrafiltration*, S. Sourirajan and T. Matsura (eds.), ACS Symposium Series Number 281, American Chemical Society, Washington, DC, 231–246.
44. Kanazek, R.A., Gullino, P.M., Kohler, P.O., and Dedrick, R.C. 1972. Cell culture on artificial capillaries, an approach to tissue growth. *In Vitro Sci.* 178: 65.
45. Hu, W.S. and Dodge, T.C. 1985. Cultivation of mammalian cells in bioreactors. *Biotechnol. Prog.* 1: 209.
46. Ali, J.K. and Rippin, D.W.T. 1995. Comparing mono and bimetallic noble metal catalysts in a catalytic membrane reactor for methyl-cyclohexane dehydrogenation. *Ind. Eng. Chem. Res.* 34: 722.
47. Hsieh, H.P. 1989. Inorganic membrane reactors: A review, *Membrane Reactor Technology*, R. Govind and N. Itoh (eds.), AIChE Symposium Series Number 268, Vol. 85, AIChE, New York, 53–67.
48. Miller, S.J., Rezac, M.E., and Koros, W.J. July 1995. Dehydrogenation using dehydrogenation catalyst and polymer-porous solid composite membrane, US Patent 5,430,218.

49. Teraoka, Y., Zhang, H., Furukawa, H., and Yamazoe, N. 1985. Oxygen permeation through perovskite-type oxides. *Chem. Lett.* 11: 1743.
50. Teraoka, Y., Nobunaga, T., and Yamazoe, N. 1988. The effect of cation substitution on the oxygen semipermeability of perovskite-type oxides. *Chem. Lett.* 503.
51. Lin, Y.S. 2001. Microporous and dense inorganic membranes: Current status and prospective. *Sep. Purif. Tech.* 25: 39.
52. Thorogood, R.M., Srinvasan, S., Yee, T.F., and Drake, M.P. 1996. Composite mixed conductor membrane for producing oxygen, US Patent 5,240,480 (August 1993); US Patents 5,240,473 (August 1993), 5,261,932 (November 1993), 5,269,822 (December 1993), and 5,516,359 (May 1996).
53. Lopez, J.L. and Matson, S.L. 1997. A multiphase/extractive enzyme membrane reactor for production of diltiazem chiral intermediate. *J. Membr. Sci.* 125: 189.
54. Matson, S.L. January 1989. Method for resolution of stereoisomers in multiphase and extractive membrane reactors, US Patent 4,800,162.
55. Shen, Y.-X., Saboe, P.O., Sines, I.T., Erbakan, M., and Kumar, M. 2014. Biomimetic membranes: A review. *J. Membr. Sci.* 454: 359–381.
56. Ghosh, A.K. 2003. Ultrafiltration membrane, PhD thesis, Bombay University, Mumbai, India.
57. Kluiters, S.C.A. 2004. *Status Review on Membrane System for Hydrogen Separation*, Intermediate Report EU Project MIGREYD NNE5-2001-670, December 2004
58. Hwang, S.-T. and Kammermeyer, K. 1975. *Membranes in Separations*, John Wiley & Sons, New York.
59. Druin, M.L., Loft, J.T., and Plovan, S.G. 1974. Novel open-celled microporous film, US Patent 3,801,404.
60. Gore, R.W. 1976. Process for producing porous products, US Patent 3,962,153.
61. Schneider, K. and van Gassel, T.J. 1984. Membrane distillation. *Chem. Ing. Tech.* 56: 514.
62. Fleischer, R. L., Price, P.B., and Walker, R.M. 1969. Nuclear tracks in solids. *Sci. Am.* 220: 30.
63. Riedel, C. and Spohr, R. 1980. Transmission properties of nuclear track filters. *J. Membr. Sci.* 7: 225.
64. Porter, M.C. 1974. A novel membrane filter for the laboratory. *Am. Lab.* 6: 63.
65. Kesting, R.E. 1971. *Synthetic Polymeric Membranes*. McGraw-Hill, New York.
66. Strathmann, H. 1982. Production of microporous media by phase inversion processes, *Material Science of Synthetic Membranes*, D.R. Lloyd (ed.), ACS Symposium Series 269, American Chemical Society, Washington, DC, 165–195.
67. Zsigmondy, R. 1992. Filter and method of producing same, US Patent 1,421,341.
68. Pall, D. 1982. Process for preparing hydrophilic polyamide membrane filter media and product, US Patent 4,340,479.
69. MacLean, D.W. 1958. The MF Millipore filter. *J. New Engl. Water Works Assoc.* 72: 272.
70. Michaels, A.S. 1976. Synthetic polymeric membranes: Practical applications—Past, present and future. *Pure Appl. Chem.* 46: 193–204.
71. Kesting, R.E. 1955. Phase inversion membranes, *Material Science of Synthetic Membranes*, D.R. Lloyd (ed.), ACS Symposium Series 269, American Chemical Society, Washington, DC, 131–164.
72. Loeb, S. and Sourirajan, S. 1964. High flow porous membranes for separating water from saline solutions, US Patent 3,133,132.

73. Merten, U. (ed.). 1966. *Desalination by Reverse Osmosis*, MIT Press, Cambridge, MA.
74. Manjikian, S. 1967. Desalination membranes from organic casting solutions. *Ind. Eng. Chem. Prod. Res. Dev.* 6: 23.
75. Strathmann, H., Scheible, P., and Baker, R.W. 1971. A rationale for the preparation of Loeb-Sourirajan-type cellulose acetate membranes. *J. Appl. Polym. Sci.* 15: 811–828.
76. Strathmann, H., Kock, K., Amar, P., and Baker, R.W. 1975. The formation mechanism of asymmetric membranes. *Desalination* 16: 179–302.
77. Broens, L., Altena, F.W., and Smolders, C.A. 1980. Asymmetric membrane structures as a result of phase separation phenomena. *Desalination* 32: 33.
78. Frommer, M.A., Feiner, I., Kedem, O., and Block, R. 1970. The mechanism for formation of "skinned" membranes: II. Equilibrium properties and osmotic flows determining membrane structure. *Desalination* 7: 393.
79. Altena, F.W. and Smolders, CA. 1982. Calculation of liquid-liquid phase separation in a ternary system of a polymer in a mixture of a solvent and a nonsolvent. *Macromolecules* 15: 1491.
80. Porter, M.C. and Consultant Pleasanton (eds.). 1990. *Handbook of Industrial Membrane Technology*, California Noyes Publications, Park Ridge, NJ.
81. Mulder, M. 1996. *Basic Principles of Membrane Technology*, 2nd Edition, Kluwer Academic Publishers, Dordrecht, the Netherlands.
82. Lonsdale, H.K., Merten, U., and Riley, R.L. 1965. Transport properties of cellulose acetate osmotic membranes. *J. Appl. Polym. Sci.* 9: 1341–1362.
83. Stannett, V., Szwarc, M., Ghargava, R.L., Meyer, J.A., Myers, A.W., and Rogers, C.E. 1962. *Permeability of Plastic Films and Coated Paper to Gases and Vapors*, Tappi Monograph Series No. 23, Technical Association of the Pulp and Paper Industry, New York.
84. Brubaker, D.W. and Kammermeyer, K. 1952. *Ind. Engr. Chem.* 44: 1465–1474.
85. Stern, S.A. and Walawender, W.P. 1969. Analysis of membrane separation parameters. *J. Sep. Sci.* 4: 129.
86. Stannett, V.T., Koros, W.J., Paul, D.R., Lonsdale, H.K., and Baker, R.W. 1979. Recent advances in membrane science and technology. *Adv. Polymer Sci.* 32: 99–151.
87. Crank, J. and Park, G.S. (eds.). 1968. *Diffusion in Polymers*, Academic Press, New York.
88. Stern, S.A. 1976. The separation of gases by selective permeation, *Membrane Separation Processes*, P. Meares (ed.), Elsevier, Amsterdam, the Netherlands, 295–326.
89. Ward III, W.J., Browall, W.R., and Salemme, R.M. 1976. Ultrathin silicone/polycarbonate membranes for gas separation processes. *J. Memb. Sci.* 1: 99–108.
90. Richter, J.W. and Hoehn, H.H. 1971. Selective aromatic nitrogen-containing polymeric membranes, US Patent 3,567,632.
91. McLain, E.A. and Mahon, H.I. 1969. Permselective hollow fibers and method of making, US Patent 3,423,491.
92. Hunter, J.B. 1956. Silver–Palladium film for separation and purification of hydrogen, US Patent 2,773,561.
93. Tosti, S., Bettinali, L., Castelli, S., Sarto, F., Scaglione, S., and Violante, V. 2002. Sputtered, electroless, and rolled palladium–ceramic membranes. *J. Membr. Sci.* 196: 241–249.
94. Ward III, W.J. 1970. Analytical and experimental studies of facilitated transport. *AIChE J.* 16: 405–410.
95. Cussler, E.L. 1971. Membranes which pump. *AIChE J.* 17: 1300–1303.

96. Li, N.N. 1968. Separating hydrocarbons with liquid membranes, US Patent 3,410,794.
97. Li, N.N. 1971. Permeation through liquid surfactant membranes. *AIChE J* 17: 459.
98. Babcock, W.C., Baker, R.W., LaChapelle, E.D., and Smith, K.L. 1980. Coupled transport membranes II: The mechanism of uranium transport with a tertiary amine. *J. Membr. Sci.* 7: 71–87.
99. Largman, T. and Sifniades, S. 1978. Recovery of copper (II) from aqueous solutions by means of supported liquid membranes. *Hydrometallurgy* 3: 153–162.
100. Baker, R.W., Roman, J.C., Smith, K.L., and Lonsdale, H.K. 1982. *Ind. Heating.* 16.
101. Kunin, R. 1958. *Ion-Exchange Resins,* Wiley, New York.
102. Bergsma, F. and Kruissink, C.A. 1961. Ion-exchange membranes. *Fortschr. Hochpolym. Forsch* 21: 307–362.
103. Meares, P. 1983. Trends in ion-exchange membrane science and technology, *Ion-Exchange Membranes,* D.S. Flett (ed.), E. Horwood, Chichester.
104. Spiegler, K.S. and Laird, A.D.K. (eds.). 1980. *Principles of Desalination,* 2nd Edition, Academic Press, New York.
105. Juda, W. and McRae, W.A. 1953. Ion exchange material and method of making and using, US Patent 2,636,851.
106. Ogami, K. and Kanda, C. 1958. Japan. Patent 1897.
107. Atsugi, Ichikawa, T.M., and Yamada, M. 1958. Japan Patent 2293.
108. Onogram, K. and Mizutani, Y. 1959. Japan Patent 91.
109. Evers, W.L. 1948. US Patent 2,518,420.
110. Bodamer, G.W. 1952. US Patent 2,597,438.
111. Dekb'rb'sy, F. and Schorr, Y. 1963. US Patent 891,562.
112. Zschocke, P. and Quellmalz, D. 1985. Novel ion exchange membranes based on an aromatic polyethersulfone. *J. Membr. Sci.* 22: 325–332.
113. Bodamer, G.W. 1954. US Patent 2,681,320.
114. Riley, R.L., Lonsdale, H.K., Lyons, C.R., and Merten, U. 1967. Preparation of ultrathin reverse osmosis membranes and the attainment of theoretical salt rejection. *J. Appl. Polym. Sci.* 11: 2143–2158.
115. Riley, R.L., Fox, R.L., Lyons, C.R., Milstead, C.E., Seroy, M.W., and Tagami, M. 1976. Spiral-wound poly(ether/amide) thin film composite membrane systems. *Desalination* 19: 113–126.
116. Cadotte, J.E., King, R.S., Majerle, R.J., and Petersen, R. J. 1981. Interfacial synthesis in the preparation of reverse osmosis membranes. *J. Macromol. Sci. Chem.* A15: 727–755.
117. Yasuda, H. and Marsh, H.C. 1976. Preparation of composite reverse osmosis membranes by plasma polymerization of organic compounds: IV. Influence of plasma–polymer (substrate) interaction. *J. Appl. Polym. Sci.* 20: 543.
118. Cadotte, J.E. 1977. US Patent 4,039,440.
119. Henis, J.M.S. and Tripodi, M.K. 1980. A novel approach to gas separations using composite hollow fiber membranes. *Separ. Sci. Technol.* 15: 1059–1068.
120. Yasuda, H. and Lamaze, C.E. 1973. Preparation of reverse osmosis membranes by plasma polymerization of organic compounds. *J. Appl. Polym. Sci.* 17: 201.
121. Rozelle, L.T., Cadotte, J.E., Cobian, K.E., and Kopp, C.V. 1977. Nonpolysaccharide membranes for reverse osmosis: NS-100 Membranes, *Reverse Osmosis and Synthetic Membranes,* S. Sourirajan (ed.), National Research Council Canada, Ottawa, Canada, 249–261.
122. Brinker, C.J. and Scherer, G. 1990. *Sol-Gel Science,* Academic Press, New York.

123. Guizard, C. 1996. Sol-gel chemistry and its application to porous membrane processing, *Fundamentals of Inorganic Membrane Science & Technology*, A.J. Burggraaf and L. Cot (eds.), Elsevier, Amsterdam, the Netherlands.
124. Baum, B., Holley Jr., W., and White, R.A. 1976. Hollow fibres in reverse osmosis, dialysis, and ultrafiltration, *Membrane Separation Processes*, P. Meares (ed.), Elsevier, Amsterdam, the Netherlands, 187–228.
125. Moch Jr., I. 1995. Hollow fiber membranes, *Encyclopedia of Chemical Technology*, Mary Howe-Grant (ed.), 4th Edition, Vol. 13, John Wiley-InterScience, New York, 312.
126. Stern, S.A., Sinclaire, T.F., Gareis, P.J., Vahldieck, N.P., and Mohr, P.H. 1965. Helium recovery by permeation. *Ind. Eng. Chem.* 57: 49–60.
127. Sudak, R.G. 1990. Reverse osmosis, *Handbook of Industrial Membrane Technology*, M.C. Porter (ed.), Noyes Publications, Park Ridge, NJ, 260–306.
128. Riley, R.L. 1991. Reverse osmosis, *Membrane Separation Systems*, R.W. Baker, E.L. Cussler, W. Eykamp, W.J. Koros, R.L. Riley, and H. Strathmann (eds.), Noyes Data Corporation, Park Ridge, NJ, 276–328.
129. Amjad, Z. (ed.). 1993. *Reverse Osmosis*, Van Nostrand Reinhold, New York.
130. Parekh, B. (ed.). 1988. *Reverse Osmosis Technology*, Marcel Dekker, New York.
131. Petersen, R.J. 1993. Composite reverse osmosis and nanofiltration membranes. *J. Membr. Sci.* 83: 81.
132. Reid, C.E. and Breton, E.J. 1959. Water and ion flow across cellulosic membranes. *J. Appl. Polym. Sci.* 1: 133.
133. Rosenbaum, S., Mahon, H.I., and Cotton, O. 1967. Permeation of water and sodium chloride through cellulose acetate. *J. Appl. Polym. Sci.* 11: 2041.
134. Lonsdale, H.K. 1966. Properties of cellulose acetate membranes, *Desalination by Reverse Osmosis*, M. Merten (ed.), MIT Press, Cambridge, MA, 93–160.
135. Endoh, R., Tanaka, T., Kurihara, M., and Ikeda, K. 1977. New polymeric materials for reverse osmosis membranes. *Desalination* 21: 35.
136. McKinney, R. and Rhodes, J.H. 1971. Aromatic polyamide membranes for reverse osmosis separations. *Macromolecules* 4: 633.
137. Richter, J.W. and Hoehn, H.H. March 1971. Selective aromatic nitrogen-containing polymeric membranes, US Patent 3,567,632.
138. Cadotte, J.E. 1985. Evaluation of composite reverse osmosis membrane, *Materials Science of Synthetic Membranes*, D.R. Lloyd (ed.), ACS Symposium Series Number 269, American Chemical Society, Washington, DC.
139. Larson, R.E., Cadotte, J.E., and Petersen, R.J. The FT-30 seawater reverse osmosis membrane-element test results. *Desalination* 38, 473.
140. Cadotte, J.E. July, 1981. Interfacially synthesized reverse osmosis membrane, US Patent 4,277,344.
141. Zhong, S., Widjojo, N., Chung, T.-S., Weber, M., and Maletzko, C. 2012. Positively charged nanofiltration (NF) membranes via UV grafting on sulfonated polyphenylenesulfone (sPPSU) for effective removal of textile dyes from wastewater. *J. Membr. Sci.* 417–418: 52–60.
142. Wang, T., Yang, Y., Zheng, J., Zhang, Q., and Zhang, S. 2013. A novel highly permeable positively charged nanofiltration membrane based on an anoporous hyper-crosslinked polyamide barrier layer. *J. Membr. Sci.* 448: 180–189.
143. Lin, S.W., Sicairos, S.P., and Navarro, R. M. F. 2007. Preparation, characterization and salt rejection of negatively charged polyamide nanofiltration membranes. *J. Mex. Chem. Soc.* 51: 129–135.

144. Childress, A. and Chemelimelech, M. 2000. Relating nanofiltration membrane performance to membrane charge (electrokinetic) characteristics. *Environ. Sci. Technol.* 34: 3710–3716.
145. Alefeld, G. and Völkl, J. (eds.). 1978. *Hydrogen in Metals I: Basic Properties*, Springer-Verlag, Heidelberg, Germany.
146. Hunter, J.B. December, 1956. Silver-palladium film for separation and purification of hydrogen, US Patent 2,773,561.
147. Buxbaum, R.E. and Kinney, A.B. 1996. Hydrogen transport through tubular membranes of palladium-coated tantalum and niobium. *Ind. Eng. Chem. Res.* 35: 530–537.
148. Edlund, D.J. and McCarthy, J. 1995. The relationship between intermetallic diffusion and flux decline in composite-metal membranes: Implications for achieving long membrane lifetime. *J. Membr. Sci.* 107: 147–153.
149. Merten, U. and Gantzel, P.K. 1968. Method and apparatus for gas separation by diffusion, US Patent 3,415,038.
150. Henis, J.M.S. and Tripodi, M.K. 1980. Multicomponent membranes for gas separations, US Patent 4,230,436.
151. Makino, H., Kusuki, Y., Harada, T., and Shimazaki, H. 1983. US Patent 4,378,400.
152. Nakamura, A. 1985. *Energy Resour.* 6(3): 67.
153. Nakamura, A. and Hotta, M. May 1985. *Kagaku Keizai*, 13–19.
154. Paul, D.R. and Morel, G. 1981. *Kirk-Othmer Encyclopedia of Chemical Technology*, 3rd Edition, Vol. 15, John Wiley & Sons, New York, 118.
155. Caro, J., Noack, M., Kolsch, P., and Schäfer, R. 2000. Zeolite membranes: State of their development and perspective. *Micropor. Mesopor. Mater.* 38: 3–24.
156. Bruschke, H.E.A. 1988. State of art of pervaporation, *Proceedings of the 3rd International Conference on Pervaporation Processes in the Chemical Industry*, R. Bakish (ed.), Bakish Materials Corporation, Englewood, NJ, 2–11.
157. Watanabe, K. and Kyo, S. 1992. Pervaporation performance of hollow-fiber chitosan-polyacrylonitrile composite membrane in dehydration of ethanol. *J. Chem. Eng. Jpn* 25: 17–21.
158. Wenzlaff, A., Boddeker, K.W., and Hattenbach, K. 1985. Pervaporation of water–ethanol through ion exchange membranes. *J. Membr. Sci.* 22: 333–344.
159. Cabasso, I. and Liu, Z.-Z. 1985. The permselectivity of ion-exchange membranes for non-electrolyte liquid mixtures: I. Separation of alcohol/water mixtures with Nafion hollow fibers. *J. Membr. Sci.* 24: 101–119.
160. Athayde, A.L., Baker, R.W., Daniels, R., Le, M.H., and Ly, J.H. 1997. Pervaporation for wastewater treatment. *ChemTech* 27: 34–39.
161. Bennett, M., Brisdon, B.J., England, R., and Feld, R.W. 1997. Performance of PDMS and organofunctionalized PDMS membranes for the pervaporation recovery of organics from aqueous streams. *J. Membr. Sci.* 137: 63–88.
162. Chen, M.S.K., Eng, R.M., Glazer, J.L., and Wensley, C.G. September 1988. Pervaporation process for separating alcohols from ethers, US Patent 4,774,365.
163. Schucker, R.C. Highly aromatic polyurea/urethane membranes and their use of the separation of aromatics from non-aromatics, US Patents 5,063,186, 5,055,632, and 4,983,338 and many others.
164. Wijmans, J.G. and Baker, R.W. 1993. A simple predictive treatment of the permeation process in pervaporation. *J. Membr. Sci.* 79: 101–113.

165. Ellinghorst, G., Steinhauser, H., and Hubner, A. 1992. Improvement of pervaporation plant by choice of pva or plasma polymerized membranes, *Proceedings of the 6th International Conference on Pervaporation Processes in the Chemical Industry*, R. Bakish (ed.), Bakish Materials Corporation, Englewood, NJ, 484–493.
166. Boddeker, K.W. and Bengtson, G. 1990. Pervaporation of low volatility aromatics from water. *J. Membr. Sci.* 53: 143–158.
167. Boddeker, K.W. and Bengtson, G. 1991. Selective pervaporation of organics from water, *Pervaporation Membrane Separation Processes*, R.Y.M. Huang (ed.), Elsevier, Amsterdam, the Netherlands, 437–460.
168. Meckl, K. and Lichtenthaler, R.N. 1992. Hybrid-processes including pervaporation for the removal of organic compounds from process and waste water, *Proceedings of the 6th International Conference on Pervaporation Processes in the Chemical Industry*, R. Bakish (ed.), Bakish Materials Corporation, Englewood, NJ.
169. Eisenberg, A. and Yeager, H.L. (eds.). 1982. *Perfluorinated Ionomer Membranes*, ACS Symposium Series Number 180, American Chemical Society, Washington, DC.
170. Gierke, T.D. 1977. Ionic clustering in Nafion perfluorosulfonic acid membranes and its relationship to hydroxyl rejection and chlor-alkali current efficiency, *Paper Presented at the Electrochemical Society Fall Meeting*, Atlanta, GA.
171. Seko, M., Miyauchi, H., and Omura, J. 1983. Ion exchange membrane application for electrodialysis, electroreduction, and electrohydrodimerisation, *Ion Exchange Membranes*, D.S. Flett (ed.), Ellis Horwood, Chichester, 121–136.
172. Sata, T., Motani, K., and Ohaski, Y. 1983. Perfluorinated ion exchange membrane, neosepta-F and its properties, *Ion Exchange Membrane*, D.S. Flett (ed.), Ellis Horwood, Chichester, 137–150.
173. Bader, H., Dorn, K., Hupfer, B., and Ringsdorf, H. 1985. Polymeric monolayers and liposomes as models for biomembranes, *Advances in Polymer Science: Polymer Membranes*, M. Gordon (ed.), Springer-Verlag, Berlin, Germany.
174. Shimidzu, T. and Yoshikawa, M. 1983. Photo-induced carrier mediated transport of alkali metal salts. *J. Membr. Sci.* 13: 1–13.

2

Synthesis of Nanocomposite Membranes

2.1 Introduction

Nanocomposites are composites in which at least one of the phases shows dimensions in the nanometer range. These are high-performance materials and exhibit unusual property combinations and unique design possibilities. Nanocomposite membranes have emerged as suitable alternatives to overcome limitations of conventional membranes. However, there are challenges with respect to the control of elemental composition and stoichiometry in the nanocluster phase while making the nanocomposite materials. Nanocomposites are emerging as the promising material due to their superior properties compared to conventional composites.

Changes in properties occur when the particle size is less than a particular level, called *the critical size*. As dimensions reach the nanometer level, interactions at phase interfaces become largely improved, and this is important to enhance materials properties. The surface area-to-volume ratio of reinforcement materials employed in the preparation of nanocomposites is crucial to the understanding of their structure–property relationships. Common particle geometries and their respective surface area-to-volume ratios[1] are shown in Figure 2.1. For the fiber and layered material, the surface area-to-volume ratio is dominated, especially for nanomaterials, by the first term in the equation. The second term ($2/l$ and $4/l$) has a small influence (and is often omitted) compared to the first term. Therefore, a change in particle diameter, layer thickness, or fibrous material diameter from the micrometer to nanometer range will affect the surface area-to-volume ratio by 3 orders of magnitude.

Carbon nanotubes (CNTs) and their use to make composites exhibiting some of the unique CNT-related mechanical, thermal, and electrical properties have added a new and interesting dimension in this area. The possibility of spinning CNTs into composite products has provided further inroads for the processing and applications of CNT-containing nanomaterials.

As in the case of microcomposites, nanocomposite materials can be classified, as per their matrix materials, in different categories[2] such as (1) ceramic

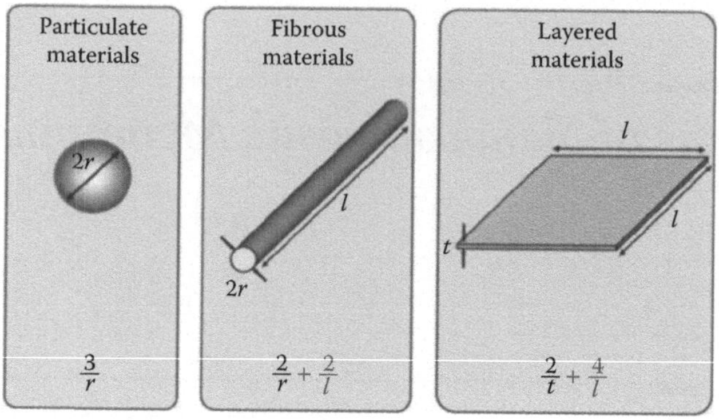

FIGURE 2.1
Common particle geometries and their respective surface area-to-volume ratios. (Reprinted from *J. Compos. Sci. Technol.*, 61, Thostenson, E. et al., Review nanocomposites in context, 491–516, Figure 3, Copyright 2005, with permission from Elsevier.)

matrix nanocomposites (CMNCs), (2) metal matrix nanocomposites (MMNCs), and (3) polymer matrix nanocomposites (PMNCs).

Nanocomposite systems, reinforced with CNTs, have been studied since 1990s. There has been a steady and continuous increase in the number of publications on the subject, including reviews from time to time.[3–16]

In CMNCs, the incorporation of high-strength nanofibers into ceramic matrices leads to preparation of advanced nanocomposites with high toughness and superior characteristics compared to the ceramic materials.[17–19]

MMNCs refer to materials consisting of a ductile metal or alloy matrix in which nanosized reinforcement material is implanted. These materials combine the properties of metals and ceramics, that is, ductility and toughness with high strength and modulus. Thus, MMNCs are suitable for production of materials with high strength in shear/compression processes and high service temperature capabilities. They show immense potential for application in areas such as aerospace and automotive industries.[20]

PMNCs have been developed to improve the separation properties of polymer membranes because they possess the properties of both organic and inorganic membranes such as good hydrophilicity, selectivity, permeability, mechanical strength, and thermal and chemical stability.

The nanostructure, the degree of organization, and the properties that can be obtained with nanocomposite materials certainly depend on the chemical nature of their components, but they also rely on the synergy between them. As a consequence, the nature of the interface or the nature of the links and interactions exchanged by the organic and inorganic components has been used to categorize these hybrids.[21] The preparations and structures of polymer–inorganic nanocomposite membranes, their applicability to gas separation, and the separation mechanisms have been reviewed by dividing the polymeric membranes in two types according to their structure: type I, polymer and inorganic phases bonded by van der Waals force or hydrogen bonds and type II, polymer and inorganic phases bonded by covalent bonds.[22,23]

Type I membrane corresponds to all the systems where no covalent or ionic–covalent bonds are present between the organic and inorganic components. In such materials, the various components only undergo weak interactions such as hydrogen bonding, van der Waals contacts, π–π interactions, or electrostatic forces. In type II membrane, at least a fraction of the organic and inorganic components are linked through strong chemical bonds, such as ionic–covalent bonds. Thus, a key point for the design of new hybrids is the tuning of the nature, the extent and the accessibility of the inner interfaces.[24] Because of the differences between the polymer and inorganic phase properties and the strong aggregation tendency of the nanofillers, fabricating an ideal nanocomposite mixed matrix membrane (Figure 2.2) is difficult.[25] Due to the weak polymer–particle adhesion, nanocomposite membranes frequently show defects at the polymer–particle interface. There are three major categories of interfacial defects: (1) interfacial voids or sieves-in-a-cage (Figure 2.3a), (2) rigidified polymer layer around the particles (Figure 2.3b), and (3) particle pore blockage.[26,27]

Rigidified polymer layer formation and interfacial void formation are due to stresses which arise during membrane formation, a result of solvent evaporation or removal. Rigidified polymer layer formation near the particle is

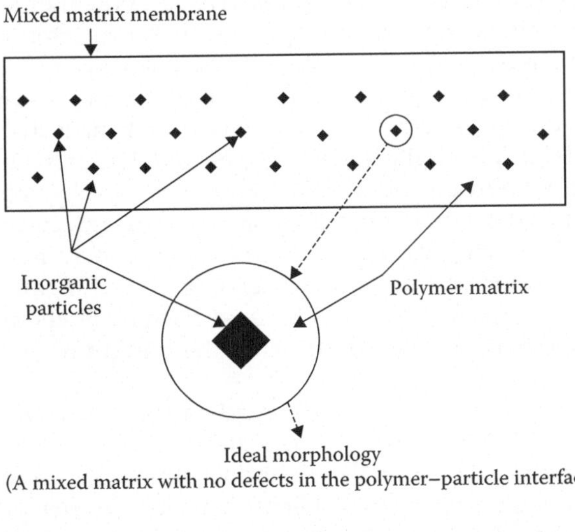

FIGURE 2.2
Schematic diagram of an ideal mixed matrix membrane. (Reprinted from *Sep. Purif. Technol.*, 75, Aroon, M.A. et al., Performance studies of mixed matrix membranes for gas separation: A review, 229–242, Figures 4 and 5, Copyright 2010, with permission from Elsevier.)

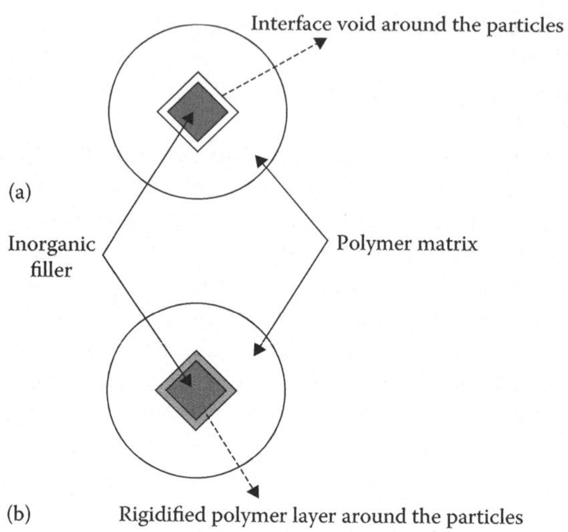

FIGURE 2.3
Interfacial void (a) and rigidified polymer layer (b) in the polymer–particle interface. (Reprinted from *Sep. Purif. Technol.*, 75, Aroon, M.A. et al., Performance studies of mixed matrix membranes for gas separation: A review, 229–242, Figures 4 and 5, Copyright 2010, with permission from Elsevier.)

related to uniform stress around the particles. In this case, polymer chain mobility in the vicinity of the particle surface is less than in the polymer bulk. If the stress directions are not uniform around the inorganic particles, interfacial voids will be formed at the particle–polymer interface. The interfacial defects are hypothesized to form as a result of nucleation of nonsolvent and/or a polymer lean phase around the inorganic phase during the phase separation process. It appears to be necessary that nucleation of solvents and nonsolvents at the zeolite surface be restricted. One such approach is by increasing the hydrophobicity of the zeolite surface by capping surface hydroxyls with hydrophobic organic molecules.[28] Aggregation/dispersion behavior control, which is the first process for the preparation of new functional materials incorporating nanoparticles, is very difficult for nanoparticles less than 100 nm in diameter due to surface interactions.[29] Examples of surface interactions between particles in the liquid phase[30] are given in Table 2.1.

The surface interaction can take place when the nanoparticles in polymeric solution fulfil the conditions as stated. Though the mechanism is known, the factors that enhance or induce the agglomeration remain unclear. This causes difficulty in dispersing the nanoparticles during membrane preparation. However, an increment in concentration of nanoparticles, ionic strength, and pH of the solution may also induce agglomeration between nanoparticles.[31,32]

TABLE 2.1

Surface Interaction between Particles in the Liquid Phase

Surface Interaction	Generation Mechanism
van der Waals interaction	Short-range electromagnetic force between molecules and/or atoms
Overlap of electrical double layer	Electrical interaction from the overlap of electrical double layers around particles in solution
Steric interaction of adsorbed polymer	Short-range interaction due to the overlap of adsorbed polymer layer on particles
Bridge force	Formation of a bridge of polymer binder and/or surfactant between particles
Hydration force	Overlap of hydrogen-bonded water molecules on a hydrophilic surface of the particles
Depletion	Negative adsorption of solute and polymer with less affinity for the surface than the solvent

2.2 Raw Materials

The nanostructured raw materials can be in the form of nanoparticles, nanotubes, nanofibers, and nanoplates. Depending upon the application area and requirement of a specific nano candidate, the raw material is synthesized by suitable methods.

2.2.1 Nanoparticles

There are two general approaches for the synthesis of nanomaterials and the fabrication of nanostructures: a top-down (comminution and dispersion) approach or bottom-up (nucleation and growth) approach. The decision to adopt a particular method depends on several factors, including desirable properties.

The top-down approach by attrition/milling uses macroscopic initial structures that can be externally controlled in the processing of nanostructures. Typical examples are etching through the mask, ball milling, and plastic deformation. It begins with a pattern on a larger size and then reduced to nanosize. Crushing and grinding have typically been treated as low-technology operations. Even soft organic matter can be ground by first freezing it in liquid nitrogen. The main disadvantages are polydispersity of the final particles and the introduction of defects. The product may get contaminated by the material used to make the grinding machinery. The probability of contamination increases as the particle size decreases. Crushing and grinding are common industrial processes, but the advent of nanotechnology has given rise to novel, well-controlled methods of achieving monodisperse nanoparticle generation by comminution and dispersion. One such process is electroerosion dispersion (EED); in which granulated metal is ground into a fine powder by electrical discharges—typically a few hundred volts are discharged in a microsecond. The plasma temperature in the discharge filament is 4,000–10,000 K, sufficient to melt any metal.

The bottom-up approach via vapor-phase synthesis such as spray pyrolysis,[33] inert gas condensation,[34] and liquid-phase fabrication such as solvothermal reaction,[35] sol–gel fabrication, and micellar structured media[36] includes the miniaturization of material components up to atomic level with further self-assembly process leading to the formation of nanostructures. During self-assembly, the physical forces operating at nanoscale are used to combine basic units into larger stable structures. Typical examples are quantum dot formation during epitaxial growth and formation of nanoparticles from colloidal dispersion. It begins with atoms or molecules and builds up to nanostructures. Nucleation and growth imply the first-order phase transition from an atomically dispersed phase to a solid condensed phase. During the first stage of the transition, fluctuations in the homogeneous, metastable parent phase result in the appearance of small quantities of the new phase. In order to synthesize nanoparticles by nucleation and growth method, first the atoms are dispersed or dissolved in a medium under conditions such that the dispersion is stable. Then, one or more of the external parameters is changed such that the bulk phase of the material now dispersed is stable. This could be accomplished by cooling the vapor of the material. The formation of the new bulk phase is a first-order phase transition involving nucleation. Two key challenges in this process are (1) to obtain particles that are as uniform (monodisperse) as possible, and (2) to be able to control the mean size. In the case of synthesis by chemical reaction, the key parameter is the rate of mixing. Two extreme situations yield the desired monodispersity: ultrarapid mixing of very concentrated solutions and ultraslow mixing of very dilute solutions.

2.2.2 Nanofibers

Nanofiber is the generic term describing nano-objects with two external dimensions in the nanoscale. A nanorod is a rigid nanofiber. A nanotube is a hollow nanofiber. A nanowire is an electrically conducting nanofiber. The nucleation method can be used to generate nuclei, followed by a growth stage to elongate them. Heterogeneous nucleation can be induced at the solid/gas interface by predepositing small catalytic clusters.[37] Upon addition of vapor, condensation on the clusters and growth perpendicular to the solid substratum takes place. This is used as an efficient way of synthesizing CNTs. A drawback of the method is that the preparation is almost inevitably contaminated with the catalyst. If uniform nanopores can be formed in a membrane by laser drilling or self-assembly, they can be used as templates for nanofiber formation. The material for the fiber should be deposited as a shell on the inner surface of the pores (if the goal is to make nanotubes), or should completely fill the pores (for nanorods). Nanofibers, especially nanorods, formed by either of the two previous methods can also be used as templates for making nanotubes of a different material.

2.2.3 Nanoplates

For a long time, thin coatings on a substrate have not been considered as nano-objects, but simply as thin films, because typically they have been more than 100 nm thick. Exceptions are Langmuir films, transferred to solid substrata using the Langmuir–Blodgett and Langmuir–Schaefer techniques; these films are a few nanometers thick. Laterally cohesive Langmuir films can be manipulated as free-standing objects. Nevertheless, the trend is to develop thinner functional surfaces[38] by coating or modifying bulk materials.

In Langmuir films and the Langmuir–Blodgett and Langmuir–Schaefer techniques, the precursors are molecules of general formula XP, where X is typically an apolar chain (e.g., an alkyl chain), called the *tail*, and P is a polar *head* group such as oligoethylene oxide. When spread on water, they mostly remain at the water/air interface, where they can be compressed to form two-dimensional (2D) liquid-like and solid-like arrays. The Langmuir–Blodgett technique refers to the transfer of the floating monomolecular films to solid substrata by vertically dipping them into and out of the bath. In the Langmuir–Schaefer technique, the substratum is pushed horizontally through the floating monolayer. Very stable multilayer films[39] can be assembled by making P a chelator for multivalent metal ions, which bridge lateral neighbors and/or successive layers (assembled head–head and tail–tail). Lateral stability can be increased by UV irradiation of films with an unsaturated alkyl chain (photopolymerization).

2.2.4 Graphene

Different methods have been reported for synthesis of graphene.[40] The graphene lamellae stacked to make bulk graphite are known to be weakly bound to each other. Individual sheets of graphene can actually be peeled them off graphite using an adhesive tape. Alternatively, a crystal of silicon carbide can be heated under vacuum to 1300°C; the silicon evaporates and the remaining carbon slowly reorganizes to form some graphene.

2.2.5 Carbon Nanotubes

The CNT is a seamless tube made by rolling up graphene. It is known that carbon filaments are formed by passing hydrocarbons over hot metal surfaces, especially iron and nickel. The actual nature of CNTs has been, however, established by Iijima in 1991. Multiwalled CNT consists of several concentric tubes of graphene nested inside each other. The three methods for producing CNTs[41] are the laser furnace, the carbon arc, that is, vaporizing graphitic electrodes, and plasma-enhanced chemical vapor deposition (CVD). CNTs are often closed at one or both ends by a hemisphere of fullerene. They tend to be aggravated into bundles. Normally, CNT of 20 nm diameter and 20 μm length is fabricated, which is not useful for several applications. The major challenges in CNTs are dispersing them in a liquid, reducing their length, and manipulating them into a desired position.

2.3 Processing Methods

Despite their nano dimensions, most of the processing techniques of the nanocomposites remain almost the same as in microcomposites.

Nanocomposites can be prepared by *in situ* synthesis of inorganic particles or by dispersion of fillers in a polymeric matrix.[42] Selection of the preparation technique is critical to obtain nanomaterials with suitable properties.[43] The synthesis of polymer nanocomposites usually applies bottom-up or top-down methodologies (Figure 2.4). In the bottom-up approach, precursors are used to construct and grow, from the nanometric level to well-organized structures. The building block approach is used, where nano-objects are combined to get the desirable material. It has an advantage compared to *in situ* nanoparticle formation, because at least one structural unit is well defined and usually does not have significant structural changes during the matrix formation. Chemical processes, such as sol–gel, CVD, template synthesis, or spray pyrolysis, are normally employed as bottom-up methodologies.[44] In the top-down approach, bulk material is broken down into smaller pieces using physical methods, as the dispersion layered silicates in polymer matrices.

The procedure of the mixed matrix membrane making is similar to that of the ordinary polymer membrane making. The first step is preparing a homogeneous solution of polymer and particles. For this purpose, the following methods for mixed matrix dope preparation are used:

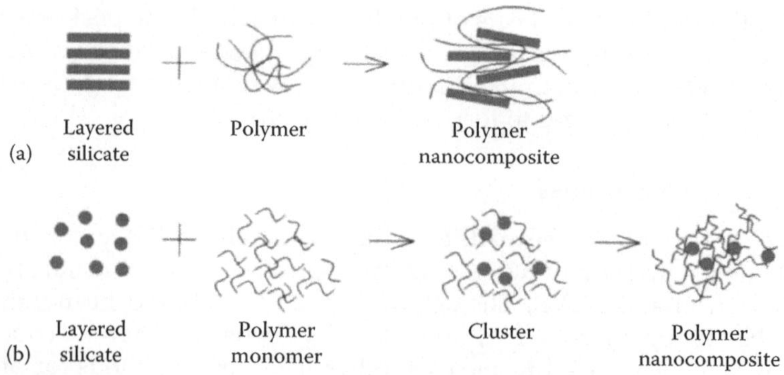

FIGURE 2.4
Schematic representation of (a) top-down and (b) bottom-up approaches.

1. Particles are dispersed into the solvent and stirred for a predetermined period of time and then the polymer is added (Figure 2.5a).
2. The polymer is dissolved in the solvent and stirred; a predetermined mass of inorganic particles is then added to the polymer solution (Figure 2.5b).
3. Particles are dispersed into the solvent and stirred for a predetermined period of time and the polymer is dissolved in a solvent separately. The particle suspension is then added to the polymeric solution (Figure 2.5c).

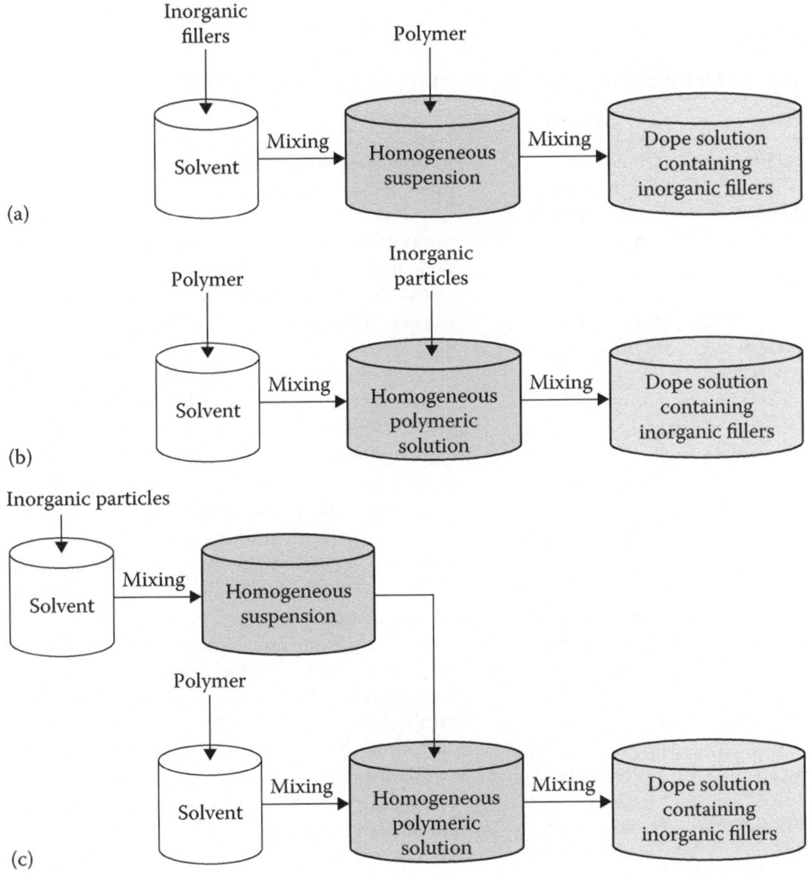

FIGURE 2.5
Different methods for mixed matrix dope preparation. (a) Particles are dispersed in solvent and stirred, subsequently polymer is added and mixed. (b) Polymer is dissolved in solvent and stirred, subsequently inorganic particle is added. (c) Particles are dispersed into solvent and stirred. Polymer is dissolved in a solvent separately. Subsequently, the particle suspension is added to the polymeric solution.

Some of the important methods for the preparation of polymer nanocomposites, including layered materials and those containing CNTs, are (1) intercalation of the polymer or pre-polymer from solution, (2) *in situ* intercalative polymerization, (3) melt intercalation, (4) direct mixture of polymer and particulates, (5) template synthesis, (6) *in situ* polymerization, and (7) sol–gel process. Table 2.2 gives the processing methods for polymer-based nanocomposite systems. Their advantages and limitations are listed in Table 2.3. Processing of CNT-reinforced polymer nanocomposites is carried out by different methods such as direct mixing, solution mixing, melt mixing, and *in situ* polymerization (Table 2.4).

TABLE 2.2

Processing Methods for Polymer-Based Nanocomposite Systems

Process	System	Procedure
Intercalation/prepolymer from solution	Clay with high-density polyethylene, etc.	Employed for layered reinforcing material in which the polymer may intercalate; mostly for layered silicates, with intercalation of the polymer or pre-polymer from solution; use of a solvent in which the polymer or pre-polymer is soluble and the silicate layers are swellable
In situ intercalative polymerization	Montmorillonate with epoxy, etc.	Encasing of the layered silicate within the liquid monomer or a monomer solution leading to formation of polymer between the intercalated sheets; polymerization by heat or radiation, by diffusion of a suitable initiator, or by a catalyst fixed through cation exchange inside the interlayer, before the swelling step
Melt intercalation	Montmorillonate with PS/polypropylene, etc.	Annealing of a mixture of the polymer and the layered host above the softening point of the polymer, statically or under shear; diffusion of polymer chains from the bulk polymer melt into the galleries between the host layers during annealing
Template synthesis	Hectorite with poly-acrylonitrile, etc.	*In situ* formation of the layered structure of the inorganic material in an aqueous solution containing the polymer; the water-soluble polymer acts as a template for the formation of layers; widely used for the synthesis of layered double hydroxide nanocomposites, but less developed for layered silicates
Mixing and *in situ* polymerization	Epoxy vinyl ester/Fe_3O_4, epoxy vinyl ester/γ-Fe_2O_3, polyacrylic acid/Ag, polyacrylic acid/Ni, etc.	Mixing of either polymer or monomer with reinforcing materials; dispersion of inorganic particles into a precursor of the polymeric matrix (monomer); polymerization of the mixture by addition of an appropriate catalyst; processing of this material by conventional molding technologies Use of ultrasonics for dispersion in epoxy systems; exposition of accelerator based gamma induced systems to 60 Co γ-ray to promote simultaneous polymerization and metal nanoparticle formation

(*Continued*)

TABLE 2.2 (*Continued*)

Processing Methods for Polymer-Based Nanocomposite Systems

Process	System	Procedure
Sol–gel process	Polyimide/SiO_2, 2-hydroxyethyl acrylate/SiO_2; polycarbonate/ SiO_2, poly(amide-imide)/TiO_2	Embedding of organic molecules and monomers on sol–gel matrices; introduction of organic groups by formation of chemical bonds leading to *in situ* formation of sol–gel matrix within the polymer and/ or simultaneous generation of inorganic/organic networks.

TABLE 2.3

Advantages and Limitations of Polymer-Based Nanocomposite Processing Methods

Process	Advantages	Limitations
Intercalation/ prepolymer from solution	Synthesis of intercalated nanocomposites based on polymers with low or even no polarity. Preparation of homogeneous dispersions of the filler	Large amount of solvents
In situ intercalative polymerization	Easy procedure, based on the dispersion of the filler in the polymer precursors	Difficult control of intragallery polymerization/ limited applications
Melt intercalation	Environmentally benign; use of polymers not suited for other processes; compatible with industrial polymer processes	Limited applications to polyolefins
Template synthesis	Large-scale production; easy procedure	Limited applications mainly in water-soluble polymers/ contaminated by side products
Sol–gel process	Simple, low processing temperature; versatile; high chemical homogeneity; rigorous stoichiometry control; high-purity products; formation of 3D polymers containing metal–oxygen bonds; single or multiple matrices; applicable specifically for the production of composite materials with liquids or with viscous fluids that are derived from alkoxides	Greater shrinkage and lower amount of voids, compared to the mixing method

Membranes utilizing nanoporous one-dimensional (1D) and 2D materials are emerging as attractive means for applications in molecular separations and related areas. Such nanotubular and nanolayered materials include CNTs, metal oxide nanotubes, layered zeolites, porous layered oxides, layered aluminophosphates, and porous graphenes. By virtue of

TABLE 2.4
Processing Methods for Polymer–CNT Nanocomposite Systems

Process	System	Procedure	Remarks
Direct mixing	Thermoset resins	Dispersion of CNTs; curing	
Solution mixing	Thermoplastic resins (PS/epoxy)	Dispersion of 0.2%–1% CNTs, (100 nm diameter, 10 μm long); removal of solvent or precipitation of polymer; curing	Modification of polymer behavior; synergistic effect; shape memory nanocomposites
Melt mixing	Polymers, N6	Mechanical mixing of CNTs with prepolymer melt followed by extrusion, injection, or compression molding	Use of 0.2%–2.0% multiwalled CNT, twin screw mixer
In situ polymerization	Polyaniline-CNT, epoxy-CNT, poly(ether ester)	Use of ultrasonics for dispersion in monomer/matrix; curing	Preparation of the polymer with CNT; good chemical bonding
Others	PP-CNT, etc.	Solid-state mechanochemical pulverization; blending + sonication; melt blending	0–10 wt.% CNTs

their unique shape, size, and structure, these materials possess transport properties, which are advantageous for membrane and thin-film applications. These materials also have a different chemistry compared to conventional porous 3D materials, due to large, chemically active, external surface area. This feature also necessitates the development of innovative strategies to process these materials into membranes and thin films with high performance. Porous 1D nanotubular[45–47] and 2D layered[48–50] materials have been suggested as leading to membranes with unique transport properties that could overcome the limitations of isotropic inorganic fillers in nanocomposite membranes. Figure 2.6[51] illustrates the schematic diagram of composite membranes containing 1D and 2D materials. In Figure 2.6a, the nanotube materials are shown as vertically aligned and completely spanning the thickness of the membrane. By realizing such an ideal structure through an appropriate membrane-making process, favorable properties of nanotubes such as fast molecular transport can be efficiently achieved and utilized. The potentially tunable functionality of the nanotube mouths, as well as inner and outer surfaces, can be used to explore the range of separation applications and compatibility with various types of matrix materials. Figure 2.6b shows a schematic diagram of a membrane incorporating high-aspect ratio layered (2D) materials. In comparison with isotropic fillers, the high aspect ratio of the layers or *flakes* requires longer and tortuous paths for the larger molecule that

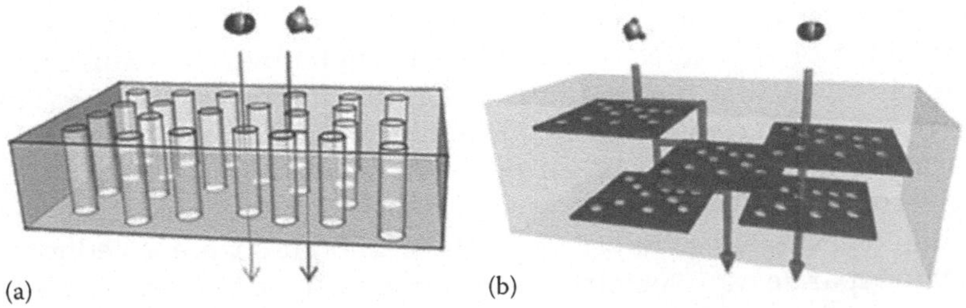

FIGURE 2.6
Schematic diagrams of composite membranes incorporating (a) 1D nanotubular materials and (b) 2D nanoplatelets. (Reprinted from *J. Membr. Sci.*, 441, Kim, W.-G. et al., Nanoporous layered silicate AMH3/cellulose acetate nanocomposite membranes for gas separations, 129–136, Figure 1, Copyright 2013, with permission from Elsevier.)

cannot easily pass through the pores, and therefore effectively decrease the permeability of the larger molecule.[52] At the same time, the nanoscopic thickness of the flakes allows fast permeation of the smaller molecule. Membranes containing nanoscopically thin flakes can allow high separation performance with a much lower loading of the flakes than required for isotropic fillers. From the viewpoint of membrane fabrication and scale-up, the nanoscopically thin layered flakes appear favorable because they can be incorporated in ultrathin membranes such as the skin layers of hollow-fiber membranes.[53]

Organic–inorganic interfacial defects can affect the overall membrane properties. The polymer–inorganic nanocomposite membranes can be prepared by the sol–gel process, phase inversion method, which is also called the *in situ* blending method, and the *in situ* or interfacial polymerization.

2.3.1 Phase Inversion Method

The phase inversion process is one of the most common methods for the preparation of polymeric membranes. The casting solution is cast as a thin film and immersed into a nonsolvent coagulation bath. Due to the solvent and nonsolvent exchange, phase separation of the casting solution is induced to form a polymer-rich phase and polymer-lean phase, and the membrane is therefore obtained.[54] Studies on composite ultrafiltration membranes based on poly(vinylidene fluoride) (PVDF) and alumina (Al_2O_3) materials by the phase inversion method and characteristics such as the membrane hydrophilicity porosity, protein retention, and surface morphologies have been reported.[55] The enhancement in permeation flux of the membrane was attributed to a surface hydrophilicity due to the added hydrophilic inorganic nanosized Al_2O_3 particles. Studies have been carried out on nanocomposite membranes containing polyethersulfone (PES) and organically modified montmorillonite (OMMT).[56] The membranes were prepared by a combination of solution dispersion and wet-phase inversion methods. The effect of OMMT addition on the properties and performance of fabricated nanofiltration membranes was investigated. It was observed that the addition of OMMT improved the mechanical properties and thermal stability of the membranes. The pure water flux, permeation, and rejection of nitrophenols were significantly improved. Hybrid composite membranes were prepared[57] by incorporating nano-TiO_2 into regenerated cellulose (RC). The membranes were tested for their ability to separate caprolactam– water mixtures by pervaporation. Among all the prepared membranes, an RC-TiO_2 inorganic–polymer hybrid membrane containing 5 wt.% TiO_2 exhibited good pervaporation performance with a flux of 1787.3 g m^{-2} h^{-1} and a separation factor of 55091.7 at 328 K for 50 wt.% caprolactam. Addition of nano-silica in polysulfone (PS) resulting into PS/silica (SiO_2) nanocomposite membrane prepared by phase inversion method provides thermal stability and enhancement of the gas permeability of PS.[58]

2.3.2 Sol–Gel Method

The sol–gel technique is used for the preparation of polymer–inorganic nanocomposite membranes with highly homogeneous and controlled morphology. In this method, organic monomers, oligomers, or polymers and inorganic nanoparticle precursors are mixed together in the solution. The inorganic precursors then hydrolyze and condense into well-dispersed nanoparticles in the polymer matrix. The concentrations of organic and inorganic components are easy to control in the solution; the reaction conditions are moderate—usually room temperature and ambient pressure—which allows the formation of inorganic frameworks under mild conditions and the incorporation of minerals into polymers, resulting in an increased chemical, mechanical, and thermal stability.[59] A hybrid ultrafiltration membrane of PS with titanium dioxide (TiO_2) as the inorganic phase shows an improvement in porosity and thermal stability.[60] Inorganic–organic polymer (hybrid PEO–[$Si(OCH_3)_3]_2$) was synthesized[61] by coupling *N*-(3-[trimethoxysilyl]propyl)ethylenediamine (A-1120) to end-capped PEO-400. Based on this hybrid, a series of positively charged membranes were prepared by the sol–gel process. From the coating time control and the concentration of sol, a desired hybrid membrane for nanofiltration can be obtained. The effect of silica particles prepared by the sol–gel technique on the gas transport properties of a nanocomposite polyimide film shows higher permeability (N_2, O_2, CO_2, H_2, and CH_4) coefficients and hydrogen permselectivities than the polyimide.[62] Nanocomposite membranes for gas separations are traditionally composed of zeolites dispersed in a polymer matrix in order to improve permeability and selectivity. A limited number of possible zeolite structures limit the potential of zeolites for use in nanocomposites. Metal organic frameworks (MOFs), a new class of microporous materials, are an attractive alternative to the use of zeolites in nanocomposite membranes. The advantage of MOFs over well-known nanoporous materials is the ability to tune their physical and chemical properties during synthesis by changing the combination of metals and organic linkers. Rational combination of different metal ions and organic linkers in the synthesis of MOFs results in materials with various pore sizes and connectivities, offering theoretically an infinite number of possible structures. An MOF of copper and terephthalic acid (CuTPA) can be synthesized, characterized, and used to make nanocomposite membranes.[63] The particles are dispersed in poly(vinyl acetate) giving defect-free zeolite–polymer nanocomposite membranes.[64] Significant increase in permeability and the desired penetrant selectivity enhancement are achieved for pure helium, carbon dioxide, oxygen, nitrogen, and methane permeation experiments. Polymer–inorganic nanocomposite membranes prepared by cross-linking polyvinyl alcohol (PVA), maleic acid (MA), and silica by aqueous sol–gel route reveal that silica nanoparticles (<10 nm) significantly reduce the swelling of the membrane.[65] Pervaporation under various operating conditions has been carried out to evaluate the separation performance of aqueous salt solutions through the hybrid PVA/MA/silica membrane. It is seen that salt rejection remains high (up to 99.9%) under all operating conditions, indicating that the salt rejection performance of the hybrid PVA/MA/silica membrane is independent of the operating conditions due to the nonvolatile nature of sodium chloride. It is possible to prepare negatively charged inorganic–organic hybrid membranes[66] by the sol–gel and oxidation processes of 3-(mercaptopropyl)trimethoxysilane (MPTS). Due to the formation of an inorganic Si–O–Si network during the sol–gel process, the membranes can withstand as high as 250°C temperature displaying extraordinary thermal stability compared to polymeric membrane.

2.3.3 *In Situ*/Interfacial Polymerization

In this method, the nanoparticles are well mixed with organic monomers and the monomers are polymerized. There are often some functional groups such as hydroxyl or carboxyl on the surface of the inorganic particles, which can generate radicals, cations, or anions under high-energy radiation, plasma, or other circumstances to initiate the polymerization of the monomers on their surface. The interfacial polymerization of polyamide is a common method of preparing thin-film composite (TFC) membrane. In case of interfacial polymerization of thin-film nanocomposite (TFN) membranes,[67] nanoparticles are embedded throughout the polyamide thin-film layer of an interfacial composite membrane (Figure 2.7). Synthesized zeolite nanoparticles, characterized by a superhydrophilic and negatively charged three-dimensional (3D) molecular sieve pore network, are used as the dispersed nanophase. Super-hydrophilic, negatively charged, molecular

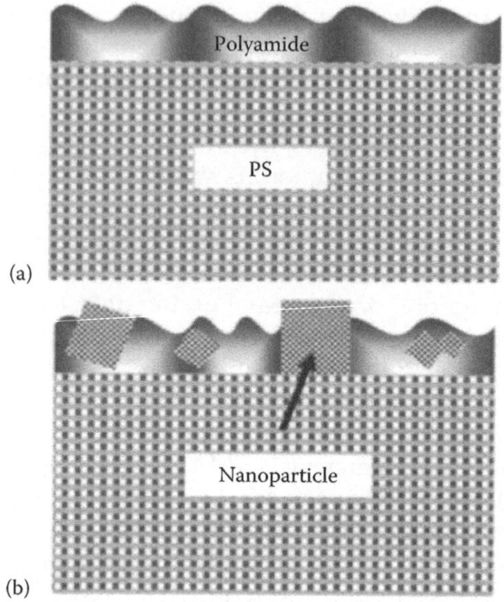

FIGURE 2.7
Conceptual illustration of (a) TFC and (b) TFN membrane structures. (Reprinted from *J. Membr. Sci.*, 294, Jeong, B.-H. et al., Interfacial polymerization of thin film nanocomposites: A new concept for reverse osmosis membranes, 1–7, Figure 1, Copyright 2007, with permission from Elsevier.)

sieve zeolite nanoparticles provide preferential flow paths for water permeation while maintaining high solute rejection through combination of steric and Donnan exclusion.

2.4 Challenges

Polymeric membranes modified by adding nanoparticles possess superior properties of both organic and inorganic membranes such as good hydrophilicity, selectivity, permeability, mechanical strength, and thermal and chemical stability. The challenges are proper dispersion of the fillers, because of the difference between the polymer and inorganic phase properties. Without proper dispersion and distribution of the fillers, the high surface area is compromised. There is a need to develop better polymer–inorganic nanocomposite hybrid membranes for different applications in separation processes.

References

1. Thostenson, E., Li, C., and Chou, T. 2005. Review nanocomposites in context. *Compos. Sci. Technol.* 65: 491–516.
2. Camargo, P.H.C., Satyanarayana, K.G., and Wypych, F. 2009. Nanocomposites: Synthesis, structure, properties and new application opportunities. *Mater. Res.* 12: 1–39.
3. Aruna, S.T. and Rajam, K.S. 2003. Synthesis, characterisation and properties of Ni/PSZ and Ni/YSZ nanocomposites. *Scr. Mater.* 48: 507–512.
4. Giannelis, E.P. 1996. Polymer layered silicate nanocomposites. *Adv. Mater.* 8: 29–35.
5. Sternitzke, M. 1997. Review: Structural ceramic nanocomposites. *J. Eur. Ceram. Soc.* 17: 1061–1082.
6. Peigney, A., Laurent, C.H., Flahaut, E., and Rousset, A. 2000. Carbon nanotubes in novel ceramic matrix nanocomposites. *Ceram. Int.* 26: 677–683.
7. Alexandre, M. and Dubois, P. 2000. Polymer-layered silicate nanocomposites: Preparation, properties and uses of a new class of materials. *Mater. Sci. Eng.* 28: 1–63.
8. Thostenson, E.T., Ren, Z., and Chou, T.W. 2001. Advances in the science and technology of carbon nanotubes and their composites: A review. *Compos. Sci. Technol.* 61: 1899–1912.
9. Pokropivnyi, V.V. 2002. Two-dimensional nanocomposites: Photonic crystals and nanomembranes (review). II: Properties and applications. *Powder Metall. Met. Ceram.* 41: 369–381.
10. Kickelbick, G. 2003. Concepts for the incorporation of inorganic building blocks into organic polymers on a nanoscale. *Prog. Polym. Sci.* 28: 83–114.
11. Fischer, H. 2003. Polymer nanocomposites: From fundamental research to specific applications. *Mater. Sci. Eng.* 23: 763–772.
12. Andrews, R. and Weisenberger, M.C. 2004. Carbon nanotube polymer composites. *Curr. Opin. Solid State Mater. Sci.* 8: 31–37.
13. Wang, C., Guo, Z.X., Fu, S., Wu, W., and Zhu, D. 2004. Polymers containing fullerene or carbon nanotube structures. *Prog. Polym. Sci.* 29: 1079–1141.
14. Jordan, J., Jacob, K.I., Tannenbaum, R., Sharaf, M.A., and Jasiuk, I. 2005. Experimental trends in polymer nanocomposites: A review. *Mater. Sci. Eng. A.* 393: 1–11.
15. Choi, S.M. and Awaji, H. 2005. Nanocomposites: A new material design concept. *Sci. Technol. Adv. Mater.* 6: 2–10.
16. Xie, X.L., Mai, Y.W., and Zhou, X.P. 2005. Dispersion and alignment of carbon nanotubes in polymer matrix: A review. *Mater. Sci. Eng. R.* 49: 89–112.
17. Nakahira, A. and Niihara, K. 1992. Structural ceramics-ceramic nanocomposites by sintering method: Roles of nano-size particles. *J. Ceram. Soc. Jpn.* 100: 448–453.
18. Ferroni, L.P., Pezzotti, G., Isshiki, T., and Kleebe, H.J. 2001. Determination of amorphous interfacial phases in Al2O3/SiC nanocomposites by computer-aided high-resolution electron microscopy. *Acta Mater.* 49: 2109–2113.
19. She, J., Inoue, T., Suzuki, M., Sodeoka, S., and Ueno, K. 2000. Mechanical properties and fracture behavior of fibrous Al2O3/SiC ceramics. *J. Eur. Ceram. Soc.* 20: 1877–1881.

20. Tjong, S.C. and Wang, G.S. High-cycle fatigue properties of Al-based composites reinforced with in situ TiB2 and Al2O3 particulates. *Mater. Sci. Eng. A* 386: 48–53.
21. Sanchez, C. and Ribot, F. 1994. Design of hybrid organic inorganic materials synthesized via sol gel chemistry. *New J. Chem.* 18: 1007–1047.
22. Cong, H., Radosz, M., Towler, B.F., and Shen, Y. 2007. Polymer–inorganic nanocomposite membranes for gas separation. *Sep. Purif. Technol.* 55: 281–291.
23. Guizard, C., Bac, A., Barboiu, M., and Hovnanian, N. 2001. Hybrid organic-inorganic membranes with specific transport properties—Applications in separation and sensors technologies. *Sep. Purif. Technol.* 25: 167–180.
24. Sanchez, C., Soler-Illia, G.J.A.A., Ribot, F., and Grosso, D. 2003. Design of functional nano-structured materials through the use of controlled hybrid organic–inorganic interfaces. *C. R. Chimie* 6: 1131–1151.
25. Aroon, M.A., Ismail, A.F., Matsuura, T., and Rahmati, M.M.M. 2010. Performance studies of mixed matrix membranes for gas separation: A review. *Sep. Purif. Technol.* 75: 229–242.
26. Chung, T.S., Ying Jiang, L., and Kulprathipanja, L.S. 2007. Mixed matrix membranes (MMMs) comprising organic polymers with dispersed inorganic fillers for gas separation. *Prog. Polym. Sci.* 32: 483–507.
27. Moore, T.T. and Koros, W.J. 2005. Non-ideal effects in organic-inorganic materials for gas separation membranes. *J. Mol. Struct.* 739: 87–98.
28. Husain, S. and Koros, W. J. 2007. Mixed matrix hollow fiber membranes made with modified HSSZ-13 zeolite in polyetherimide polymer matrix for gas separation. *J. Membr. Sci.* 288: 195–207.
29. Ng, L.Y., Mohammad, A.W., Leo, C.P., and Hilal, N. 2010. Polymeric membranes incorporated with metal/metal oxide nanoparticles: A comprehensive review. *Desalination.* 308: 15–33.
30. Schaep, J., Vandeeasteele, C., Leysen, R., and Doyen, W. 1998. Salt retention of Zirfon® membranes. *Sep. Purif. Technol.* 14: 127–131.
31. Gilbert, B., Ono, R.K., Ching, K.A., and Kim, C.S. 2009. The effects of nanoparticle aggregation processes on aggregate structure and metal uptake. *J. Colloid. Interface Sci.* 339: 285–295.
32. Yu, L.-Y., Shen, H.-M., and Xu, Z.-L. 2009. PVDF-TiO_2 composite hollow fiber ultrafiltration membranes prepared by TiO_2 sol-gel method and blending method. *J. Appl. Polym. Sci.* 113: 1763–1772.
33. Gröhn, A.J., Pratsinis, S.E., Sánchez-Ferrer, A., Mezzenga, R., and Wegner, K. 2014. Scale-up of nanoparticle synthesis by flame spray pyrolysis: The high-temperature particle residence time. *Ind. Eng. Chem. Res.* 53: 10734–10742.
34. Lee, D.-W., Tolochko, O.V., Turaev, F.R., Kim, D., and Kim, B.-K. 2009. Synthesis and characterization of WS2 nanoparticles by chemical vapor condensation. *J. Nanosci. Nanotechnol.* 9: 1–6.
35. Singh, A., Kumar, R., Malhotra, N., and Suman. 2012. Preparation of ZnO nanoparticles by solvothermal process. *Int. J. Sci. Emerg. Technol. Latest Trends* 4: 49–53.
36. Mohapatra, M. and Anand, S. 2010. Synthesis and applications of nano-structured iron oxides/hydroxides a review. *Int. J. Eng. Sci. Technol.* 2: 127–146.
37. Singh, C., Quested, T., Boothroyd, C.B., Thomas, P., Kinloch, I.A., Abou-Kandil, A.I., and Windle, A.H. 2002. Synthesis and characterization of carbon

nanofibers produced by the floating catalyst method. *J. Phys. Chem. B* 106: 10915–10922.
38. Yu, H., Zhang, Q., Liu, H., Dahl, M., Joo, J.B., Li, N., Wang, L., and Yin, Y. 2014. Thermal synthesis of silver nanoplates revisited: A modified photochemical process. *ACS Nano* 8: 10252–10261.
39. Robitaille, L. and Leclerc, M. 1994. Synthesis, characterization, and Langmuir-Blodgett films of fluorinated polythiophenes. *Macromolecules* 27: 1847–1851.
40. Avouris, P. and Dimitrakopoulos, C. 2012. Graphene: Synthesis and applications. *Mater. Today* 15: 86–97.
41. Prasek, J., Drbohlavova, J., Chomoucka, J., Hubalek, J., Jasek, O., Adam, V., and Kizek, R. 2011. Methods for carbon nanotubes synthesis—Review. *J. Mater. Chem.* 21: 15872–15884.
42. Kickelbick, G. 2007. *Hybrid Materials: Synthesis, Characterization, and Applications*, Wiley-VCH, Weinheim, Germany.
43. Ruiz-Hitzky, E., Aranda, P., Darder, M., and Ogawa, M. 2011. Hybrid and biohybrid silicate based materials: Molecular vs block-assembling bottom-up processes. *Chem. Soc. Rev.* 40: 801–828.
44. Ariga, K. and Nalwa, H.S. 2009. *Bottom-up Nanofabrication: Supramolecules-I*, American Scientific Publications, Stevenson Ranch, CA.
45. Cong, H., Zhang, J., Radosz, M., and Shen, Y. 2007. Carbon nanotube composite membranes of brominated poly(2,6-diphenyl-1,4-phenyleneoxide) for gas separation. *J. Membr. Sci.* 294: 178–185.
46. Ismail, A.F., Goh, P.S., Sanip, S.M., and Aziz, M. 2009. Transport and separation properties of carbon nanotube-mixed matrix membrane. *Sep. Purif. Technol.* 70: 12–26.
47. Kang, D.Y., Tong, H.M., Zhang, J., Choudhury, R.P., Sholl, D.S., Beckham, H.W., Jones, C.W., and Nair, S. 2012. Single-walled aluminosilicate nanotube/poly (vinylalcohol) nanocomposite membranes. *ACS Appl. Mater. Interfaces.* 4: 965–976.
48. Jeong, H.K., Krych, W., Ramanan, H., Nair, S., Marand, E., and Tsapatsis, M. 2004. Fabrication of polymer/selective-flake nanocomposite membranes and their use in gas separation. *Chem. Mater.* 16: 3838–3845.
49. Galve, A., Sieffert, D., Vispe, E., Téllez, C., Coronas, J., and Staudt, C. 2011. Copolyimide mixed matrix membranes with oriented microporous titanosilicate JDF-L1sheet particles. *J. Membr. Sci.* 370: 131–140.
50. Zornoza, B., Gorgojo, P., Casado, C., Téllez, C., and Coronas, J. 2011. Mixed matrix membranes for gas separation with special nanoporous fillers. *Desalination Water Treat.* 27: 42–47.
51. Kim, W.-G., Lee, J.S., Bucknall, D.G., Koros, W.J., and Nair, S. 2013. Nanoporous layered silicate AMH-3/cellulose acetate nanocomposite membranes for gas separations. *J. Membr. Sci.* 441: 129–136.
52. Ray, S.S. and Okamoto, M. 2003. Polymer/layered silicate nanocomposites: A review from preparation to processing. *Prog. Polym. Sci.* 28: 1539–1641.
53. Johnson, J.R. and Koros, W.J. 2009. Utilization of nanoplatelets inorganic-inorganic hybrids eparation materials: Separation advantages and formation challenges. *J. Taiwan Inst. Chem. Eng.* 40: 268–275.
54. Dong, C., He, G., Li, H., Zhao, R., Han, Y., and Deng, Y. 2012. Antifouling enhancement of poly(vinylidene fluoride) microfiltration membrane by adding $Mg(OH)_2$ nanoparticles. *J. Membr. Sci.* 387: 40–47.

55. Yan, L., Li, Y.S., and Xiang, C. B. 2005. Preparation of poly(vinylidene fluoride) (PVDF) ultrafiltration membrane modified by nano-sized alumina (Al_2O_3) and its antifouling research. *Polymer* 46: 7701–7706.
56. Ghaemi, N., Madaeni, S.S., Alizadeh, A., Rajabi, H., and Daraei, P. 2011. Preparation, characterization and performance of polyethersulfone/organically modified montmorillonite nanocomposite membranes in removal of pesticides. *J. Membr. Sci.* 382: 135–147.
57. Zhu, T., Lin, Y., Luo, Y., Hu, X., Lin, W., Yu, P., and Huang, C. 2012. Preparation and characterization of TiO_2—Regenerated cellulose inorganic–polymer hybrid membranes for dehydration of caprolactam. *Carbohydr. Polym.* 87: 901–909.
58. Ahn, J., Chunga, W.J., Pinnau, I., and Guiver, M.D. 2008. Polysulfone/silica nanoparticle mixed-matrix membranes for gas separation. *J. Membr. Sci.* 314: 123–133.
59. Cho, J.W. and Sul, K.I. 2001. Characterization and properties of hybrid composites prepared from poly (vinylidene fluoride–tetrafluoroethylene) and SiO_2. *Polymer* 42: 727–35.
60. Yang, Y. and Wang, P. 2006. Preparation and characterizations of a new PS/ TiO_2 hybrid membranes by sol–gel process. *Polymer* 47: 2683–2688.
61. Wu, C., Xu, T., and Yang, W. 2003. Fundamental studies of a new hybrid (inorganic–organic) positively charged membrane: Membrane preparation and characterizations. *J. Membr. Sci.* 216: 269–278.
62. Joly, A., Goizet, S., Schrotter, J.C., Sanchez, J., and Escoubes, M. 1997. Sol-gel polyimide-silica composite membrane: Gas transport properties. *J. Membr. Sci.* 130: 63–74.
63. Banerjee, R., Furukawa, H., Britt, D., Knobler, C., O'Keeffe, M., and Yaghi, O.M. 2009. Control of pore size and functionality in isoreticular zeolitic imidazolate frameworks and their carbon dioxide selective capture properties. *J. Am. Chem. Soc.* 131: 3875–3877.
64. Zimmerman, C.M., Singh, A., and Koros, W.J. 1997. Tailoring mixed matrix composite membranes for gas separations. *J. Membr. Sci.* 137: 145–154.
65. Xie, Z., Ng, D., Hoang, M., Duong, T., and Gray, S. 2011. Separation of aqueous salt solution by pervaporation through hybrid organic–inorganic membrane: Effect of operating conditions. *Desalination* 273: 220–225.
66. Wu, C., Xu, T., and Yang, W. 2003. A new inorganic–organic negatively charged membrane: Membrane preparation and characterizations. *J. Membr. Sci.* 224: 117–125.
67. Jeong, B.-H., Hoek, E.M.V., Yan, Y., Subramani, A., Huang, X., Hurwitz, G., Ghosh, A.K., and Jawor, A. 2007. Interfacial polymerization of thin film nanocomposites: A new concept for reverse osmosis membranes. *J. Membr. Sci.* 294: 1–7.

3

Characterization of Nanocomposite Membranes

3.1 Introduction

Any membrane developed under a standard set of conditions, such as temperature and relative humidity needs to be characterized. Any small change in the conditions of membrane preparation may lead to a significant change in membrane structure and membrane morphology, which in turn may affect the membrane performance. The ultimate goal is to get membrane morphology to achieve desired performance. Membrane characterization is carried out to evaluate the structural membrane properties, such as pore size and pore size distribution, and the separation properties.

The methods of characterization for conventional membranes remain equally applicable for the nanocomposite membranes. The micrographic and spectroscopic methods become more important in terms of characterizing the nanomaterials embedded in the host matrix (polymer/ceramic).

The characterization methods can be classified[1] into the following categories:

- Physical methods to determine pore size and pore size distribution
- Micrographic methods to have photographical images
- Spectroscopic methods to know the membrane structure in its molecular level
- Drop shape analysis (contact angle measurements) to ensure hydrophilicity/hydrophobicity of membrane surface
- Zeta potential measurement to ensure surface charge on membrane surface
- Methods to obtain bulk properties of membranes such as thermal and mechanical properties

3.2 Methods for Measurement of Pore Size and Pore Size Distribution

The following methods are used for the measurement of pore size and pore size distribution:

- Bubble gas transport method
- Mercury intrusion porosimetry
- Adsorption–desorption method (Barrett–Joyner–Halenda [BJH] method)
- Liquid–solid equilibrium method (thermoporometry)
- Gas–liquid equilibrium method (permporometry)
- Permeability method
- Solute rejection method

3.2.1 Bubble Gas Transport Method

The bubble gas transport method is based on the measurement of the pressure to blow air through a water-filled porous membrane. The method so developed has been used to characterize membranes and is called the bubble point method. The principle of the bubble point method is shown schematically in Figure 3.1. It can be seen from Figure 3.1 that the liquid on the top of the membrane wets the membrane surface. The bottom of the membrane is in contact with air and the air pressure is gradually increased until bubbles of air penetrate through the membrane.

An air bubble penetrates through a pore of radius r_p when the transmembrane pressure difference $\Delta P = P_1 - P_2$ given by the following Laplace equation is reached:

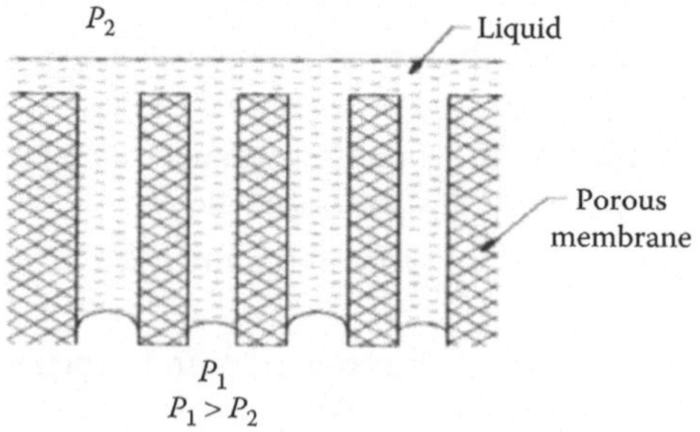

FIGURE 3.1
Principle of the bubble point method.

$$r_p = \frac{2\gamma \cos\theta}{\Delta P} \tag{3.1}$$

where:
γ is the surface tension at the liquid/air interface
θ is the contact angle

Penetration will first occur through the largest pores, and because the pressure difference is known, the pore radius can be calculated from Equation 3.1. It is also possible to obtain pore size distribution by performing this technique by a stepwise increase of pressure. The method has been improved for both liquid–gas interfaces and liquid–liquid ones allowing the evaluation of pore sizes for a range of porous materials. This method is an easy, fast, and inexpensive method to determine the maximum pore size and pore size distribution of membranes.[2] Though only active pores are measured, the disadvantage is that different results are obtained with different liquids. Liquid displacement is similar to gas bubble transport method, where a liquid is pressurized to displace the liquid in the pores. The disadvantage is that swelling is probable because of stagnant liquid in the pores, which can affect the pore size.

3.2.2 Mercury Intrusion Porosimetry

Mercury intrusion porosimetry is used extensively for the characterization of various aspects of porous media, including porous membranes and powders. It is applicable to pores ranging from 30 to 900°A in diameter. It involves placing the sample in a special sample cup (penetrometer), surrounding the sample with mercury. Mercury is a nonwetting liquid to most materials and resists entering voids, doing so only when pressure is applied. The pressure at which mercury enters a pore is inversely proportional to the size of the opening to the void. As mercury is forced to enter pores within the sample material, it is depleted from a capillary stem reservoir connected to the sample cup. The incremental volume depleted after each pressure change is determined by measuring the change in capacitance of the stem. This intrusion volume is recorded with the corresponding pressure or pore size. By this technique, both pore size and pore size distribution can be determined. The relationship of pressure and pore size is given by the Laplace equation (3.1). As mercury does not wet the membrane (because its contact angle is greater than 90° and cos θ will have a negative value), Equation 3.1 is modified as follows:

$$r_p = -\left(\frac{2\gamma \cos\theta}{P}\right) \tag{3.2}$$

The contact angle of mercury with polymeric material is often 141.3° and the surface tension at the Hg/air interface is 0.48 N/m. Hence, Equation 3.2 can be written as

$$r_p = \frac{7492}{P} \tag{3.3}$$

where:
r_p is expressed in nanometer (nm)
P is expressed in bar

The volume of mercury can be determined very accurately, which can result in the precise estimation of pore size distribution. In the above equation, it is assumed that the membranes have capillary pores. As in general, it is not always the case and morphology constant must be introduced. Also, very high pressure should be avoided as it may damage the porous structure and lead to an erroneous pore size distribution. Some of the disadvantages of this technique are as follows: (1) It needs high pressure that could damage the surface and (2) it measures all the pores present in the structure, including dead-end pores. The technique is not widely used due to the limitations.

3.2.3 Adsorption–Desorption Method

Gas adsorption is a popular and commonly used method for characterization of surface and structural properties of porous materials allowing the determination of their surface area, pore volume, pore size distribution, and adsorption energy distribution. Nitrogen is often used for the adsorbent gas, but other adsorbents such as argon and benzene are also used. According to this method, adsorption isotherm (the amount of adsorbed gas vs. the relative pressure) is drawn and the data are analyzed by assuming capillary condensation. The relative pressure is defined as the ratio of the adsorbent pressure to saturation vapor pressure of the adsorbent. The vapor pressure, p, of the adsorbent liquid in the pore of radius r_p is given by the following Kelvin equation:

$$\ln\left(\frac{p}{p_0}\right) = -\left(\frac{2\gamma V\cos\theta}{r_p RT}\right) \tag{3.4}$$

where:
p_0 is the saturation vapor pressure
γ is the surface tension of the adsorbent liquid
V is the molar volume of the adsorbent liquid
R is the universal gas constant
T is the absolute temperature
θ is the contact angle

Assuming $\theta = 0°$, the above equation becomes for the liquid nitrogen

$$r_p = \frac{-4.1}{\ln(p/p_0)} \tag{3.5}$$

Assuming further all the pores whose radii are smaller than r_p are filled at a given relative pressure p/p_0, the cumulative pore volume curve versus r_p can be drawn. Often, the thickness of adsorbed layer t is added to r_p to calculate more precise pore radius. Ceramic membranes often give better results because of their uniform structure and the membrane is less susceptible to capillary forces. Dead-end pores that do not contribute toward transport are also measured with this technique.

3.2.4 Liquid–Solid Equilibrium Method (Thermoporometry)

Thermoporometry is based on the calorimetric measurement of a solid–liquid transition in a porous material and can be applied to determine the pore size in porous membranes.[3–5] The temperature at which the liquid in the pore freezes depends on the pore size. As the pore size decreases, the freezing point decreases. Each pore gets its own specific freezing point. For cylindrical pores containing water,

$$r_p = 0.68 - \left(\frac{32.33}{\Delta T}\right) \tag{3.6}$$

where:

r_p is the pore radius
ΔT is the extent of undercooling

As the pore radius decreases, the extent of undercooling increases. The heat effect of the liquid–solid transition is measured by means of a differential scanning calorimeter (DSC). All pores are measured with this technique including the dead-end ones.

3.2.5 Gas–Liquid Equilibrium Method (Permporometry)

Permporometry, unlike thermoporometry, characterizes only the active pores.[6,7] It is based on the blockage of the pores by means of a condensable gas, linked with the simultaneous measurement of gas flux through the membrane. This blockage is based on the same principle of capillary condensation as adsorption–desorption hysteresis. It is important that the vapor should be inert and should not swell the membrane; otherwise, the pore size will be affected. It is important to note that permporometry measures the active pores, whereas adsorption–desorption and thermoporometry measure active, dead-end and even smaller pores in the sublayer. In case the membrane is asymmetric in nature, where the transport is governed by the thin top layer, this technique can give the information on pore size and pore size distribution of the top layer.

3.2.6 Permeability Method

Assuming the pores to be capillary in nature, the pore size can be determined by measuring the flux through a membrane at a constant pressure using Hagen–Poiseuille equation:

$$J = \left(\frac{\varepsilon r^2 \Delta P}{8 \eta \tau \Delta x} \right) \tag{3.7}$$

where:

J is the flux through the membrane at a driving force of $\Delta P/\Delta x$, with ΔP being the pressure difference

Δx is the membrane thickness

Other parameters are the pore radius (r), the liquid viscosity (η), the surface porosity of the membrane (ε), and the tortuosity factor (τ). The pore size distribution can be obtained by varying the pressure, that is, by a combination of bubble point and permeability methods. It is not necessary that the liquid should wet the membrane.

The Hagen–Poiseuille equation assumes the pores to be cylindrical, which may not be the case. In that case, Kozeny–Carman equation can be used. It is assumed in this relation that the pores are interstices between close-packed spheres. The flux is given by

$$J = \frac{\varepsilon^3 \Delta P}{\left[K \eta S^2 (1-\varepsilon)^2 \Delta x \right]} \tag{3.8}$$

where:

K is the membrane constant, called the Kozeny–Carman constant, which is dependent on the pore shape and tortuosity

ε is the porosity

S is the specific surface area

The permeability method is widely used for both microfiltration and ultrafiltration membranes. This method is entirely based upon the assumptions taken toward the pore geometry which is largely unknown, and hence, the experimental results are difficult to interpret.

3.2.7 Solute Rejection Method

This is the method frequently used for industrial assessment of membranes. Usually, membrane manufacturers use the concept of *cutoff* to characterize the ulrafiltration membranes. Cutoff is defined as the molecular weight that is 90% rejected by the membrane. The membrane has a cutoff value of 40,000, which implies that all solutes with a molecular weight greater than 40,000 are rejected by the membrane by more than 90%. However, it is not possible to define the separation characteristics of a membrane by this single parameter, that is, molecular weight of solute, because the parameters such as shape, flexibility of macromolecular solute, and its interaction with the membrane material are all important to be taken into account. However, concentration polarization and membrane fouling can affect the separation performance of the membrane. Cutoff values are often expressed in different ways under different test conditions (pressure, geometry of test cell, type and concentration of solute, cross-flow velocity, and more importantly molecular-weight distribution of solute), which makes it difficult to hold a comparison of the results obtained.

3.3 Micrographic Methods

Micrographic methods are more pronounced where the structure of nanocomposites usually consists of the matrix material containing the nanosized reinforcement components in the form of particles, whiskers, fibers, nanotubes, and so on. These characterization techniques help to understand the correlation between structure and performance of a particular nanocomposite membrane.

- Scanning electron microscopy (SEM)
- Transmission electron microscopy (TEM)
- Atomic force microscopy (AFM)

3.3.1 Scanning Electron Microscopy

A scanning electron microscope (SEM) is a type of electron microscope that images a sample by scanning it with a high-energy beam of electrons. The electrons interact with the atoms that make up the sample-producing signals that contain information about the sample's surface topography, composition, and other properties such as electrical conductivity.

The types of signals produced by a SEM include secondary electrons, back-scattered electrons (BSE), characteristic X-rays, light (cathodoluminescence), and specimen current and transmitted electrons. Secondary electron detectors are common in all SEMs. The signals result from interactions of the electron beam with atoms at or near the surface of the sample. In the most common or standard detection mode, secondary electron imaging, the SEM can produce very high-resolution images of a sample surface, revealing details of about 5–20 nm in size. Due to the very narrow electron beam, SEM micrographs have a large depth of field yielding a characteristic three-dimensional (3D) appearance useful for understanding the surface structure of a sample. A wide range of magnification is possible, from about 10 times (about equivalent to that of a powerful hand lens) to more than 500,000 times, about 250 times the magnification limit of the best light microscopes. BSEs are beam electrons that are reflected from the sample by elastic scattering. They are often used in analytical SEM along with the spectra made from the characteristic X-rays. Because the intensity of the BSE signal is strongly related to the atomic number (Z) of the specimen, BSE images can provide information about the distribution of different elements in the sample.

In a typical SEM, an electron beam is thermionically emitted from an electron gun fitted with a tungsten filament cathode. Tungsten is normally used in thermionic electron guns because it has the highest melting point and lowest vapor pressure of all metals, thereby allowing it to be heated for electron emission. Other types of electron emitters include lanthanum hexaboride (LaB_6) cathodes, which can be used in a standard tungsten filament SEM if the vacuum system is upgraded, and field emission guns, which may be of the cold cathode type using tungsten single crystal emitters or the thermally assisted Schottky type, using emitters of zirconium oxide.

The electron beam, which typically has an energy ranging from 0.5 to 40 keV, is focused by one or two condenser lenses to a spot about 0.4–5 nm in diameter. The beam passes through pairs of scanning coils or pairs of deflector plates in the electron column, typically in the final lens, which deflect the beam in the x and y axes so that it scans in a raster fashion over a rectangular area of the sample surface. The raster scanning of the CRT display is synchronized with that of the beam on the specimen in the microscope, and the resulting image is therefore a distribution map of the intensity of the signal being emitted from the scanned area of the specimen. The image can be captured from a high-resolution cathode ray tube. It can be digitally captured and displayed on a computer monitor and saved to a computer's hard disk. Depending on the instrument, the resolution can fall somewhere between less than 1 and 20 nm. SEM gives information about topography, that is, surface features, and its textures as well as morphology, that is, shape and size of the particles. It also furnishes information about the elements and compounds that the object is composed of. In other words, SEM furnishes direct relationship between features and material properties, structure and material properties as well as composition and material properties.

The method that is suitable for the direct estimation of surface porosity is electron microscopy. A major drawback of electron microscopy is that microscopic analysis is very local and that the resolution is insufficient to study fine porous structures.[8] Also, the method becomes very laborious when a reasonable level of precision has to be reached because it is necessary to count and measure a large number of pores. The processing of the data is time consuming, although computer-aided image analysis can be used. Top layer thickness is one of the parameters frequently estimated from electron microscopic pictures. Asymmetry in composite membranes can also be readily observed. SEM has been used to study polysulfone/sulfonated polyether ether ketone (SPEEK) membranes.[9] Changes in pore structure resulting from different casting conditions and composition are shown in Figure 3.2. Obtaining cross-sectional samples usually involves razor blades and can result in compression and tearing.[10] The direct freeze fracture method, where a membrane is made brittle in liquid nitrogen and is then broken, can be used. The preparation of a thin film, as is required in transmission electron microscope (TEM), is not required for SEM. The disadvantage is that the membrane pore structure can be affected by drying and coating, so SEM-observed porosity may not exactly correspond to *in situ* membrane porosity.

3.3.2 Transmission Electron Microscopy

In TEM, an image is produced by passing electrons under high vacuum through a sample. The electron beam typically has an energy ranging from 100 to 300 keV. The image is formed from the interaction of the electrons transmitted through the specimen, which is magnified and focused onto an imaging device, such as a fluorescent screen, or on a layer of photographic film, or to be detected by a sensor. Near-atomic resolution is attainable using TEM; however, this technique is limited to thin samples which are at most several tens of nanometers in thickness. Therefore, TEM is restricted to imaging thin-film samples rather than fully constructed membranes. As an example, a TEM image of a graft copolymer is shown in Figure 3.3.[11] It may be concluded that the differences in electron densities between the two components of this amphiphilic graft copolymer were sufficient to be observed as image contrast. Dark regions represent the hydrophobic crystalline regions of the main chains, whereas lighter regions represent the hydrophilic side chains.

In terms of sample preparation, the vacuum environment requires dry samples. Also, coating may be necessary to prevent sample destruction by the electron beam.

3.3.3 Atomic Force Microscopy

Surface roughness of the membranes can be measured with atomic-level resolution using AFM. AFM is a nonoptical surface imaging technique that approaches atomic resolution. The atomic force microscope (AFM) consists of a cantilever with a sharp tip (probe) at its end that is used to scan the specimen surface.[12] The cantilever is typically silicon or silicon nitride with a tip

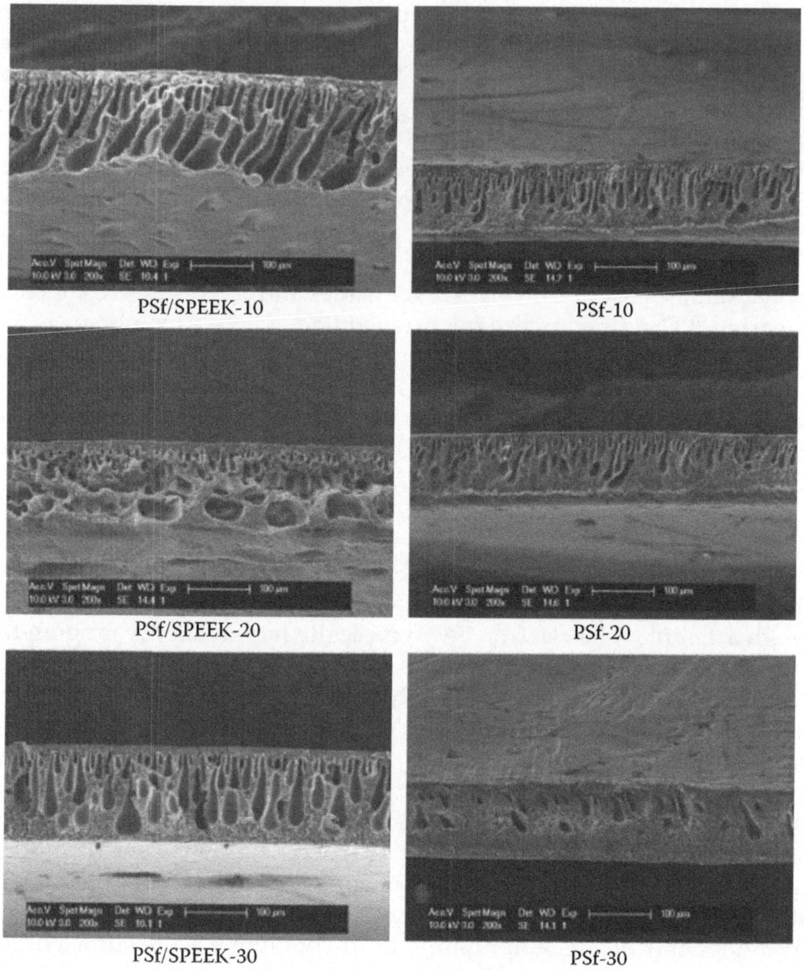

FIGURE 3.2
Scanning electron microscope images (200×) of cross sections of polymer blend membranes. (Reprinted from *J. Membr. Sci.*, 324, Li, X., De Feyter, S., and Vankelecom, I.F.J., Poly(sulfone)/sulfonated poly(ether ether ketone) blend membranes: Morphology study and application in the filtration of alcohol based feeds, 67–75, Copyright 2008, Figure 1, with permission from Elsevier.)

radius of curvature on the order of nanometers. When the tip is brought into proximity of a sample surface, forces between the tip and the sample lead to a deflection of the cantilever according to Hooke's law. Depending on the situation, forces that are measured in AFM include mechanical contact force, van der Waals forces, capillary forces, chemical bonding, electrostatic forces, magnetic forces, and solvation forces. Along with force, additional parameters may simultaneously be measured through the use of specialized types of probe

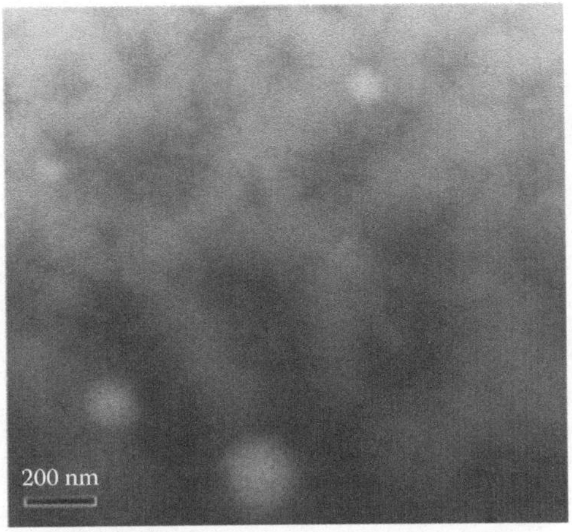

FIGURE 3.3
Transmission electron microscope image of a graft copolymer film. (Reprinted from *J. Membr. Sci.*, 325, Kim, Y., Park, J.T., Koh, J.H., Roh, D.K., and Kim, J.H., Anhydrous proton conducting membranes based on crosslinked graft copolymer electrolytes, 319–325, Copyright 2008, Figure 3, with permission from Elsevier.)

(scanning thermal microscopy, scanning Joule expansion microscopy, photothermal microspectroscopy, etc.). The deflection is measured using a laser spot reflected from the top surface of the cantilever into an array of photodiodes. Other methods that are used include optical interferometry, capacitive sensing, or piezoresistive AFM cantilevers. These cantilevers are fabricated with piezoresistive elements that act as a strain gauge. Using a Wheatstone bridge, strain in the AFM cantilever due to deflection can be measured, but this method is not as sensitive as laser deflection or interferometry.

If the tip was scanned at a constant height, a risk would exist that the tip collides with the surface, causing damage. Hence, in most cases, a feedback mechanism is employed to adjust the tip-to-sample distance to maintain a constant force between the tip and the sample. Traditionally, the sample is mounted on a piezoelectric tube that can move the sample in the z direction for maintaining a constant force, and the x and y directions for scanning the sample. Alternatively, a *tripod* configuration of three piezo crystals may be employed, with each responsible for scanning in the x, y, and z directions. This eliminates some of the distortion effects seen with a tube scanner. In newer designs, the tip is mounted on a vertical piezo scanner, whereas the sample is being scanned in X and Y using another piezo block. The resulting map of the area $z = f(x, y)$ represents the topography of the sample.

The AFM can be operated in a number of modes, depending on the application. In general, possible imaging modes are divided into static (also called *contact*) modes and a variety of dynamic (or noncontact) modes where the cantilever is vibrated. The membrane surface can be scanned in vacuum, air,

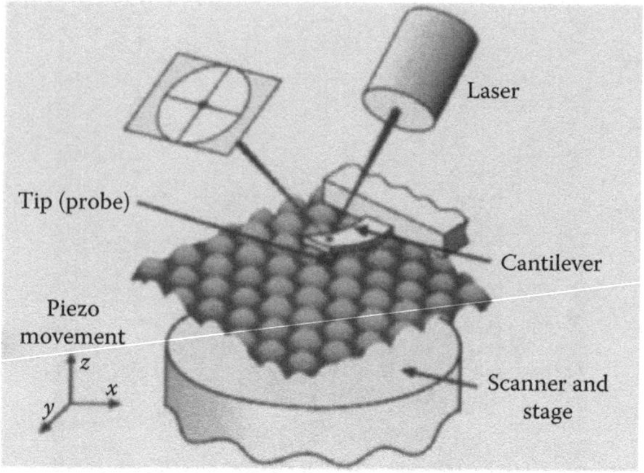

FIGURE 3.4
Schematic illustration of the operating principles associated with the atomic force microscope. (Reprinted from *J. Membr. Sci.*, 315, Wyart, Y., Georges, G., Demie, C., Amra, C., and Moulin, P., Membrane characterization by microscopic methods: Multiscale structure, 82–92, Copyright 2008, Figure 2, with permission from Elsevier.)

or water, and no sample preparation is necessary. An example illustration of an AFM experimental setup is given in Figure 3.4. Based on the sample profile, $h(x, y)$, it is possible to calculate mean and root mean square (RMS) roughness (*Ra* and *Rrms*, respectively) and 10-point mean roughness (*Rz*) parameters. There is relationship between *Ra* and the water flux and separation performance of hollow-fiber UF membranes.[13] As *Ra* decreases, flux decreases and rejection increases.

3.4 Spectroscopic Methods

Under this category, the following techniques are widely used:

- Infrared (IR) and Fourier transform IR (FTIR) spectroscopy
- X-ray photoelectron spectroscopy (XPS)
- Energy-dispersive X-ray spectroscopy (EDS or EDX)
- Positronium annihilation lifetime spectroscopy (PALS)

3.4.1 Attenuated Total Reflection FTIR Spectroscopy

The principle of attenuated total reflection FTIR (ATR-FTIR) is based on the total internal reflection phenomenon. A block of material with greater optical density than the sample is put into close contact with the sample surface.

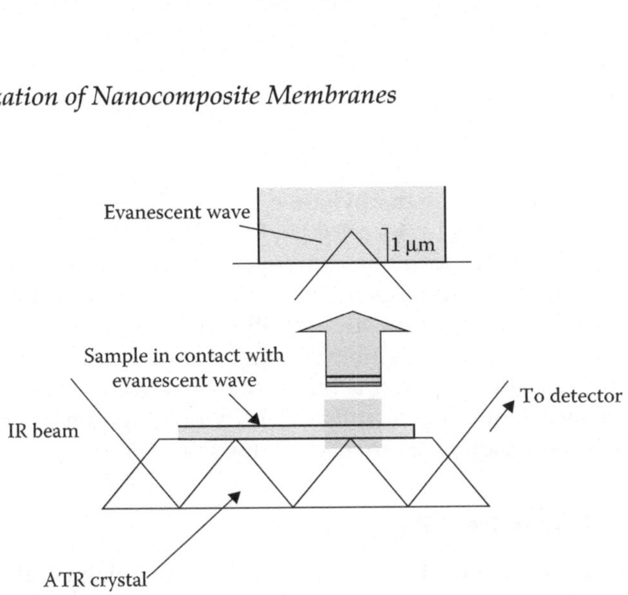

FIGURE 3.5
Principle of ATR-FTIR.

The incident radiation angle must be greater than the critical angle value. The evanescent field decays exponentially into the sample. The principle of ATR-FTIR is shown in Figure 3.5.[14] ATR-FTIR spectroscopy allows for IR analysis of surfaces. The IR spectrum can provide determination of vibrational frequencies and transition intensities of most molecules (with the exception of diatomics such as N_2 and O_2), including characteristic functional group frequencies. Knowledge of vibrational frequencies of functional groups (or reference spectra) allows for chemical identification of at least a class of compounds (e.g., aromatic amides). IR radiation typically penetrates 1 μm into the surface, which is a disadvantage because this is deeper than the active layer thickness of most composite membranes, so the top layer is not necessarily isolated. The penetration depth can be decreased by careful selection of crystal and incident angle. Alternatively, it is possible to perform ATR-FTIR measurements of both sides of an asymmetric membrane and then to essentially subtract out the support layer spectral regions.

Vibrational frequencies cannot be easily converted into complete chemical structure of the material, but can provide more molecular structure information than XPS.

3.4.2 X-Ray Photoelectron Spectroscopy

In XPS, the sample surface is subjected to X-ray radiation capable of removing electrons from the inner shells of the atoms (except H and He). The amount of emitted electrons is recorded as a function of binding energy. Surface depths of 0.5–10 nm can be probed, depending on the incident beam angle. The inner shell energies are characteristic for a given element, but sensitive to the external chemical environment. For carbon, this chemical shift can amount to a few electronvolts. For the carbonyl group (C=O), the binding energy is 285 eV, and for the methylene group (CH_2), the binding energy is 287.8 eV. Routine detection limits are 0.1%. From the emitted electron intensities, surface atomic concentrations can be found out through empirical formulae. XPS results for a variety of membranes are reported in Table 3.1.[15] The advantage of XPS is the requirement of shallow probing depth (i.e., it obtains information about the near membrane surface rather than the membrane interior). It has the ability to characterize the elemental composition of the membrane surface. However, chemical identity of the polymer (or additives) cannot be deduced based on elemental composition alone.

3.4.3 Energy-Dispersive X-Ray

EDS or EDX is an analytical technique used for the elemental analysis or chemical characterization of a sample. It is one of the variants of X-ray fluorescence spectroscopy which relies on the investigation of a sample through interactions between electromagnetic radiation and matter, analyzing X-rays emitted by the matter in response to being hit with charged particles. It is based on the principle that each element has a unique atomic structure allowing X-rays that are characteristic of an element's atomic structure to be identified uniquely from one another.

To stimulate the emission of characteristic X-rays from a specimen, a high-energy beam of charged particles such as electrons or protons, or a beam of X-rays, is focused into the sample. At rest, an atom within the sample contains ground-state (or unexcited) electrons in discrete energy levels or electron shells bound to the nucleus. The incident beam may excite an electron in an inner shell, ejecting it from the shell while creating an electron hole. An electron from an outer, higher energy shell then fills the hole, and the difference in energy between the higher energy shell and the lower energy shell may be released in the form of an X-ray. The number and energy of the X-rays emitted from a specimen can be measured by an energy-dispersive spectrometer. As the energy of the X-rays is characteristic of the difference in energy between

TABLE 3.1

Measured Atomic Concentration Percentages (%) of C_{1s}, O_{1s}, N_{1s} and S_{2p} Obtained by XPS for Commercially Available and Experimental Membranes

	NF270	Desal51HL	Desal5DL	NTR7450	N30F	NFPES10	D71	N13
C_{1s}	72.0	75.9	72.4	72.1	79.9	77.3	77.0	79.7
O_{1s}	17.0	12.6	17.6	19.2	14.4	16.8	16.3	14.4
N_{1s}	10.9	10.7	9.5	5.2	1.8	2.4	2.6	2.3
S_{2p}	0.1	0.8	0.5	3.5	3.9	3.5	4.1	3.6

Source: Boussu, K., Baerdemaeker, J.D., Dauwe, C., Weber, M., Lynn, K.G., Depla, D., Aldea, S., Vankelecom, I.F.J., Vandecasteele, C., and Bruggen, B.V.: Physico-chemical characterization of nanofiltration membranes. *Chem. Phys. Chem*. 2007. 8. 370–379. Copyright Wiley-VCH Verlag GmbH & Co. KGaA. Reproduced with permission.

the two shells, and of the atomic structure of the element from which they were emitted, this allows the elemental composition of the specimen to be measured. There are four primary components of the EDX setup: the beam source, the X-ray detector, the pulse processor, and the analyzer. A number of free-standing EDX systems exist. However, EDX systems are most commonly found on scanning electron microscopes (SEM-EDX) and electron micro-probes. SEMs are equipped with a cathode and magnetic lenses to create and focus a beam of electrons, and since the 1960s, they have been equipped with elemental analysis capabilities. A detector is used to convert X-ray energy into voltage signals; this information is sent to a pulse processor, which measures the signals and passes them onto an analyzer for data display and analysis. Typical EDX analyses of polysulfone and silver–polysulfone nanocomposite membrane are shown in Figure 3.6a and b, respectively.[16]

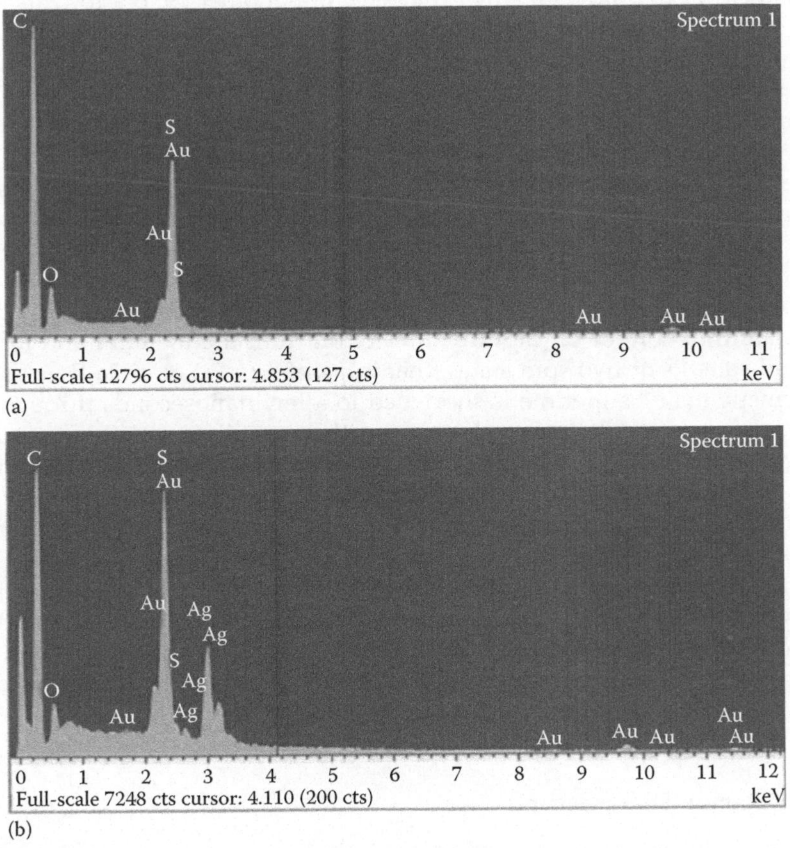

FIGURE 3.6
(a) EDX spectra of polysulfone membrane surface. (b) EDX spectra of polysulfone–silver membrane surface. (Data from Kar, S. et al., *Desalination Water Treat.*, 27, 224–230, Copyright 2011, Figures 5 and 6, Taylor & Francis Group.)

3.4.4 Positronium Annihilation Lifetime Spectroscopy

PALS measures positronium (Ps) annihilation lifetimes and intensities, which can be related to the size and amount of defect structures, such as voids or pores in the range of several angstroms to tens of nanometers. It has tremendous potential[17–19] as a powerful tool for quantifying the types and densities of defects in solids. Following the injection of positrons into a solid material, the positrons pair-annihilate at a rate depending on the density of electrons near the injection site. If there are lattice vacancy or dislocation defects (voids) near the injection site, the positrons are attracted to these areas, which have lower electron densities and thus give rise to longer positron lifetimes. This is suitable for insulating materials while not good in the bulk of metals because Ps cannot be formed in the latter. A beam of monoenergetic low-energy positrons (several kiloelectronvolts) is used for thin-film study. Thickness of the film can vary from several nanometers to a few microns. The energy of the implanted positrons is tunable, which enables PALS to depth profile thin films. By comparison, positrons typically used for bulk PALS have an implantation depth of 0.2–0.3 mm and as such are unsuitable for probing thin films.

PALS is a unique and valuable technique that characterizes the free-volume hole properties in solid polymeric systems[20,21] based on the detection of γ-radiation.[22] When positrons from a ^{22}Na radioisotope source are injected into a polymer, they lose their kinetic energy. After thermalization occurs in the sample, the positrons either (1) diffuse into the media and become annihilated as a free positron with an electron in about 0.4 ns (free positron annihilation) or (2) capture an electron from the material and form a bound-state Ps of two spin states, that is, *para*-Ps (*p*-Ps) or *ortho*-Ps (*o*-Ps). In polymers, the *o*-Ps lifetime is shortened to a few nanoseconds through the trapping in molecular free volumes and the annihilation with an electron from the inner wall of the free volume (annihilation of *o*-Ps). The lifetime of *o*-Ps has been proven to be highly correlated with the average size and distribution of hole spaces present in the solid polymers; the longer the lifetime of *o*-Ps, the larger the R, and vice versa.[23,24] The spectral intensity associated with this lifetime component is related to the number of free volume hole sites if no Ps-quenching functional groups exist in the molecular structure of polymers, which is expected to provide a better understanding of the flux enhancement and eventually could aid in design of new TFC membranes that display outstanding performance.

3.4.5 Raman Spectroscopy

Raman spectroscopy is a spectroscopic technique based on inelastic scattering of monochromatic light, usually from a laser source. Inelastic scattering means that the frequency of photons in monochromatic light changes upon interaction with a sample. Photons of the laser light are absorbed by the sample and then reemitted. Frequency of the reemitted photons is shifted up or down in comparison with original monochromatic frequency, which is called the Raman effect. This shift provides information about vibrational, rotational, and other low-frequency transitions in molecules. Raman spectroscopy can be used to study solid, liquid, and gaseous samples. This is a complementary method to IR spectroscopy. It is an alternative technique for molecular identification. In particular, Raman spectroscopy provides information about functional groups (—C—S—, —C—C—, —C—H— etc.), which are important in membrane technology. Raman spectroscopy has been used to examine changes in polymer structure in the membrane and to characterize membrane morphology. Raman techniques require no special sample preparation and produce cleaner, narrower spectra than IR, making band interpretation more robust. However, because of the inherently weak signals and interference from fluorescence, the technique has not gained significant attention for identification of organic compounds. Surface-enhanced Raman spectroscopy is an extension of Raman spectroscopy that allows both an enhanced Raman signal and substantial quenching of fluorescence through the use of coinage metal nanostructures (silver, gold, etc.), thus offering highly specific molecular-level identification of extremely small samples and samples capable of fluorescence.[25,26] It allows nondestructive characterization of organics. Nafion membranes have been studied earlier by IR and Raman spectroscopy.[27,28]

3.4.6 Small-Angle Neutron Scattering

Small-angle neutron scattering (SANS) is a neutron scattering technique that enables the study of materials on the nanometer to micrometer length scales. The experiment consists of a well-collimated beam of neutrons being passed through a sample and detectors to count the number of neutrons scattered as a function of angle and neutron wavelength. This data is used to extract information about the shape, size, arrangement, and interactions of the components of the sample.

SANS technique finds particular use in the study of soft matter (e.g., colloids and polymers), biophysics (e.g., lipids and lipid–protein complexes), biology (e.g., solution structures of proteins), and hard condensed matter (e.g., superconductors and magnetic materials). SANS has been sued to characterize a phospholipid/alkoxysilane hybrid bilayer membrane at the interface of porous aluminum oxide films.[29]

3.4.7 Small-Angle X-Ray Scattering

The small-angle X-ray scattering (SAXS) technique is based on the detection of the elastic scattering of X-rays by samples with inhomogeneities of the electron density in the nanometer range. X-ray scattering in the small-angle range contains information about the shape and size of nano-objects in dilute solution such as macromolecules, characteristic distances of partially ordered nanostructured materials, nanopore sizes, and other low-resolution structure features. SAXS provides structural parameters of macromolecules with sizes between 5 and 25 nm.

In small-angle scattering technique, a sample is irradiated with a well-collimated beam of radiation such as X-rays, neutrons, or light; the resulting intensity is measured as a function of angle between the incoming beam and the scattered beam; and then the structure that caused the observed pattern is determined. Scattering patterns are caused by the interference of secondary waves that are emitted from various structures when irradiated, that is, electrons for X-rays and light, or nuclei for neutrons. Scattering of X-rays is caused by differences in electron density. Scattering of neutrons is caused by differences in scattering power of different nuclei. Scattering of light is caused by differences in refractive index. The larger the diffraction angle, the smaller the length scale probed; wide-angle X-ray scattering is used to determine crystal structure on the atomic length scale, whereas SAXS or SANS is used to explore the microstructure on the colloidal length scale. Unlike an electron micrograph, SAXS patterns do not give morphological information directly. The result of a SAXS experiment is essentially the intensity of the Fourier transform of the electron density and must be interpreted in order to determine morphology. One fundamental problem with any scattering experiment is that two different morphologies can, in theory, give identical scattering patterns. Generally, one cannot reconstruct the exact microstructure uniquely from a SAXS pattern because in a scattering experiment only the scattered radiation intensity can be measured and all phase information is lost. Therefore, one cannot be absolutely sure that a scattering pattern is due to a particular morphology. Nanopores are characterized in track-etched polymer membranes using SAXS.[30] Improved performance of phosphonated carbon nanotube–polybenzimidazole composite membranes in proton-exchange membrane fuel cells is supported by the SAXS studies.[31]

3.4.8 Nuclear Magnetic Resonance

Nuclear magnetic resonance (NMR) spectroscopy is a powerful and theoretically complex analytical tool. It is based on the fact that when a population of magnetic nuclei is placed in an external magnetic field, the nuclei become aligned in a predictable and finite number of orientations. For ^{1}H there are two orientations: In one orientation, the protons are aligned with the external magnetic field (north pole of the nucleus aligned with the south pole of the magnet and south pole of the nucleus with the north pole of the magnet) and in the other orientation, the nuclei are aligned against the field (north with north and south with south). The alignment with the field is also called the *alpha* orientation and the alignment against the field is called the *beta* orientation. The NMR spectroscopy experiment involves using energy in the form of electromagnetic radiation to pump the excess alpha-oriented nuclei into the beta state. When the energy is removed, the energized nuclei relax back to the alpha state. The fluctuation of the magnetic field associated with this relaxation process is called resonance, and this resonance can be detected and converted into the peaks, which is seen in an NMR spectrum. Membrane proteins are important targets for structure determination. Protein structure determination[32,33] has played a key role in biomedical research for more than 60 years NMR spectroscopy is particularly well suited for determining the 3D structures of proteins and for describing their local and global motions in lipid environments, where they maintain their native conformation.

3.4.9 Electron Spin Resonance

Electron spin resonance (ESR) is also known as the electron paramagnetic resonance (EPR). Paramagnetic molecules have unpaired electrons, and while changing the magnetic field, these will absorb energy at particular values of the field. This absorption is due to a change in direction of the magnetic moment resulting from the electron's spin. The absorption spectra that are produced can give information about the structure of free radicals and complex ions. ESR works on the same principle as NMR, except that microwave (rather than radiowave) frequencies are employed and spin transitions of unpaired electrons rather than nuclei are recorded. Unlike NMR spectra, where absorption is recorded directly, ESR spectrometers plot the first derivative of the absorption curve. ESR permits observation of any substance having unpaired electrons. It is sensitive to local environment. Polymers themselves contain paramagnetic free radicals. The free radicals do take part in the transportation of gases through the membrane.[34,35]

3.5 Contact Angle Goniometer

Contact angle goniometer is used to measure the static contact angle, advancing and receding contact angles, and surface tension. The first contact angle goniometer was designed by William Zisman. The original manual contact angle goniometer used an eyepiece with microscope. The current generations of contact angle instruments use cameras and software to capture and analyze the drop shape suitable for dynamic and advanced studies. A gonioreflectometer is used to measure the reflectivity of a surface at a variety of angles.

The contact angle, θ, is the angle formed by a liquid at the three-phase boundary where the liquid, gas, and solid intersect. The contact angle depends on the interfacial tensions between the gas and liquid, liquid and solid, and gas and solid. Young's relation expresses the contact angle analytically.

$$\gamma_{SG} = \gamma_{SL} + \gamma_{LG}\mathrm{Cos}\,\theta \tag{3.9}$$

where:

γ_{SG} is the interfacial tension between the solid and the gas
γ_{SL} is the interfacial tension between the solid and the liquid
γ_{LG} is the interfacial tension between the liquid and the gas

Contact angle goniometer measures a droplet's contact angle by assuming that the droplet fits the geometry of a sphere, an ellipsoid, or the Young–Laplace equation.

Contact angle is the ratio of cohesion force to adhesion force. Cohesion is the force between the liquid molecules that hold the liquid together. Adhesion is the force between the liquid molecules and the solid molecules. If the contact angle is near zero, meaning that the liquid droplet spreads completely on the solid surface, adhesive forces are dominating. If the contact angle is very high, meaning that the liquid droplet beads up on the solid surface as water does on a freshly waxed car, cohesive forces are dominating.

Although static contact angle gives static information about the interfacial tensions between the solid, the liquid, and the gas; advancing and receding contact angles give some information about the dynamic interaction of the liquid, solid, and gas. An advancing contact angle is determined by pushing a droplet out of a pipette onto a solid. When the liquid initially comes into contact with the solid, it forms some contact angle. As the pipette injects more liquid through the pipette, the droplet increases in volume and the contact angle increases, but its three-phase boundary remains stationary until it suddenly jumps outward. The contact angle the droplet has immediately before jumping outward is termed the advancing contact angle. The droplet will decrease in volume and the contact angle decreases, but its three-phase boundary remains stationary until it suddenly jumps inward. The contact angle the droplet has immediately before jumping inward is termed the receding contact angle. The receding contact angle is measured by sucking the liquid back out of the droplet. The difference between advancing and receding contact angles is termed as contact angle hysteresis, which can be used to characterize surface heterogeneity, roughness, and mobility.

The contact angle (θ) that is formed at the three-phase interface between solid, liquid, and gas/vapor phases is used to elicit information regarding membrane surface energy properties. The information that is gathered from contact angle analysis may be used to calculate specific surface energy properties (van der Waals, Lewis acid–base) for a detailed interfacial analysis, as well as for qualitatively assessment of the wettability, or hydrophobicity/hydrophilicity, of membrane surface. The greatest challenge with goniometric contact angle measurements is contact angle hysteresis, which is the difference in the measured contact angle depending on whether it is an advancing or receding measurement. The important issue for contact angle measurement reproducibility is the effect of drop and bubble size on the contact angle. The advancing angle should be more reproducible, as it depends on the bubble/drop size to a lesser degree, on both smooth and imperfect surfaces.[36,37] However, information on surface quality that can be gained from receding angle/bubble size relationship will be lost if only advancing angle is measured. Multiple linear regression has been used to study the influence of surface hydrophobicity, surface roughness, surface charge, molecular-weight cutoff, permeability, and porosity of the top layer on nanofiltration membrane performance.[38] Contact angle, volume fraction of small pores, and membrane charge are the significant variables in predicting relative flux. Contact angle measurements are useful in membrane integrity testing.

3.6 Zeta Potential Measurements

Zeta potential (ζ) is the potential difference between the bulk of solution and the sheer (slipping) plane of the interfacial double layer. It is a function of surface and solution chemistry (pH, ionic composition, and ionic strength) at the solid–liquid interface and is an important membrane characteristic for assessing membrane fouling potential and developing chemical cleaning protocols. Membrane zeta potential is typically determined from streaming potential measurements. A streaming potential is generated when an electrolyte solution flows through a thin channel or porous media (e.g., a sand column) and is related to zeta potential by the Helmholtz–Smoluchowski equation:

$$\frac{E}{P} = \frac{\varepsilon \varepsilon_0 \zeta}{\lambda \eta} \tag{3.10}$$

where:

- E is the streaming potential due to electrolyte flow through a capillary channel
- P is the applied pressure driving the flow
- ζ is the zeta potential
- λ is the electrolyte conductivity
- η is the viscosity of the electrolyte solution
- ε is the relative permittivity of the solution (dimensionless)
- ε_0 is the vacuum permittivity (fundamental constant)

The values of E, P, and λ are measured by the streaming potential analyzer, whereas ε and η are calculated based on temperature measurement.

In experimental investigations of membrane charge, streaming potential measurements are used to calculate zeta potential. Streaming potential is the potential induced when an electrolyte solution is pumped across a stationary charged surface. It can be used to calculate zeta potential using the Helmholtz–Smoluchowski equation, which relates the pressure dependence of streaming potential to the properties of the solution, that is, conductivity and viscosity. Unlike earlier streaming potential investigations, now the streaming potential is measured over a range of solution pH.[39–41] Surface charge as a function of pH is crucial for understanding the acid–base properties of membrane surface functional groups.

Although streaming potential measurements are the most frequently used method for evaluating charge properties, there are concerns with respect to the effect of membrane roughness on the measurement, the relationship between the measured zeta potential and the double layer structure, the inherent assumptions of the Helmholtz–Smoluchowski equation (i.e., laminar flow), and the lack of a calibration standard. However, zeta potential from streaming potential measurement is one of the few techniques capable of describing the charge properties of membranes.

3.7 Differential Scanning Calorimetry

The DSC is used for thermal analysis of nanomaterials and polymers. The information is used to understand amorphous and crystalline behavior, polymorph and eutectic transitions, curing and degree of cure, and many other material properties used to design, manufacture, and test products. A sample of known mass is heated or cooled and the change in its heat capacity is monitored. The DSC measures the amount of energy absorbed or released by a sample when it is heated or cooled, providing quantitative and qualitative data on endothermic (heat absorption) and exothermic (heat evolution) processes. It is used to determine the melting behavior of complex organic materials; both temperatures and enthalpies of melting can be used to determine purity of a material. It is also widely used to measure material glass transition temperatures or softening temperatures of plastic or glassy materials, which change depending upon the temperature history of the polymer or the amount and type of fill material, among other effects. It can determine the crystalline-to-amorphous transition temperatures in polymers and the energy associated with the transition. The biggest advantage of DSC is the ease and speed with which it can be used to see transitions in materials. A schematic DSC curve for a semicrystalline polymer is shown in Figure 3.7 illustrating the possible heat effects. Such DSC curves allow the glass transition temperature and the degree of crystallinity to be obtained. Both first-order and second-order transitions are observed in Figure 3.7.

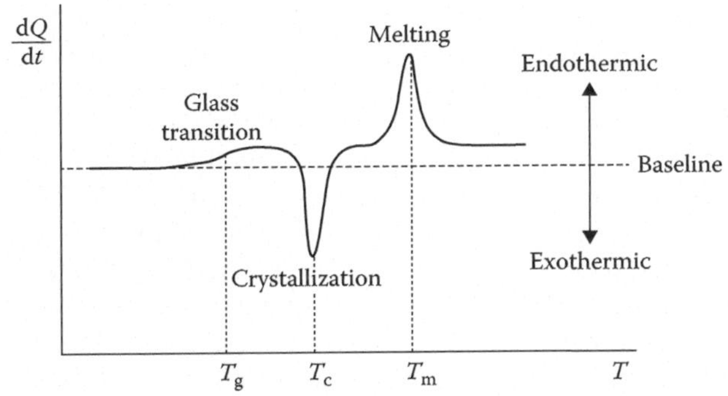

FIGURE 3.7
Schematic DSC curve for a semicrystalline polymer.

The former transitions, referred to as crystallization and melting, give narrow peaks (the peak area being proportional to the amount of crystalline material present). The glass transition temperature corresponds to second-order transition. Studies have been carried out to see the effects of nanoparticles on the glass transition temperature of the prepared nanocomposite membranes by DSC, where the glass transition temperature of nanocomposite membranes was found to increase with nanoparticle incorporation.[42–44]

3.8 Tensile Strength Measurements

Tensile strength measurements are done by universal testing machine. By pulling the material, it is determined that how the material will react to forces being applied in tension. By continuing to pull on the material until it breaks, a good, complete tensile profile of the material can be obtained. A curve will result, showing how it reacted to the forces being applied. The point of failure is of much interest and is typically called its *ultimate strength.*

For most tensile testing of materials, in the initial part of the test, when a material deforms elastically, the amount of deformation depends on the size of the material, but the strain for a given stress is always the same and the two are related by Hooke's law (stress is directly proportional to strain):

$$E = \left(\frac{\sigma}{\varepsilon}\right) \tag{3.11}$$

where:

σ is the stress
E is the modulus of elasticity
ε is the strain

E is the slope of the line in this region where stress (σ) is proportional to strain (ε) and is called the *modulus of elasticity* or *Young's modulus*. The modulus of elasticity is a measure of the stiffness of the material. If a specimen is loaded within this linear region, the material will return to its exact same condition if the load is removed. At the point that the curve is no longer linear and deviates from the straight-line relationship, Hooke's law no longer applies and some permanent deformation occurs in the specimen. This point is called the *elastic or proportional limit*. From this point on in the tensile test, the material reacts plastically to any further increase in load or stress. It does not return to its original, unstressed condition if the load is removed. A parameter called *yield strength* of a material is defined as the stress applied to the material at which plastic deformation starts to occur although the material is loaded. The amount of stretch or elongation the specimen undergoes during tensile testing can also be found out. This can be expressed as an absolute measurement in the change in length or as a relative measurement called *strain*. Strain can be expressed as the ratio of the change in length to the original length:

$$\varepsilon = \frac{\Delta l}{\Delta l_0} = \frac{(l_1 - l_0)}{l_0} \tag{3.12}$$

where:

Δl is the change of the length
l_1 is the length after elongation
l_0 is the original (initial) length

The *ultimate tensile strength* of a material is the maximum load the specimen sustains during the test. It may or may not equate to the strength at break. This depends on the type of material under testing.

As membranes vary in chemical and material composition, test methods are prescribed in various standards, which specify different specimen shapes, sizes, grip lengths, and loading speeds. Some of these *standard* test methods, frequently referred to in testing of roofing membranes or similar materials, are presented in Table 3.2.[45] It is seen that the temperature, speed of loading, gauge length, and strain variation within a specimen affect the tensile tests of a membrane sample.

Nanoparticles have great potential in improving the mechanical properties of the membrane. In particular, PVDF flat sheet membrane with 0.54 wt.% of SBA-15 (a mesoporous silica material with a highly ordered two-dimensional hexagonal mesostructure and thick uniform silica walls) increases tensile strength from 0.151 to 0.183 MPa, whereas that loaded with 0.36 wt.% increased elongation at break from 22.6% to 49.4%.[46] Likewise, PVDF nanocomposite membranes incorporated with ZnO nanoparticles exhibits increased tensile strength and elongation at break.[47] PVDF/TiO_2 hollow-fiber membranes

TABLE 3.2
ASTM Standards Applicable to Properties of Roofing Membrane Material Where Tensile Testing Is Involved

ASTM Standard	ASTM Designation	Shape	Specimen Width (mm)	Gauge or Grip Length (mm)	Rate of Grip Separation (mm/min)	Approximate Operating Time (s)
D146	Bitumen-saturated felts and woven fabrics for roofing and water proofing	Rectangular	25	100	51	15
D412	Rubber properties in tension	Dumbbell	6	33	510	30
D624	Rubber property—tear resistance	Curly and notched	19	58	450–550	25
D638	Tensile properties of plastic	Dumbbell	6	25	50–500	30–300
D751	Coated fabrics	Rectangular (grab)	100	75	300	30
D882	Tensile properties of thin plastic sheeting	Rectangular	5–25	50–125	12.5–500	15–120
D1004	Initial tear resistance of plastic film and sheeting	Curly and notched	19	25	51	20
D2523	Load–strain properties of roofing membranes	Dumbbell	25	76	1.3–500	3–1500
D2707	Hard rubber in tension	Dumbbell	12.5	75	5–10	90
D4073	Tensile tear strength of bituminous roofing membranes	Rectangular and notched	75	100	2.5	30

Note: ASTM, American Society for Testing and Materials.

prepared by either TiO_2 sol–gel or blending method shows 30% increase in tensile strength.[48] However, elongation at break decreases from 162% to 120% likely due to the rigidity of the inorganic particles. The improvement in tensile strength is attributed to the reduced macro-void formation observed in the nanocomposite membranes. The decreased ductility could be owing to the increased cross-linking arising from the nanoparticle inclusion rather than the brittleness of the particles.

3.9 Methods to Characterize Proton-Exchange Membranes

In addition to the above-mentioned different characterization techniques, specific methods are adopted for characterization and testing of proton-exchange membranes.

3.9.1 Water Uptake Measurement

In case of proton-exchange membranes, the ion-exchange capacity is linked with the water uptake capacity of the membrane, due to migration methods presented in proton-exchange membranes, which are mainly Grotthuss and vehicular mechanisms. In the Grotthuss mechanism, protons *jump* from an H^+ donor site to any receiving water molecule in the vicinity, forming an H_3O^+ complex, whereas in the vehicular mechanism, protons are transferred through hydronium ions. The presence of water is necessary for proton conduction in both mechanisms.[49] For the water uptake measurement, a sample of the dry membrane prepared is weighed[50] and immersed in distilled water for 24 h, then excess water is removed with absorbent paper and the wet sample is weighed. The water uptake is calculated according to the following equation:

$$\%\text{water uptake} = \frac{w_w - w_d}{w_d} \times 100 \quad (3.13)$$

where:

w_w is the weight of wet sample
w_d is the weight of dry sample

3.9.2 Ion-Exchange Capacity Measurement

The ion-exchange capacity is measured by a classic titration method. Initially, the membranes is converted to its proton form by soaking them in 1 M HCl solution for 24 h.[51] Subsequently, the membranes are washed with distilled water and immersed in a 1 M solution of NaCl for 24 h, to perform the exchange of H^+ protons for Na^+ ions. Then the solution with H^+ protons is titrated with a 0.01 M NaOH solution using phenolphthalein indicator. The ion-exchange capacity was calculated according to the following equation:

$$\text{IEC}\left(\text{meq/g dry sample}\right) = \left(\frac{VM}{W}\right) \tag{3.14}$$

where:

IEC is the ion-exchange capacity
V is the volume of the titrant solution when equilibrium point is reached
M is the concentration of titrant solution
W is the weight of the dry sample

3.9.3 Electrochemical Impedance Spectroscopy

Electrochemical impedance spectroscopy (EIS) is a powerful diagnostic tool that can be used to improve the performance of the fuel cells. It is frequently used for the determination of proton and ionic conductivity of proton-exchange membranes. There are three fundamental sources of voltage loss in fuel cells: (1) charge transfer activation or *kinetic* losses, (2) ion and electron transport or *ohmic* losses, and (3) concentration or *mass transfer* losses. The EIS technique is used to measure the frequency dependence of the impedance of a fuel cell by applying a small sinusoidal AC potential (or current) as a perturbation signal to the fuel cell and measuring the current (or potential) response. The application of EIS in proton-exchange membrane fuel cell studies involve the following: (1) to provide microscopic information about the fuel cell system, which can help in fuel cell structure optimization and the selection of the most appropriate operating conditions; (2) to allow modeling of the system with an appropriate equivalent circuit and consequently to obtain the electrochemical parameters of the system; (3) to differentiate the individual contributions of each component, such as the membrane and the gas diffusion electrode, to fuel cell performance, which can assist in identifying problems within the fuel cell components; and (4) to identify individual contributions to the total impedance of a PEM fuel cell from different electrode processes such as interfacial charge transfer and mass transport in both the catalyst layer and the backing diffusion layer.[52–54]

References

1. Khulbe, K.C., Feng, C.Y., and Matsuura, T. 2002. Membrane characterization, *Water and wastewater treatment technologies: Encyclopedia of Life Support Systems*, EOLSS Publishers, Oxford, UK, 131.
2. Mulder, M. 1996. *Basic Principles of Membrane Technology*, 2nd Edition, Kluwer Academic, Dordrecht, the Netherlands.

3. Brun, M., Lallemand, A., Quinson, J.F., and Eyraud, Ch. 1977. A new method for the simultaneous determination of the size and shape of pores: Thermoporometry. *Therm. Acta.* 21: 59–88.
4. Quinson, J.F., Mameri, N., Guihard, L., and Bariou, B. 1991. The study of the swelling of an ultrafiltration membrane under the influence of solvents by thermoporometry and measurement of permeability. *J. Membr. Sci.* 58: 191–200.
5. Cuperus, F.P., Bargeman, D., and Smolder, C.A. 1992. Critical points in the analysis of membrane pore structures by thermoporometry. *J. Membr. Sci.* 66: 45–53.
6. Mey-Marom, A. and Katz, M.G. 1986. Measurement of active pore size distribution of microporous membranes - a new approach. *J. Membr. Sci.* 27: 119–130.
7. Cuperus, F.P., Bargeman, D., and Smolder, C.A. 1992. Permporometry: The determination of the size distribution of active pores in UF membranes. *J. Membr. Sci.* 71: 57–67.
8. Cuperus, F.P. and Smolders, C.A. 1991. Characterization of UF membranes membrane characteristics and characterization techniques. *Adv. Colloid Interface Sci.* 34: 135–173.
9. Li, X., De Feyter, S., and Vankelecom, I.F.J. 2008. Poly(sulfone)/sulfonated poly(ether ether ketone) blend membranes: Morphology study and application in the filtration of alcohol based feeds. *J. Membr. Sci. 324*: 67–75.
10. Childress, A.E., Brant, J.A., Rempala, P., Phipps, D.W Jr., and Kwan, P. 2012. *Evaluation of Membrane Characterization Methods,* Subject Area: Water quality, Web report #4102, Water Research Foundation, Denver, CO.
11. Kim, Y., Park, J.T., Koh, J.H., Roh, D.K., and Kim, J.H. 2008. Anhydrous proton conducting membranes based on crosslinked graft copolymer electrolytes. *J. Membr. Sci.* 325: 319–325.
12. Wyart, Y., Georges, G., Demie, C., Amra, C., and Moulin, P. 2008. Membrane characterization by microscopic methods: Multiscale structure. *J. Membr. Sci.* 315: 82–92.
13. Chung, T.S., Qin, J.J., Huan, A., and Toh, K.C. 2002. Visualization of the effect of die shear rate on the outer surface morphology of ultrafiltration membranes by AFM. *J. Membr. Sci.* 196: 251–266.
14. Hofstetter, H. 2007. FT-IR spectroscopy attenuated total reflectance ATR. http://www.niu.edu/analyticallab/ftir/ATR_lecture.pdf.
15. Boussu, K., Baerdemaeker, J.D., Dauwe, C., Weber, M., Lynn, K.G., Depla, D., Aldea, S., Vankelecom, I.F.J., Vandecasteele, C., and Bruggen, B.V. 2007. Physicochemical characterization of nanofiltration membranes. *Chem. Phys. Chem.* 8: 370–379.
16. Kar, S., Subramanian, M., Ghosh, A.K., Bindal, R.C., Prabhakar, S., Nuwad, J., Pillai, C.G.S., Chattopadhyay, S., and Tewari, P.K. 2011. Potential of nanoparticles for water purification: A case-study on anti-biofouling behaviour of metal based polymeric nanocomposite membrane. *Desalination Water Treat.* 27: 224–230.
17. Positronannihilation.net, 2011. Positron lifetime spectroscopy. http://www.positronannihilation.net/index_files/Positron%20Lifetime%20Spectroscopy.pdf.
18. Tsai, K. and McNulty, D. 2012. Positron lifetime spectroscopy (PALS). http://www2.cose.isu.edu/~mcnudust/publication/reports/Wyle_PALS_quarterly_Apr2012.pdf.
19. Krause-Rehberg, R. 2013. Fundamentals of positron annihilation spectroscopy and its application in semiconductors. http://positron.physik.uni-halle.de/talks/ICPA-15_ISPS_lecture_RKR.pdf.

20. Kim, S.H., Kwak, S.Y., and Suzuki, T. 2005. Positron annihilation spectroscopic evidence to demonstrate the flux-enhancement mechanism in morphology-controlled thin-film-composite (TFC) membrane. *Environ. Sci. Technol.* 39: 1764–1770.
21. Sawada, S., Kawasuso, A., Maekawa, A., and Yabuuchi, Y.M. 2010. Free-volume structure of fluoropolymer-based radiation grafted electrolyte membranes investigated by positron annihilation lifetime spectroscopy. *J. Phys. Conf. Ser.* 225: 012048.
22. Jean, Y.C., Mellon, P.E., and Schrader, D.M. 2003. *Principles and Applications of Positron and Positronium Chemistry*, World Scientific, Singapore.
23. Tao, S.J. 1972. Positronium annihilation in molecular substances. *J. Phys. Chem.* 56: 5499.
24. Nakanishi, H. and Jean, Y.C. 1988. Positrons and positronium in liquids, *Positron and Positronium Chemistry*, D.M. Schrader and Y.C. Jean (eds.), Elsevier, Amsterdam, the Netherlands, 159–192.
25. Moscovits, M. 1985. Surface-enhanced spectroscopy. *Rev. Mod. Phys.* 57: 783–826.
26. Moskovits, M. 2006. Surface-enhanced Raman spectroscopy: A brief perspective, *Surface-Enhanced Raman Scattering*, K. Kneipp, M. Moskovits, and H. Kneipp (eds.), Springer, Berlin, Germany, 1–17.
27. Ovstrovskii, D.I., Torell, L.M., Paronen, M., Hietala, S., and Sundholm, F. 1997. Water sorption properties of and the state of water in PVDF-based proton conducting membranes. *Solid State Ion.* 97: 315–321.
28. Innocent, C., Huguet, P., Bribes, J.L., Pourcelly, G., and Kameche, M. 2001. Characterisation of cation exchange membrane in hydro-organic media by electrochemistry and Raman spectroscopy. *Phys. Chem. Chem. Phys.* 3: 1481–1485.
29. Marchal, D., Bourdillon, C., and Deme, B. 2001. Small-angle neutron scattering by highly oriented hybrid bilayer membranes confined in anisotropic porous alumina. *Langmuir* 17: 8313–8320.
30. Cornelius, T.W., Schiedt, B., Severin, D., Pepy, G., Toulemonde, M., Apel, P.Y., Boesecke, P., and Trautmann, C. 2010. Nanopores in track-etched polymer membranes characterized by small-angle x-ray scattering. *Nanotechnology* 21: 155702.
31. Kannan, R., Kagalwala, H.N., Chaudhari, H.D., Kharul, U.K., Kurungota, S., and Pillai, V.K. 2011. Improved performance of phosphonated carbon nanotube–polybenzimidazole composite membranes in proton exchange membrane fuel cells. *J. Mater. Chem.* 21: 7223–7231.
32. Shahid, S.A., Bardiaux, B., Franks, W.T., Krabben, L., Habeck, M., Rossum, B.-J., and Linke, D. 2012. Membrane-protein structure determination by solid-state NMR spectroscopy of microcrystals. *Nat. Methods* 9: 1212–1217.
33. Opella, S.J. 2013. Structure determination of membrane proteins by nuclear magnetic resonance spectroscopy. *Annu. Rev. Anal. Chem. (Palo Alto Calif)* 6: 305–328.
34. Khulbe, K.C., Chowdhury, G., Matsuura, T., and Lamarche, G. 1997. Characterization of PPO (poly(phenylene oxide)) powder and membranes by ESR technique. *J. Membr. Sci.* 123: 9–15.
35. Altschuler, S.A. and Kozyrev, B.M. 1964. *Electron Para-Magnetic Resonance*, C.P. Poole Jr., (ed.), Academic Press, New York, 312–313.
36. Drelich, J. 1997. The effect of drop (bubble) size on contact-angle at solid-surfaces. *J. Adhes.* 63: 31–51.

37. Drelich, J., Miller, J.D., and Good, R.J. 1996. The effect of drop (bubble) size on advancing and receding contact angles for heterogeneous and rough solid surfaces as observed with sessile-drop and captive-bubble techniques. *J. Colloid Interface Sci.* 179: 37–50.
38. Boussu, K., Vandecasteele, C., and van der Bruggen, B. 2008. Relation between membrane characteristics and performance in nanofiltration. *J. Membr. Sci.* 310: 51–65.
39. Childress, A.E. and Elimelech, M. 2000. Relating nanofiltration membrane performance to membrane charge (electrokinetic) characteristics. *Environ. Sci. Technol.* 34: 3710–3716.
40. Hagmeyer, G. and Gimbel, R. 1999. Modelling the rejection of nanofiltration membranes using zeta potential measurements. *Sep. Purif. Technol.* 15: 19–30.
41. Tay, J.H., Liu, J.L., and Sun, D.D.L. 2002. Effect of solution physico-chemistry on the charge property of nanofiltration membranes. *Water Res.* 36: 585–598.
42. Mohagheghian, M., Sadeghi, M., Chenar, M.P., and Naghsh, M. 2014. Gas separation properties of polyvinylchloride (PVC)-silica nanocomposite membrane. *Korean J. Chem. Eng.* 31: 2041–2050.
43. Khosravi, A., Sadeghi, M., Banadkohi, H.Z., and Talakesh, M.M. 2014. Polyurethane-silica nanocomposite membranes for separation of propane/methane and ethane/methane. *Ind. Eng. Chem. Res.* 53: 2011–2021.
44. Tohidiana, M., Ghaffariana, S.R., Nourib, M., Jaafarniac, E., and Haghighid, A.H. Polyelectrolyte nanocomposite membranes using imidazole-functionalized nanosilica for fuel cell applications. *J. Macromol. Sci. Part B Phys.*, 54: 17–31.
45. Dutt, O.M. and Ashton, H.E. PVC roofing membranes: Factors affecting tensile tests. http://docserver.nrca.net/technical/298.pdf.
46. Liao, C., Zhao, J., Yu, P., Tong, H., and Luo, Y. 2010. Synthesis and characterization of SBA-15/poly (vinylidene fluoride) (PVDF) hybrid membrane. *Desalination* 260: 147–152.
47. Liang, S., Xiao, K., Mo, Y., and Huang, X. 2012. A novel ZnO nanoparticle blended polyvinylidene fluoride membrane for anti-irreversible fouling. *J. Membr. Sci.* 394–395: 184–192.
48. Yu, L.-Y., Shen, H.-M., and Xu, Z.-L. 2009. PVDF-TiO_2 composite hollow fiber ultrafiltration membranes prepared by TiO_2 sol-gel method and blending method. *J. Appl. Polym. Sci.* 113: 1763–1772.
49. Amjadi, M., Rowshanzamir, S., Peighambardoust, S., Hosseini, M., and Eikani, M. 2010. Investigation of physical properties and cell performance of Nafion/TiO_2 nanocomposite membranes for high temperature PEM fuel cells. *Int. J. Hydrogen Energy* 35: 9252–9260.
50. Shahi, V. 2007. Highly charged proton-exchange membrane: Sulfonated poly(ether sulfone)-silica polyelectrolyte composite membranes for fuel cells. *Solid State Ion.* 177: 3395–3404.
51. Balster, J., Krupenko, O., Pünt, I., Stamatialis, D., and Wessling, M. 2005. Preparation and characterisation of monovalent ion selective cation exchange membranes based on sulphonated poly(ether ether ketone). *J. Membr. Sci.* 263: 137–145.
52. Asghari, S., Mokmeli, A., and Samavati, M. 2010. Study of PEM fuel cell performance by electrochemical impedance spectroscopy. *Int. J. Hydrogen Energy* 35: 9283–9290.
53. Brunetto, C., Moschetto, A., and Tina, G. 2009. PEM fuel cell testing by electrochemical impedance spectroscopy. *Electr. Power Syst. Res.* 79: 17–26.
54. Yuan, X., Wang, H., Sun, J.C., and Zhang, J. 2007. AC impedance technique in PEM fuel cell diagnosis—A review. *Int. J. Hydrogen Energy* 32:4365–4380.

4

Nanocomposite Membranes in Water Treatment

4.1 Introduction

Water resources are becoming increasingly scarce worldwide. Global water consumption is increasing at more than double the rate of the world's population growth. Population growth, pollution, and climate change, which are all accelerating, are likely to combine to produce a drastic decline in per capita water availability in the coming decades. Water pollution is a bigger environmental threat than climate change. Water bodies in many countries are seriously polluted. This does not mean that climate change is less important an issue. Adverse impact of climate change will make the water issues more alarming in time to come. History reveals that water has been used, misused, and abused throughout. We are yet to accept that water is a limited resource. Agriculture accounts for 70% of all global water use. It is worthwhile to mention that countries in general refrain to charge farmers full operation and maintenance (O&M) costs for irrigation water, let alone investment costs.[1] Even for domestic water, people in very few cities in the world pay the real cost. Water issues are complex. There is a requirement for the availability of sufficient quantity of appropriate quality of water for different end uses. There is need for technological innovations to make the water technologies affordable, acceptable, and accessible for the masses. Technological interventions in water management will play an important role in dealing with the crisis. There is need to make the effluent treatment as an industry-driven process through technological innovations. There are challenges and opportunities.

More than a billion people lack access to safe drinking water.[2] One-third of the world's population live in water-stressed countries, and this is expected to rise to two-thirds by 2025. Lack of clean drinking water and sanitation kills thousands of children every day. Many children are missing school because neither their homes nor their schools have adequate drinking water facilities. Therefore, sustainable water management, including water treatment, is a critical aspect of addressing the challenges of poverty, equity, and related issues. The UN Millennium Development Goal of ensuring environmental sustainability commits governments to *reduce by half the proportion of people without sustainable access to safe drinking water* by 2020,[2] a goal closely linked with the separate goal of access to sanitation and basic hygiene education. Nanocomposite membrane-based water treatment is expected to play an increasingly important role in areas such as drinking water treatment, brackish and seawater desalination, and wastewater treatment and reuse. Existing membranes for water treatment, typically polymeric in nature, are still restricted by several challenges, including the trade-off relationship between permeability and selectivity and low resistance to fouling. Nanocomposite membranes, a new class of membranes made by combining polymeric materials with nanomaterials, are emerging as a promising solution to these challenges. The advanced nanocomposite membranes could be designed to meet specific water treatment applications by tuning their structure and physicochemical properties (e.g., hydrophilicity, porosity, charge density, and thermal and mechanical stability) and introducing unique functionalities (e.g., antibacterial, photocatalytic or adsorptive capabilities).

Nanocomposite membranes can be classified[3] into four categories: (1) conventional nanocomposite, (2) thin-film nanocomposite (TFN), (3) thin-film composite (TFC) with nanocomposite substrate, and (4) surface-located nanocomposite. The typical structures of these membranes are illustrated in Figure 4.1.

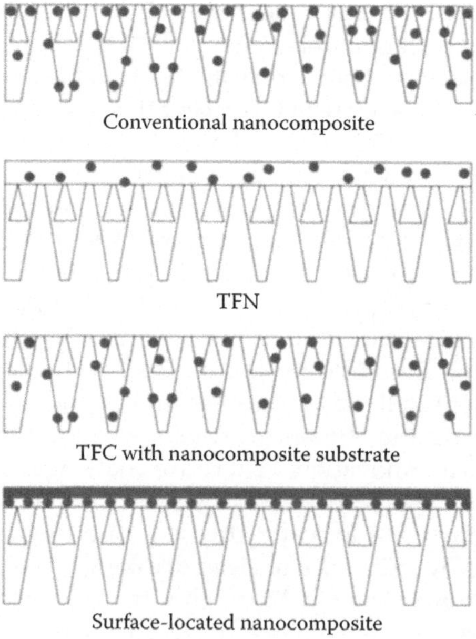

FIGURE 4.1
Classification of nanocomposite membranes. (Reprinted from *J. Membr. Sci.*, 479, Yin, J. and Deng, B., Polymer matrix nanocomposite membranes for water treatment, 256–275, Copyright 2015, Figures 1, 2, 4, and 5, with permission from Elsevier.)

4.2 Conventional Nanocomposites

Mixed matrix membranes have the advantage of ease of fabrication of organic polymeric membranes as well as the mechanical strength and functional properties of inorganic materials. The concept of making nanocomposite membranes has been originally developed to overcome the Robeson upper boundary in the field of gas separation in the 1990s,[4,5] where highly selective zeolites were incorporated into polymers to improve both permeability and selectivity.[6,7] Mixed matrix membranes present an opportunity for tunable water treatment membranes as well, through increased selectivity, targeted functionalities, and improved thermal, chemical, and mechanical stability. The interplay between enhanced properties and defect formation (beyond a certain loading of nanofiller) must be balanced to derive positive benefits without compromising the integrity of the membrane.

Preparation of nanocomposite membranes is mostly based on phase inversion (PI) method in which nanofillers are dispersed in polymer solution prior to the PI process and can be prepared in either flat sheet or hollow-fiber configurations (Figure 4.2). This type of membrane is mainly used in microfiltration (MF) or ultrafiltration (UF) processes due to its typical porous structure.

4.2.1 Carbon Nanotube Reinforcement

The potential of application of carbon nanotube (CNT)-embedded membranes is tremendous in the area of water purification technology. A substantial amount of work has been carried out with impregnation of CNTs in

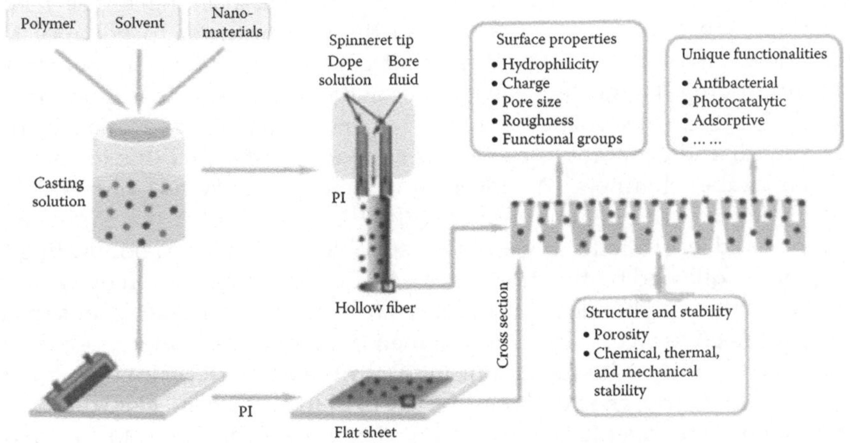

FIGURE 4.2
Schematic of fabrication of nanocomposite membrane by PI method in case of flat sheet and hollow-fiber configuration.

polymer host matrix as one of the reinforcement for incorporation of better properties such as anti-biofouling and better strength.[8–23]

Multiwalled CNT (MWCNT)/cellulose acetate (CA) nanocomposite membranes are prepared by PI[24,25] method. CNTs are functionalized by oxidation purification in a strong acidic medium to enhance their dispersion within the polymer matrix. They are randomly oriented and uniformly dispersed within the membranes. A decrease in the number of macrovoids with an increase in CNT content is observed. The pore sizes decrease with the increase in CNT content. Nanocomposite membrane permeation rates and salt retention rates have been investigated using a 1000 ppm NaCl solution. Permeation rates are found to improve by 54% with a minimal decrease in salt retention (−6%) for the membranes with the lowest CNT content. Further addition of CNTs causes a reduction in permeation rates, which is attributed to the decreased porosity and surface area. Membrane systems have been evaluated[26] for the removal of the extractable organic fraction from oil sands process-affected water (OSPW). Experiments have been performed using *in situ* integrated membrane units consisting of low pressure-driven membrane (LPM) and high pressure-driven membrane (HPM) modules. LPM has been prepared with a polysulfone (PS) PI casting process. HPM is made by polyamide (PA) TFC interfacial polymerization (IP). To change membrane physicochemical properties, each membrane has been developed with and without multiwalled CNTs (MWCNTs). The MWCNTs have been modified with strong acid to enhance dispersion in an organic solvent. The results show that acid-modified MWCNTs developed surface functional groups which increased their hydrophilicity, thus increasing the rejection of hydrophobic pollutants, increasing OSPW permeate flux, and significantly reducing membrane fouling.

The incorporation of CNTs disrupts the polymer packing in the membrane and creates nanoscale cavities in the coating layer to enhance the water flow.[27] Another significant merit of CNT incorporation in membrane is for improving the mechanical stability of the membrane. Incorporation of MWCNTs into an electrospun polyvinyl alcohol leads to improved mechanical strength and durability[28] as well as enhanced water flux due to frictionless nanotube channels. A novel functionalized MWCNT-immobilized polyethyleneimine–polyamide-imide (PEI–PAI) hollow-fiber membrane was designed and fabricated using an easily scalable method.[29] PAI hollow fiber is spun by PI, followed by functionalized MWCNT immobilization by vacuum filtration before a chemical posttreatment using PEI to obtain a positively charged selective layer. The resulting membranes give 44% enhancement in water permeability without significant compromise on the salt permeability compared to membranes without MWCNTs in forward osmosis (FO).

PS membrane blended with carboxylated MWCNTs by PI process[30] increases the pure water flux by increasing MWCNTs content up to 1.5 wt.%. The pure water flux decreases with further loading of MWCNTs. The flux increases due to hydrophilic surface and large surface pores resulting from the addition of MWCNTs. PEG6000 and MWCNTs used for the preparation of chitosan porous membranes[31] with 10 wt.% MWCNT loading gives 4.6 times higher water flux than pure chitosan membranes. The higher water flux is observed due to the formation of MWCNT nanochannels in chitosan pores. Moreover, the tensile strength of the membranes increases with MWCNT addition.

Hydroxyl-functionalized MWCNTs blended with polyacrylonitrile (PAN) to prepare UF membranes[32] by a PI process leads to increase in water flux of the membranes by 63% at 0.5 wt.% loading of MWCNTs compared to neat PAN membranes. The water flux decreases with further increasing the concentration of MWCNTs, but at 2 wt.% loading it was still higher compared to pure PAN membranes. The surface hydrophilicity of the membranes is enhanced by the addition of MWCNTs, as observed by contact angle measurements. The increased hydrophilicity leads to improved water flux. The tensile strength of the membrane at 2 wt.% MWCNT loading increases compared to neat membrane.

4.2.2 Metal Oxide Reinforcement

The incorporation of metal oxide nanomaterials (such as TiO_2,[33–67] SiO_2,[68–76] Al_2O_3,[77–81] Fe_3O_4,[82–86] MnO_2,[87] and ZnO[88]) into polymers not only tunes the structure and physicochemical properties, such as hydrophilicity, porosity, charge density, and chemical, thermal, and mechanical stability, of membranes, but also introduces the unique functionalities such as antifouling and photocatalytic characteristics into the membranes.

Fouling is a major limitation in membrane filtration process. One of the practical strategies to overcome this limitation is the use of advanced antibiofouling membrane material. Nanocomposite membranes made of mesoporous silica (MS) particles[89] in polyethersulfone (PES) by PI method offer antifouling properties. The nanocomposite membrane with 2% MS exhibits an excellent hydrophilicity, water permeability, and good antifouling performance. In addition, the introduction of the MS particles improves the thermal stability of the nanocomposite membranes. The protein adsorption on the membrane surface decreases significantly from 45.8 to 21.4 mg/cm^2 when the MS content increases from 0% to 2%. The UF experiments reveal that the incorporation of MS particles reduces membrane fouling, especially irreversible fouling. Higher MS content (4%) results in particle agglomeration. The incorporation of spray-dried nanostructured silica granules into PS nanocomposite membranes[90] gives significant enhancement of water permeability without sacrificing the separation performance. The mechanical and chemical modifications of TiO_2 nanoparticles give improvements in reducing the nanoparticle agglomeration and fouling in flat sheet membranes. The migration of nanoparticles toward the outer layer occurs in case of mechanical modifications of particle, whereas the migration and size of agglomeration reduces significantly with mechanically and chemically modified titania (TiO_2) particles.[91] Higher thermal resistance, stiffness, and lower elasticity can be observed in fibers made with chemically and mechanically modified particles. Enhancement in initial pure water flux due to lower intrinsic membrane resistance and bigger pore size is also observed, although the rejection is not compromised. Polyvinylidene fluoride (PVDF)/sulfonated PES (SPES) blend membrane was modified using TiO_2 nanoparticles.[92] Sulfonation of PES is carried out, then PVDF/SPES blend membranes are prepared with TiO_2 nanoparticles in the casting solution using PI induced by immersion precipitation technique. Polyvinylpyrrolidone (PVP) (4 wt.% concentration) was added in the casting solution as a pore former. It is observed that the average size of membrane pores in the surface and sublayer is reduced with addition of TiO_2 nanoparticles in the casting solution. It is observed through contact angle measurement that the hydrophilicity of modified membrane is enhanced by addition of TiO_2 in the casting solution. The experimental results imply that the initial flux of TiO_2-entrapped PVDF/SPES membranes is lower than the initial flux of neat PVDF/SPES membrane. The antifouling properties of membranes are improved by changing the membrane surface from hydrophobic to hydrophilic after TiO_2 addition in the casting solution. The type and size of TiO_2 nanoparticles have impact on the morphology, performance, and fouling control aspects of mixed matrix PES nanofiltration (NF) membranes.[93] Surface hydrophilicity of the TiO_2 blended membranes improves due to water affinity improvement of the membrane surface. The NF membranes made from the nanosized TiO_2 particles enhanced the pure water flux and antifouling properties when tested for whey solution. The low concentration of TiO_2 nanoparticles leads to more biofouling reduction because the aggregation of the nanoparticles does not take place at low amount. Poly(phthalazine ether sulfone ketone) (PPESK)/TiO_2 organic–inorganic composite UF membranes prepared by PI method give enhanced performance for treatment of high-temperature condensed water.[94] It is observed that the fingerlike structure in the membrane sublayer is suppressed, and the sponge-like structure begins to be developed. The mechanical strength and thermal stability of composite membrane are improved due to the existence of hydrogen bond between TiO_2 and polymer. The permeate flux is also enhanced due to the improved hydrophilicity and porosity. Compared with neat PPESK membrane, the composite membrane presents improved antifouling properties, showing lower filtration resistance and better flux recovery property during treatment of high-temperature condensed water.

4.2.3 Nanoclay Reinforcement

A common polymer material used for membranes is PVDF, which has gained popularity in water treatment for its chemical robustness. Nanoclay reinforcement in PVDF improves the mechanical properties of PVDF–nanoclay flat sheet membranes.[95,96] The wear property of polymeric and nanocomposite materials, including clay nanocomposites,[97–99] has been widely studied, but this has not been extended to the membrane field. Improvement in abrasion resistance is observed by incorporating nanoclay into nanocomposite materials and PVDF hollow-fiber membranes.[100] Novel PVDF/nanoclay hollow-fiber membranes fabricated by non-solvent-induced phase separation show significant improvement of membrane physical endurance. The incorporation of nanoclay shifts the PVDF crystalline phase from α-phase to β-phase and improves the membrane structure as well as the mechanical properties in terms of stiffness and flexibility. Tensile strength increases from 3.8 to 4.3 MPa with 5.08 wt.% Cloisite-30B loading, whereas break extension increases from 175% to 229% with 5.08 wt.% nanomer. Nanoclay gives reinforcement to nanocomposite membranes.[101–108]

4.2.4 Organic Material Reinforcement

There are several organic materials such as cyclodextrin,[109–112] polyaniline,[113–118] polypyrrole,[119] chitosan,[120] polyhedral oligomeric silsesquioxane,[121] and semi-interpenetrating network polymeric nanoparticles[122] which are used to make nanocomsposite membranes using conventional polymers such as PS, PES, PAN, CA, and ethylene vinyl alcohol for increase in hydrophilicity, improvement in adsorptive properties, and anticompaction as well as antifouling behavior of resultant membranes.

4.2.5 Dendrimer Reinforcement

Dendrimers and their derivatives are substances with diverse analytical, biomedical, and environmental applications[123–127] due to their unique molecular structure, easy functionalization, and manipulation of their terminal groups.[128–133] Dendritic polymers basically consist of a multifunctional core, high degree of repeated branching units, and high density of surface functional groups.[134] A commercial polyamidoamine (PAMAM) dendrimer can be used in separation systems for recovery of heavy metals from aqueous solution by means of chelating agents in pollution remediation processes. Particularly, aqueous heavy metal solution treated with PAMAM before passing them through an UF membrane has been proposed for water and soil remediation.[135–139] Raw PAMAM, aromatic PAMAM, and PAMAM coated with polyethylene glycol can be used in the modification of polymeric reverse osmosis (RO) membranes for remediation of copper, nickel, and chromium ions from waste water and for CO_2 separation.[140–146] Diaminobutane-based poly(propyleneimine) dendrimer functionalized with 16 thiol groups, DAB-3-(SH)16, successfully embedded in a swollen cellulosic support[147] in order to achieve an easily handle engineered membrane, shows that dendrimer inclusion improves the membrane elastic behavior. Young's modulus increases by about 20%. A significant reduction in the permeation of toxic heavy metals (Cd^{2+}, Hg^{2+}, and Pb^{2+}) is noted, indicating the possible application of dendrimer-modified membrane in electrochemical devices for water remediation.

4.2.6 Zeolite Reinforcement

Zeolites are microporous, aluminosilicate minerals commonly used as commercial adsorbents and catalysts. They belong to the members of the family of microporous solids known as *molecular sieves*. The term *molecular sieve* refers to a particular property of these materials, that is, the ability to selectively sieve molecules based on a size-exclusion process. This is due to the pore structure of molecular dimensions. The maximum size of the molecular or ionic species that can enter the pores of a zeolite is controlled by the dimensions of the channels. Zeolites are used as reinforcement materials in polymeric host matrix[148–150] to increase hydrophilicity and improve cross-linking property and molecular sieving. Zeolite/PVDF nanocomposite membranes possess enhanced mechanical stability in terms of tensile strength and break elongation[151] due to the good interaction between zeolite and PVDF matrix, where zeolite nanoparticles could act as a cross-linking agent for the polymeric chains and increase membrane rigidity. The cross-linking phenomenon is attributed to the hydrogen bonds between polymer chains and hydroxyl groups T-OH (T = Si, Al, P) on zeolite surface.[152]

4.2.7 Silver Reinforcement

As biofouling causes deterioration in membrane performance such as flux decline, increase in O&M costs, and membrane degradation, efforts are directed to develop antifouling strategies. To reduce biofouling, functional membranes containing biocides or antibacterial materials have attracted tremendous interest. Silver is one of the most widely studied biocides because of its excellent biocidal properties.[153–155] Silver nanoparticles (AgNPs) have been successfully introduced into various membrane materials such as PS,[156–158] PES[159–161], PVDC,[162,163] PA,[164,165] and chitosan.[166] The addition of AgNPs into the polymer membranes improves the membrane performance in terms of their flux and fouling resistance, attributing to an increase of hydrophilicity or change in membrane morphology. However, the chemically produced AgNPs often have problems with particle stability. The beneficial effects of added particles are often limited by aggregation and poor compatibility with the polymeric matrix. In addition, the improved durability of silver-containing membrane and the simultaneously reduced potential risks of released silver ions at high load to the environment and filtration process are still challenges for excellent membrane performance.

AgNPs can prevent the bacteria permanently attaching to the membrane surface. However, this mechanism depends on the location of AgNP into membrane matrix. The location of AgNP can change depending on the polymer type and PI parameters. PS, PES, and CA polymers are used for preparation of flat sheet bare and nanocomposite membranes. It is found that AgNPs mostly accumulate onto the top and skin layers of PES and PS nanocomposite membranes. The interaction between AgNPs and bacteria depends on the release of ionic silver from AgNPs embedded in membrane. The nanocomposite membranes that store AgNPs at surface have the best antibacterial properties. Antibacterial PS UF membranes are made[167] using different AgNPs in the casting solution. UF mixed matrix membranes are prepared using AgNPs in the PS casting solution. The crystallinity of the membrane matrix decreases by decreasing the AgNP size. More hydrophilic surface is produced for membranes in the presence of AgNPs. AgNP-containing membranes showed high antibacteriality especially with smaller AgNPs. High-concentration and high-stable biogenic AgNPs with the averaged diameter of only 6 nm was first extracted from the supernatant of *Lactobacillus fermentum*.[168] The biogenic nanocomposite PES membranes are prepared by adding different amounts of biogenic silver nanoparticles into the dope solution. The nanocomposite membranes are tested for physical properties with pure water permeability. The results demonstrate that the nanoparticles are well dispersed into PES matrix without aggregation. They slightly increase the hydrophilicity of the PES membranes and improve the water permeability. The protein adsorption on the membrane surface decreases significantly due to the increased hydrophilicity and the improved smoothness of membrane surfaces. To obtain both organic antifouling and antibacterial properties, acrylamide is grafted[168] onto a PES hollow-fiber membrane, and AgNPs were then formed within the acrylamide layer. The hydrophilicity of the membrane surface is improved by acrylamide grafting, leading to a reduction in membrane fouling. Bare PES membranes have no antibacterial activity and bacteria grow on the membrane surface, whereas PES membranes containing AgNPs indicate high antibacterial activity. Thus, polymer membranes containing AgNPs within the acrylamide gel layer have high potential for applications with organic antifouling and antibacterial properties.

Mixed matrix porous polymeric membrane with antifouling properties is prepared by the PI method[169] from a quaternary system of PS/*N,N*-dimethylformamide/PVP/nanosilver (nAg). The effect of casting mixture composition on the membrane morphology, performance, and antifouling (including anti-biofouling) properties is evaluated by microscopic, spectroscopic, and surface characterization techniques. Composition of casting mixture affects the morphology and selectivity of membranes. The incorporation of nAg into a casting mixture containing 5 wt.% PVP introduces morphological changes in the membrane structure, including increase in pore sizes, suppression of macrovoids, and thinning of the skin layer. These changes enhance the separation and antifouling properties of membranes. Nanoparticles of silver, copper, and silver–copper mixture have been impregnated to the PS[170–172] host matrix and the biofouling resistance behavior of each membrane surface has been examined. The performance of the membranes has been evaluated in terms of pure water permeability and solute rejection studies. The silver-impregnated membranes possess the best biofouling-resistant behavior. Other nanomaterials such as copper[173–175] and selenium[176] have also been explored for their potential applications to make antimicrobial membranes.

4.2.8 Graphene Oxide Reinforcement

Graphene oxide (GO) is gaining interest in the field of material research due to its high surface area, outstanding electron transport, and mechanical properties. The atomically thin carbon sheets when incorporated appropriately into polymer matrix can significantly improve the physical properties of the host polymer at extremely low dope concentration. The objective behind choosing GO as an additive lies in its hydrophilic and pH-sensitive behavior. It is also demonstrated that GO exhibits the property of negative surface charge throughout the entire pH range.[173] Due to the different types of hydrophilic functional groups present on the surface of GO, it can take up water very easily. The water uptake increases as the degree of oxidation increases.[177] GO-dispersed PS mixed matrix membranes prepared by wet PI method[178] exhibit improved salt rejection. Membrane with 2000 ppm GO loading gives 72% Na_2SO_4 rejection at 4 bar applied pressure. The salt rejection shows an increasing trend with an increase in the pH. PVDF/GO UF membranes prepared[179] by immersion precipitation PI process show a large amount of —OH groups due to the introduction of GO nanosheets, which improve the surface hydrophilicity of the modified membrane. In permeation experiment, the water flux is improved after blending GO. The flux recovery ratio (FRR) and the fouling resistance suggest that PVDF/GO UF membranes have better antifouling properties than pure PVDF due to the changes in surface hydrophilicity and membrane morphologies. GO nanosheets are employed[180] as nanofillers to improve hydrophilicity and antifouling performance of a polymer-based membrane (PES), resulting in high-performance UF membranes with substantially improved flux.

Using the immersion PI process, PVDF and GO dissolved in *N,N*-dimethylacetamide combined to prepare organic–inorganic blended UF membrane[181,182] shows that the properties and structure of the blended membrane are improved when the content of GO added in the casting solution was 0.20 wt.%. The permeability of the blended membrane increases by 96.4% with a slight change of retention; the tensile strength increases by 123%. The contact angle decreases from 79.2° to 60.7°, which implies that the antifouling ability of the membrane is improved.

Membrane bioreactor (MBR) is quite effective and advantageous compared with conventional wastewater treatment process,[183] in terms of space requirement. However, the high cost due to membrane material/replacement and maintenance due to membrane fouling restricts the application of MBRs.[184] The bio-cake layer accumulated on the membrane surface is the main cause of membrane fouling in the MBR process.[185–187] Blending modification with GO nanosheets and its application in long-term MBRs have not been extensively studied or reported. A composite MF membrane, prepared by blending PVDF and hydrophilic GO nanosheets, can be successfully used[188] in a submerged MBR system. The PVDF/GO composite membrane demonstrates sustained permeability, lower cleaning frequency, and filtration time which is 3 times longer than that of the PVDF membrane. In terms of anti-extracellular polymeric substance accumulation, the PVDF/GO composite membrane shows lower membrane resistance, particularly lower pore plugging resistance, than the PVDF membrane.

4.2.9 Hybrid Material Reinforcement

Titania (TiO_2)-coated multiwalled nanotubes (MWNTs) into the PES matrix membrane show[189] enhanced hydrophilicity and pure water flux. The resulting membrane exhibits improved fouling resistance due to the lower surface roughness and synergistic photocatalytic activity. The existence of CNTs is believed to reduce the electron/hole recombination and improve the photon efficiency. A multifunctional membrane designed by incorporating gold nanoparticles (AuNPs)/exfoliated graphite nano-platelets into PS membrane[186] shows enhanced compaction resistance and permeability as well as superior catalytic property on the reduction of 4-nitrophenol to 4-aminophenol by $NaBH_4$, where AuNPs serve as the catalyst. The structure and catalytic activity of such membranes can be controlled separately by changing the relative contents of corresponding components in the nanofiller hierarchy. Nanocomposite membranes containing silica (SiO_2)/GO hybrid materials[187–191] exhibit much better permeability, protein rejection, and fouling resistance than SiO_2/PS and GO/PS membranes. The synergistic effect of SiO_2/GO is due to its high hydrophilicity as well as its special sandwiched structure that facilitated its dispersion in the PS matrix.

Hybrid Nafion membranes are prepared[192] using various fillers, such as anatase-type TiO_2 nanoparticles, GO, and organo-modified GO (abbreviated as GOSULF) for water purification applications. The photocatalytic properties of the hybrid membranes are evaluated using the azo dye methyl orange (MO) in aqueous solutions. The two main effects contributing to the decolorization of the dye, adsorption in the membrane and degradation in solution, are investigated in detail. Among the Nafion membranes under investigation, Nafion–GOSULF, for the first time used for water purification purposes, shows the promising results in MO adsorption and photocatalytic degradation.

A partially reduced GO (rGO)/TiO_2 nanocomposite with five different molar ratios (rGO/TiO_2: 3/97, 30/70, 50/50, 70/30, and 90/10) can be synthesized[193] and characterized using X-ray diffraction and scanning electron microscopy (SEM) techniques. The PVDF mixed matrix membranes containing 0.05 wt.% of the rGO/TiO_2 nanocomposite prepared by the PI method show enhanced hydrophilicity, higher pure water flux, and higher FRR compared to the bare PVDF. A comparison of pure water flux of TiO_2- and rGO/TiO_2-containing membranes shows that dispersion of inorganic TiO_2 nanoparticles on the GO surface decreases the aggregation of the nanoparticles and improves the characteristics of the mixed matrix membrane. The blended PVDF membrane containing 0.05 wt.% rGO/TiO_2 nanocomposite with a GO-to-TiO_2 ratio of 70/30 show improved permeability and antifouling performance property of PVDF membranes makes them prone to fouling, so hydrophilic additives are frequently used to improve their hydrophilicity and antifouling performance. AgNPs prepared by using PAMAM dendrimers as templates[194] to form silver-PAMAM dendrimer nanocomposites (Ag-DENPs) have good hydrophilicity and antibacterial performance. Ag-DENPs increase the surface roughness and decrease the pore size of PVDF membranes. They provide significant improvement in PVDF membrane surface hydrophilicity. Membrane permeation and antibacterial tests carried out to characterize the antifouling performance of PVDF membranes show that FRR increases about 40% in the presence of Ag-DENPs on the PVDF membrane surface. The anti-organic fouling performance of PVDF membranes is elevated.

4.3 Thin-Film Nanocomposites

TFC membrane consists of an ultra-thin barrier layer (commonly made of PA) atop a more porous supporting layer. It is interfacially synthesized RO[195] or NF membrane and widely used to desalinate seawater/brackish water or remove heavy metals, hardness, organic micropollutants such as pesticides, and disinfection by-products. The development of FO/pressure-retarded osmosis (PRO) processes has stimulated the development of TFC membrane, as there is a potential for further energy saving or even energy generation during the seawater/brackish water desalination.

Efforts have been directed to improve water flux, solute rejection, and antifouling properties of TFC membranes such as (1) to select or modify the supporting layer so the adhesion between the barrier layer and the supporting layer could be enhanced, and (2) to optimize the barrier layer by varying the IP conditions, that is, changing monomers, applying physical coating, or chemical modification. With the advent of technologies to make nanocomposite materials, a new concept has been proposed based on dispersing nanomaterials into the ultra-thin barrier to improve membrane performance for water filtration.[196] Nanomaterials used in the past for conventional nanocomposite membrane preparation have also been explored to prepare TFN membranes, including zeolites, CNTs, silica, Ag, and TiO_2. The common method is through the *in situ* IP process between aqueous *m*-phenylenediamine (MPD) and trimesoyl chloride organic solution as

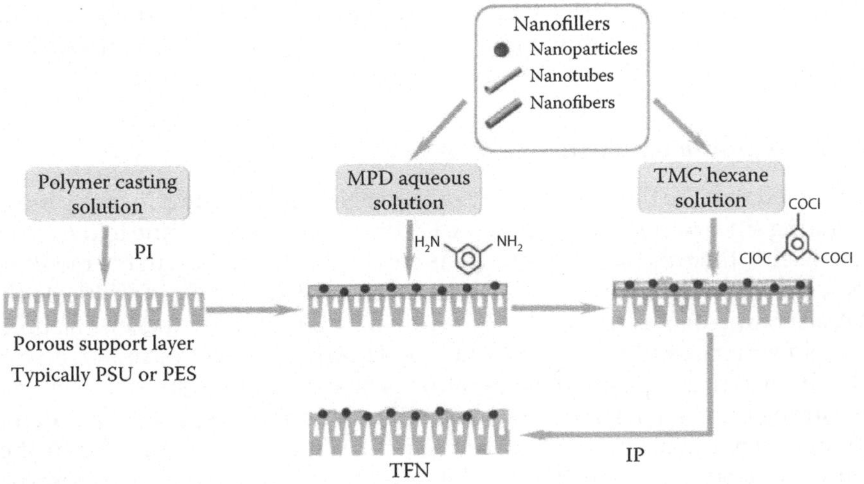

FIGURE 4.3
Schematic of fabrication of TFN membrane by IP.

presented in Figure 4.3. The nanofillers can be dispersed either in aqueous or in organic phase.

4.3.1 Improvement of Chlorine-Resistant Properties

The amide linkage in PA-TFC membranes is susceptible to attack by chlorine, leading to the undesired degradation which adversely affects the life of the membrane. Therefore, much attention has been paid to prevent the membrane from being exposed to strong oxidants, particularly chlorine.[197] Solutions proposed include coating the PA surface with chlorine-resistant materials[198] and introducing specific functional groups to the amide structure.[199]

Introducing nanomaterials into the PA structure provides a new dimension to design chlorine-resistant membranes. When MWNTs are incorporated into PA thin-film layer, the membrane shows a much improved chlorine resistance.[200,201] It is believed that the interactions between the carboxylic group of modified MWNTs and the amide bond make the membrane more stable against chlorine. Similarly, the MWNT-embedded TFN membranes have a better chlorine resistance due to the protection of amide linkage by electron-rich MWNTs.

Amide bonds are the main target of chlorine attack. Introduction of additional amide bonds or amino groups to the membrane appears to be a reasonable strategy to protect the PA cross-linking structure. Aminated zeolite[202] or hyperbranched PA-modified silica nanoparticles[203] into the PA thin-film layers show an enhanced chlorine resistance due to (1) the intermolecular hydrogen bonding between aminated nanoparticles and PA structure, which mitigated the replacement of hydrogen by chlorine and (2) the additional amide bonds or amino groups introduced by aminated nanoparticles that render the membrane more resistant to the chlorine attack.

4.3.2 Improvement in Thermal Stability

It is noted that the thermal stability of the TFN membrane improves by incorporating silica nanoparticles into the PA thin-film layer[204,205] due to the stronger electrostatic and steric interactions between silica and PA in the modified polymer network structure. The incorporation of zeolite (NaX)[206] inside the PA structure also leads to a TFN membrane with a higher thermal stability and an enhanced filtration performance. However, no significant change in the thermal decomposition temperature is observed, suggesting that there is no strong interaction between the nanofillers and polymer backbone chains. To improve the interaction between nanofiller and PA structure for further improvement of thermal stability, TiO_2 has been aminosilanized prior to IP process,[207] resulting in a membrane that indeed has a better thermal stability, in addition to improved permeability and selectivity.

4.3.3 Effects on Antifouling Properties

The introduction of hydrophilic nanofillers into the PA structure increases the surface hydrophilicity and helps to mitigate surface fouling. After incorporating hydrophilized ordered mesoporous carbons (H-OMCs) into the PA thin-film layer, membranes show an enhanced surface hydrophilicity.[204] Surface contact angle decreases from 79° to 46°. With an increasing H-OMC loading, the adsorption of bovine serum albumin (BSA) is clearly decreased, indicating that the antifouling capability to organic protein materials is enhanced by H-OMCs. It is also noted that the TFN membrane surface exhibits significantly improved hydrophilicity after incorporating *in situ* hydrophilic surface-modifying macromolecules.[208] Their long-term (200 h) fouling tests with sodium humate, silica particles, and chloroform show that the TFN membranes have a much lower flux reduction compared to the conventional TFC membranes. The reduced flux reduction is also observed when using TFN membrane prepared by MWNTs to treat feed solution containing foulants $Ca(HCO_3)_2$ or BSA.

4.3.4 Effects on Antibacterial Properties

Use of nanosize antimicrobial and biocidal materials can introduce antibiofouling capability in nanocomposite membranes. By incorporating AgNPs into the PA structure during the IP process and homogenously dispersing in the organic phase, the nanocomposite membrane can demonstrate appreciable anti-biofouling capability.[209] By incorporating silver salts (such as silver citrate hydrate, silver lactate, and silver nitrate) into the PA structure, the antimicrobial capability can be introduced in the membrane. Membranes prepared by dispersing AgNPs in MPD aqueous solution[210] show antibacterial properties. However, there is need to carry out long-term studies with feed stream containing bacteria.

Considering the dense structure and good ion rejection property of PA barrier, most of the embedded AgNPs (released as silver ions) inside the PA matrix are unlikely to be delivered to the interface between the membrane and the feed stream containing bacteria to act as the biocidal material. Although this kind of membrane may show a relatively slow leaching rate, it cannot guarantee a long-term antimicrobial activity. Accordingly, an attempt is made to prepare an anti-biofouling membrane by incorporating zeolite nanocrystals in the silver form.[211,212] Due to the significant biocidal activity of nanocrystals in silver form, it is expected that the antimicrobial property of TFN membrane would be improved. The TFN membrane exhibits higher water permeability and comparable salt rejection, but it does not show the strong biocidal activity as expected.

4.3.5 Permeability and Selectivity

In the TFC membrane, the PA thin-film layer largely controls the membrane performance in terms of permeability, selectivity, and fouling resistance. The incorporation of nanomaterials into the PA layer modifies the physicochemical properties of the membrane such as hydrophilicity, charge density, porosity, and cross-linking, and provides special water channels having potential to overcome the permeability/selectivity trade-off relationship.

Based on the solution–diffusion theory widely used to describe the mass transport in the PA-TFC membrane,[213] an increase in hydrophilicity of membranes could facilitate water solubilization and diffusion through the membrane, thus improving water permeability.[214] Most of the studies using hydrophilic nanofillers result in a TFN membrane with decreased contact angle, indicating an enhanced surface hydrophilicity. For example, the contact angle of zeolite-PA TFN membranes decreases from around 70° to 40° with increasing zeolite loading from 0% to 0.4% (w/v) in the organic phase. The contact angle for the oxidized MWNTs-PA TFN membranes decreases from around 70° to 25° with increasing MWNT loading from 0% to 0.2% (w/v) in the aqueous phase.[214] For the MS-PA TFN membranes, the contact angle decreases from around 57° to 28° with increasing silica loading from 0% to 0.1% (w/w) in the organic phase.[215] These studies exhibited an enhanced water permeability with increasing nanofiller loading.

The decrease of contact angle in the presence of embedded nanoparticles is caused by hydration and hydrophilicity. The nanoparticles may hydrate and release heat when contacting with the MPD aqueous solution. This process may affect the IP reaction between MPD and TMC, and subsequently the chemical structure of the PA thin film. If a large number of the acyl chloride groups in TMC remain on the surface without reacting with amine groups, the hydrolysis of acyl chloride may generate carboxylic acid functional groups; thus, surface hydrophilicity increases.[215] The embedded hydrophilic nanomaterials can be exposed on the membrane surface, providing more hydrophilic functional groups to membrane surface. This mechanism is similar to the one in the conventional nanocomposite membrane that improved surface hydrophilicity.

Besides the hydrophilicity, the cross-linking condition and the thickness of thin-film layer are also important factors controlling the water permeability and salt rejection.[216] In general, lower degree of cross-linking and thinner thickness of thin-film layer result in higher water permeability. Embedding nanofillers in the PA matrix can reduce the cross-linking in the thin-film layer by disturbing the reaction between amine groups and acyl chloride groups or even forming nanovoids around the interfaces between nanofiller and PA matrix.

It is noted[217] that TFN membranes are less cross-linked than the corresponding TFC membrane; however, the cross-linking extent does not show a strong correlation with water or salt permeability. It suggests that defects or molecular sieving effect might have played a major role in the membrane separation performance. In another study examining the effects of MWNTs on TFN membrane performance, it is shown[218] that the nanogaps between the external surfaces of MWNTs and the polymer matrix can provide low-resistance pathway for solvent leading to the permeability enhancement. The PA matrix is still the main contributor to the solute rejection.

In addition to causing changes for the polymer cross-linking, the incorporation of nanofillers may also provide additional channels for the transport of water but not solutes. Introduction of the zeolite-A nanoparticles into TFN membrane results in higher water flux with constant salt rejection. The permeability enhancement decreases with blocking of the pores inside the zeolite, although the permeability remains higher than that of conventional TFC membrane. The hydrophilic nanoparticles provide preferential flow paths for water molecules. Compared to the nonporous silica nanoparticles, MS nanoparticles containing highly ordered hexagonal pores give significantly higher impact on water permeability.[219] With the nonporous silica nanoparticles, the reduction in polymer cross-linking can still occur, but there will be no water permeation through internal structures of nanoparticles; therefore, although the observed water flux in the membrane containing nonporous silica is higher than the conventional TFC, it is less than that in the membrane containing MS. It indicates that the internal pores of nanofiller contribute significantly to water permeability enhancement.

In order to overcome the permeability/selectivity trade-off relationship normally observed for polymeric membranes, it is desirable in nanocomposite membrane that solute rejection should remain near the same when the water permeability is improved. The high solute rejection can be maintained by a combination of steric and Donnan exclusion.[220] Therefore, the goal of breaking the trade-off relationship mainly relies on one or more factors such as a moderate reduction on cross-linking, an enhanced membrane surface charge density, an appropriate nanovoid, and additional water channels. For example, zeolite can provide molecular sieving channels and enhanced charge density. MS can provide large water channels combined with enhanced charge density, whereas aquaporin can provide exclusive water channels.

To fully utilize the favorable properties of nanomateirals, the nanofillers of appropriate size, internal structure, and surface properties are introduced. Suitable interfacial interactions between them and polymer matrices are ensured. A typical thickness of the thin-film layer approximates several hundred nanometers. The nanofiller particle should not be too large to ruin the barrier. Small-sized nanoparticles (about 97 nm size) in the TFN membrane enhance its permeability. To improve the dispersion of nanofillers and the interaction between nanofillers and polymer matrix, it is common to modify nanofillers surface prior to the embedding process. As an example with respect to CNTs, the most common modification is to treat CNTs by concentrated acid to generate oxygen-containing functional groups so that an adequate dispersion can be achieved for the preparation of the membrane. It is also needed to provide good interactions between CNTs and polymer matrices. This is consistent with the methodology used in the conventional nanocomposite membrane preparation.

4.4 TFC with Nanocomposite Substrate

This class of membrane has been developed to investigate the effects of nanofiller on membrane compaction behavior. In this class, silica or zeolite nanoparticles were embedded into the PS substrate,[221] which is then used in the IP process to prepare TFC membrane. The prepared membrane shows a higher initial permeability and lower flux decline during the compaction compared with the original TFC membrane. The nanomaterial provides necessary mechanical support to mitigate the collapse of porous structure and resist thickness reduction due to compaction. Membranes with nanocomposite substrate undergo far less physical compaction and play an important role in maintaining high water permeability,[222] as illustrated in Figure 4.4.

This concept is implemented to mitigate internal concentration polarization (ICP) occurring inside the porous support layer, which may have adverse impact on the FO and PRO processes because it can significantly reduce the available osmotic driving force and hence lower the water flux.[223] The nanocomposite substrate has an enhanced hydrophilicity.

Incorporation of zeolite into PS substrate reduces ICP and increases water flux under either *active layer facing draw solution" or "active layer facing feed solution* condition[224] due to improved porosity, better hydrophilicity, and additional water pathways through porous nanoparticles. It demonstrates the possibility

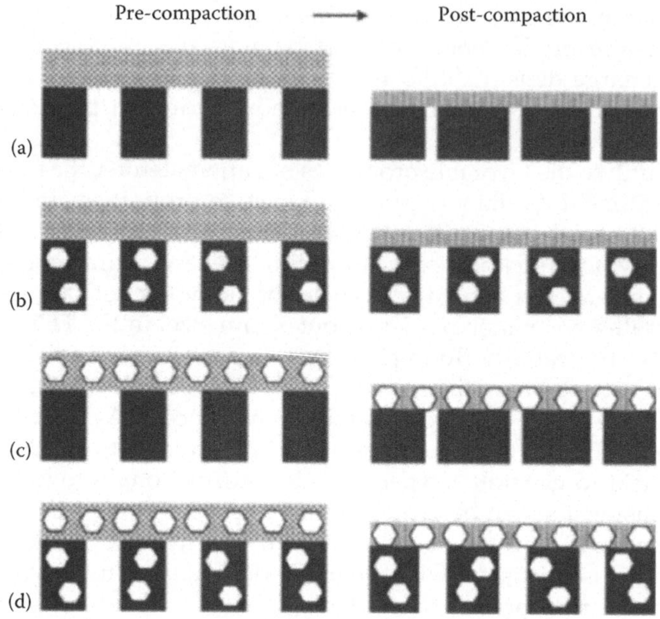

FIGURE 4.4
Schematic diagram showing the decrease in extent of compaction by incorporating nanoparticles in TFC substrate. (a) Conventional TFC membrane, (b) nanoparticle in PS substrate, (c) nanoparticle in PA film, and (d) nanoparticle in PS substrate and PA film.

of using porous nanoparticles and nanocomposite substrate to control ICP in FO operation. Subsequent studies have been carried out using MWNTs[225] and TiO_2[226–232] to mitigate ICP problem in the nanocomposite substrate for FO process. The resulting membranes have shown enhancement in performance of FO and reduction in ICP. The substrate also shows an enhanced tensile strength, which is a desirable property for practical application.

4.5 Bioinspired Membranes

Under this category, the discussions have been carried out on vertically aligned CNT, graphene, and aquaporin membranes.

4.5.1 CNT Membranes

The introduction of CNTs[233] immediately followed the synthesis of fullerenes in macroscopic quantities, and since then the research in this field has been in continuous evolution. Because of their outstanding properties,

CNTs are considered as attractive candidates in diverse nanotechnological applications, such as fillers in polymer matrices, sensors, and many others. They have great potential to offer a membrane material with improved properties of flux for water passing through CNT channels, better selectivity by side wall, and tip functionalization and antifouling characteristics for both hydrophilic and hydrophobic fouling as well as biofouling by selective chemical modification of CNT architecture. It has been reported[234–238] that CNTs have been used for removal of lead, cadmium, fluoride, and arsenate contamination from contaminated water. The alignment and functionalization of CNTs are important in deciding the sorption capacity of CNTs. A well-controlled defined macro-architecture of aligned CNTs has also been reported for removal of petroleum products and microorganisms.[239] The integrated system of CNTs and scale-up are challenging issues.[240]

4.5.1.1 Functionalization of CNTs

The functionalization of CNTs opens up excellent avenues for membranologists to exploit the potential benefits of CNTs and gain improved separation characteristics as well as high throughput. In addition, attaching organic moieties leads to better anchoring of nanotubes in host materials, and thus yield better reinforcement of composites. The main approaches for the modification of CNTs can be grouped into two categories: (1) the covalent attachment of chemical groups through reactions onto the π-conjugated skeleton of CNT and (2) the endohedral filling of their inner empty cavity.

A noteworthy breakthrough in the area of development of nanotube chemistry is the oxidation of CNT in concentrated nitric acid.[241] Such a drastic condition helps in opening of the CNT tips as well as oxidative etching along the sidewalls enabling the decoration of walls with various oxygen-containing groups (mainly carboxyl group). The incorporation of carboxyl group exposes various useful sites in CNTs for further modification as per the requirement, for example, ester or amide bond formations can take place. In addition, the formation of anhydride at the tube ends can take place through which the rings of CNTs are accessible.[242] The most important implication for introduction of carboxyl group lies in the fact that the van der Waals forces existing between the individual CNTs are reduced, and hence, the CNTs can be made water soluble by addition/substitution of new moieties (CNTs as grown are not soluble in any solvent).

However, addition reactions such as fluorination,[243–245] hydrogeneation,[246–248] and cycloaddition[249–253] help in direct coupling of functional groups onto the π-conjugated carbon framework. The fluorine atoms of fluorinated CNTs can be replaced through nucleophilic substitution reaction, and thus, functional groups of alcohols, amines, and Grignard reagent can be successfully incorporated on the CNT sidewall.[254,255]

4.5.1.2 Wetting and Filling of CNT Cavity

Wetting properties of CNT determines which liquid will fill the tube by capillary action and cover the inner surface. The Young–Laplace equation relates the pressure difference ΔP across the liquid–vapor interface in a capillary to the surface tension of the liquid (γ) and contact angle (θ) between the solid and the liquid as follows:

$$\Delta P = 2\gamma r^{-1} \cos\theta \tag{4.1}$$

where:

r is the radius of curvature of the meniscus

The contact angle θ is an indicator of the strength of the interaction between the liquid and the solid interface relative to the cohesive forces in the liquid

If θ is smaller than 90°, the contact between the liquid and the surface is said to be wetting and ΔP is positive. Therefore, the liquid will be pulled into the capillary spontaneously as there is an energy gain in the wetting process. If θ is larger than 90°, the contact angle is said to be nonwetting and ΔP will be negative. Therefore, when $\theta > 90°$, the only way to introduce liquid into a capillary is to apply pressure larger than ΔP.

The wetting properties of the CNTs measured experimentally[256] imply that there is a cutoff point in the surface tension of the liquid above which wetting no longer occurs. The capillary action of nanotubes observed[257–259] for oxides indicates that the surface tension values of compounds are below the cutoff point determined.

If the surface tension of the liquid or molten salt is sufficiently low (<200 mN/m), wetting occurs, and according to Equation 4.1, the CNTs can be filled by capillarity. However, to fill CNTs with higher surface tension materials (such as metals or metal oxides), one has to force the molten materials into open CNTs by applying large pressures followed by cooling of sample before pressure is dropped, so that the material is trapped inside. The most convenient way, however, is to use a low-surface tension solvent carrier. In such a case, the compound of interest can be dissolved in nitric acid, which would open up the tips of CNTs, and the material is carried into the empty cavity of CNTs by capillarity.

4.5.1.3 Mass Transport through CNT Channels

Extremely high aspect ratios, molecularly smooth hydrophobic graphitic walls, and nanoscale inner diameters of CNTs give rise to the unique phenomenon of ultra-efficient transport of water and gas through these ultra-narrow molecular pipes. Water and gas molecules move through nanotube pores much faster. The proposed water transport mechanism has a distinct similarity to the transport mechanisms of biological ion channels.

Thermodynamic and transport properties of confined water differ substantially from those observed in the bulk due to the reduced diameter of CNTs.[260] When simple fluids are simulated within narrow molecular sieves, it has been shown that their diffusion is dominated by concerted events in which multiple molecules move simultaneously.[261,262] These concerted events are due to strong mismatches in (1) the distance between binding sites along the pore axis and (2) the distance between adsorbed molecules that minimizes adsorbate–adsorbate interactions.[263] On the contrary, the inherent smoothness of the interior of CNTs generates exceptionally high diffusion coefficients for simple gases[264–266] as well as for water.[267–269] Water molecules move occasionally along the single wall nanotube (0.8 nm diameter and 1.35 nm long) axis via bursts of hydrogen-bonded clusters of molecules.[270]

The average flow velocity υ of an incompressible, creeping liquid, that is, Reynolds number much less than one inside a tube with a uniform cross-sectional area, is given by Darcy's law:

$$\upsilon = \gamma\left(\frac{\Delta P}{L}\right) \tag{4.2}$$

where:

ΔP is the pressure difference across the tube
L is its length
γ is the hydraulic conductivity[271]

Although Darcy's law is an empirical expression, the hydraulic conductivity of a Newtonian liquid in a circular tube subject to the no-slip boundary condition, $\gamma_{\text{no-slip}}$, can be found directly from the no-slip Poiseuille relation.[272] Liquid slip at the solid–liquid boundary, confinement induced reduction in the liquid viscosity, and subcontinuum changes to the liquid structure can all cause the actual hydraulic conductivity (γ_{actual}) to exceed the calculated hydraulic conductivity from the Poiseuille relation.[273] This increase in γ leads to the definition of a flow enhancement factor, ε, given by $\varepsilon = \gamma_{\text{actual}}/\gamma_{\text{no-slip}}$. For CNTs with diameters larger than 1.6 nm, the variation in hydraulic conductivity with CNT diameter can be understood in terms of slip at the water–CNT boundary and diameter-related changes to the water viscosity. In CNTs with smaller diameters, however, water molecules have been shown to assemble into diameter-dependent one-dimensional structures for which neither the slip length nor the effective viscosity is well defined. This confinement-induced change to the liquid structure necessitates a subcontinuum description of the liquid.[274,275]

4.5.1.4 Antimicrobial Property of CNTs

The effect of CNTs on bacteria and virus has not received enough attention, probably due to the difficulty of dispersing CNTs in water. Surfactants or polymers such as PVP or Triton-X are generally used to facilitate the dispersion.

Few studies are available on single-walled nanotubes (SWNTs) with antimicrobial activity toward gram-positive and gram-negative bacteria, and the damages inflicted are attributed to either a physical interaction with or oxidative stress on cell membrane integrity.[276,277] CNTs may therefore be useful for inhibiting microbial attachment and biofouling on surfaces. However, the degree of aggregation[278] and the bioavailability of the nanotubes[279] need to be considered to exploit the antimicrobial properties effectively.

4.5.1.5 Preparation of CNT Membranes

Keeping in view the application of CNT-based composites in the area of water purification, the *ex situ* alignment method is preferred where the CNTs are aligned in advance using CVD method. Then, they are compounded with the polymeric matrix by either *in situ* polymerization of some monomers or spin coating/dip coating of polymer solution onto the aligned CNT matrix. There are numerous challenges associated with each step of membrane making starting from the growth of CNTs to membrane performance evaluation and scale-up. Primarily, there are four approaches[280] to the synthesis of membranes based on CNTs as shown in Figure 4.5.

1. Template synthesis approach in which carbonaceous material is deposited inside the preexisting ordered porous membranes such as anodized alumina also known as the template-synthesized CNT membranes.[281] Scanning electron micrograph of the nanotubes after dissolution of the template is shown in Figure 4.5a.
2. Membranes based on the interstice between nanotubes in a vertical array of CNTs, referred as the dense-array outer-wall CNT membrane in which the fluid transport is through the interstice between the nanotubes, although some transport can occur through some open-ended tubes. SEM image demonstrating the dense array of CNTs is shown in Figure 4.5b.
3. Encapsulation of as-grown vertically aligned CNTs by a space-filling inert polymer or ceramic matrix followed by opening up the CNT tips using plasma chemistry, or the open-ended CNT membrane[282,283] in which fluid transport occurs through the inner core of the CNT. SEM image shows the cross section of the membrane with aligned CNTs in an impervious polymer matrix (Figure 4.5c).
4. Membranes composed of nanotubes as fillers in a polymer matrix, also known as mixed matrix membranes. SEM image of the composite membrane structure is shown in Figure 4.5d.

In case of CNT membrane, the pressure-driven flux is few orders of magnitude higher than conventional Newtonian flow due to atomically flat graphite planes inducing nearly ideal slip conditions.[284]

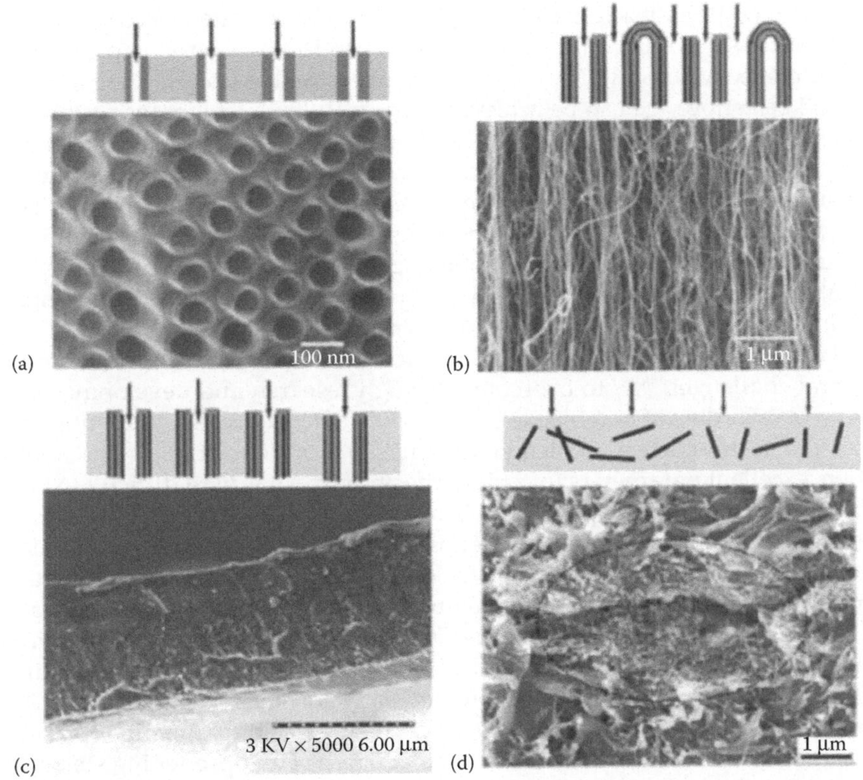

FIGURE 4.5
Different approaches to CNT membrane synthesis. (a) SEM image of nanotubes after dissolution of the template, (b) SEM image demonstrating the dense arrays of CNTs, (c) SEM image of cross section of the membrane with aligned CNTs in polymer matrix, and (d) SEM image of composite membrane structure. (Reprinted from *Comprehensive Membrane Science and Engineering*, Vol. 1, Majumder, M. and Ajayan, P.M., Carbon nanotube membranes: A new frontier in membrane science, Drioli, E. and Giorno, L., (eds.), 291–310, Copyright 2010, Figure 2, with permission from Elsevier.)

Membranes with CNT tips that were functionalized[285] with biotin show a reduction in $Ru(NH_3)_6$ by a factor of 15 when bound with streptavidin, thereby demonstrating the ability to gate molecular transport through CNT cores for potential applications in chemical separation. The embedded CNTs in host membrane matrix work on size-exclusion principle. However, the functionalization of CNT tip can introduce the required physicochemical characteristics into the membrane surface, which may lead to selective removal of contaminants based upon physicochemical interaction of species with the functional group present over the CNT tip.

Antimicrobial activity of CNTs requires direct contact between CNTs and target microorganisms.[286] Suspension of nonfunctionalized CNTs in water is extremely difficult and does not provide enough CNT–microbe contact for disinfection.

Accordingly, the antibacterial activity of CNTs can be exploited by coating CNTs on a reactor surface in contact with the pathogen-laden water. Complete retention and effective inactivation of *Escherichia coli* as well as up to 5–7 log removal of MS2 bacteriophage can be achieved using a PVDF microporous membrane coated with a thin layer of SWNTs.[287] Although the rate of bacterial inactivation by CNT is relatively low compared to conventional disinfectants, it is sufficient to prevent biofilm formation and the subsequent biofouling of membrane surface, which in turn would increase the lifetime of membrane without significant decline in throughput because of biofouling. In addition, the selective functionalization of CNT tip with hydrophilic or hydrophobic groups can help in minimizing the fouling depending upon the feed water conditions.

Though CNTs have got great potential to offer an ideal membrane, there are several challenges[288,289] to overcome through research and development. To grow 12–13 orders of magnitude (i.e., 10^{12}–10^{13}) of CNTs per square centimeter area is a technological challenge. Chemical vapor deposition methodology offers excellent parameters to achieve the objective. It is very difficult to have the yield of CNTs in a particular batch of synthesis beyond 90%. Tedious purification steps to remove the deposits and to make the CNT wall defect free make the things more complicated. It is important that the preparation methodology of nanocomposite membrane must ensure well-dispersed and well-aligned CNTs in the membrane. Functionalization of CNTs with desired functional groups may be required to have better dispersion, which is quite challenging because CNT is not soluble in any solvent. Opening of the CNT tips is a critical step, which is carried out by either acid treatment or plasma-based oxidation and is not simple to be adopted. The tip-opening step may cause thinning of CNT wall and disruption of tube integrity and subsequent failure of membrane channels. Scale-up with respect to CNT growth, CNT alignment, nanocomposite formation, and CNT tip or sidewall functionalization is quite complex involving material as well as process challenges.

4.5.2 Graphene Membranes

Among the significant recent advancements in the design and development of new nanocomposite membrane systems is the use of graphenes.[290,291] Graphenes offer a novel class of mechanically robust, ultrathin, high-flux, high-selectivity, and fouling-resistant membranes. The facile synthesis of nanoporous graphene (NPG) and GO membranes opens the door for ideal next-generation membranes as cost effective and sustainable alternative to the long-existing TFC PA membranes for water purification applications.

Graphene-based materials have attracted great interest for their potential use in desalination and water purification membranes due to their unique properties, including distinctive structural characteristics,[292] high mechanical strength,[293] and small thickness.[294] The advancement in molecular simulation of graphene family opens the door for their potential contribution in developing novel membrane desalination technologies. Graphene's unique electronic properties, high tensile strength, and impermeability to small molecules are now a well-determined fact[295–299] and these have been utilized to construct extremely thin membrane with size tunable pores (for molecular sieving) allowing for high flux. Graphene nanosheets display ideal chemical and physical properties in the desalination process. Despite its negligible thickness, membranes made of graphene exhibit adequate mechanical strength and capability of functioning under higher pressures which is superior to conventional polymeric RO membranes currently in circulation.[296]

Several simulation studies have identified NPG structures among the most promising membrane materials that can provide high water flow rates and high salt rejection as a function of nanopore morphology.[300] However, these hypotheses are based on a single layer of graphene sheet, which is difficult to assemble in the real world.[301] The transport of water through these nanoporous membranes can reach up to 66 L/cm²/day/MPa with greater than 99% salt rejection. By contrast, water transport through a conventional RO membrane approximately reaches 0.01–0.05 L/cm²/day/MPa with similar salt rejection. These values revealed the great potential for the utilization of functionalized NPG as a high-permeability desalination membrane (Figure 4.6a and b).

Figure 4.6a shows high-pressure molecular and ionic sieving across a one-atom-thick graphene sheet. Chemical functionalization of the pores with hydrogen increases water selectivity, whereas functionalization with hydroxyl groups improves the flux.[302] Performance of functionalized NPG and membranes used in existing desalination technologies is shown in Figure 4.6b. It is seen that the performance of functionalized nanoporous grapheme is superior from both water permeability and salt rejection point of view compared to conventional membranes used in desalination plants.

With current advancement, large areas of graphene have been successfully grown on plain copper foils at atmospheric pressure and roll-to-roll transfer of 750 mm graphene sheets.[303] The feasibility of graphene to be

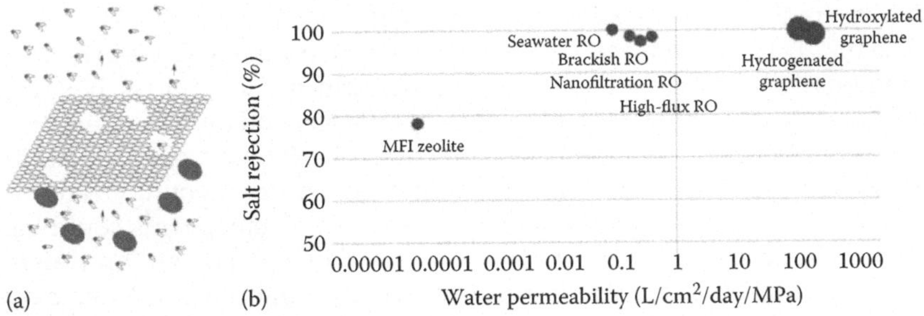

FIGURE 4.6
(a) High-pressure molecular and ionic sieving across a one-atom-thick graphene sheet. (b) Performance of functionalized NPG and membranes used in existing desalination technologies.

mass-produced has widened the application spectrum of this nanomaterial. The graphene monolayers with superior strength can allow water to flow through them while hindering the passage of other unwanted species.[304,305] Computational studies have predicted that large slip length of water molecules allows negligible friction with graphene surfaces, which results in an ultrafast water flow in the graphene nanochannel[306–309] providing much higher membrane flux compared to commercially available RO and NF membranes.[310]

4.5.3 Aquaporin Membranes

Biomimetics focuses on the basic science by fundamentally exploring the principles of biological systems, whereas bioinspiration focuses on the applied engineering by technologically implementing the principles from biological systems. Biomimetics and bioinspiration, as the complementary and interchangeable strategies for sustainable innovation and development of membrane technology, have great implications in exploring membrane materials and intensifying membrane processes. Based on their unique combination of offering high water permeability and high solute rejection, aquaporin proteins have attracted considerable interest over the past few years as functional building blocks of biomimetic membranes.[311–313]

Aquaporins are the protein channels that control water flux across biological membranes.[314] This first-characterized aquaporin is found widely in human tissues for rapid passive transport of water across cell membranes. Such transport channels exist in the cells of species in all three domains of life. Water movement in aquaporins is mediated by selective, rapid diffusion caused by osmotic gradients.[315,316] The hourglass shape of aquaporin-1 (AQP1), with selective extracellular and intracellular vestibules at each end, allows water molecules to pass rapidly in a single-file line, while excluding proteins.[317] Water molecules transport in single-file line through a narrow aquaporin channel; a constant number of molecules are assumed to occupy the channel at all times and the water molecules are assumed to move together in discrete translocations, or hops.[318]

The highly selective water permeability of aquaporin channels is an interesting possibility when considering water treatment membranes. Biological lipid bilayers containing aquaporins transport water and maintain selectivity that far surpasses all commercial RO membranes. Single aquaporins transfer water molecules at rates of $2–8 \times 10^9$ molecules per second.[315] Kaufman et al. predict that a membrane with 75% coverage of aquaporins could have a hydraulic permeability in the range of 2.5×10^{-11} m Pa^{-1} s^{-1}, an order of magnitude higher than commercial seawater RO membranes.[319] Kumar et al. include Aquaporin-Z (AqpZ) from *E. coli* bacterial cells in a polymeric membrane.[320] This aquaporin is selected based on the ability for high water permeation and high selectivity. In addition, it is easy to purify and multiply using a recombinant *E. coli* strain.

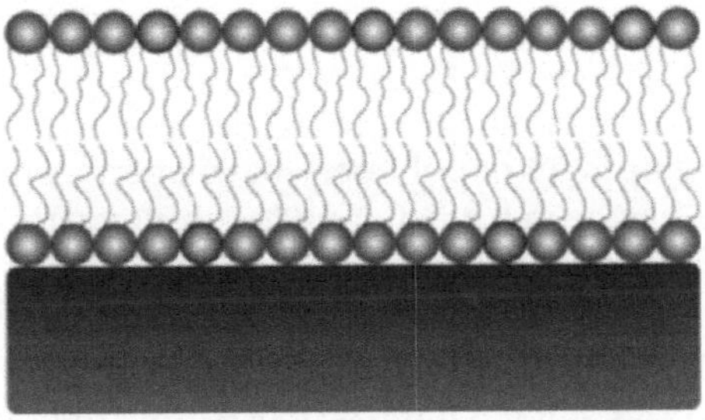

FIGURE 4.7
Conceptual cross-sectional image of a semipermeable lipid bilayer membrane cast atop a nanofiltration-type support membrane.

The transport across biological membranes is driven by osmotic pressure (or salt concentration) gradient. Conceptual cross sectional image of a semipermeable lipid bi-layer membrane cast atop a NF-type membrane is shown in Figure 4.7.

Water transport through the robust aquaporin-embedded vesicular membrane has been explored[321] in the FO process. Mathematical simulation to correlate vesicle size, vesicle permeability, and the interior solute concentration with membrane flux in FO process indicate that the water flux of the membrane is determined by both vesicle size and permeability in a pressure-retarded mode. The interior solute concentration of vesicles will impact the hydrostatic pressure of the vesicles in an FO mode.

Planar biomimetic membranes consisting of AqpZ can be made[322] upon the CA membrane substrate, functionalized with methacrylate end groups. By vesicle rupture of triblock copolymer (ABA) vesicles and UV polymerization, a selective layer upon the substrate for NF is formed. This study opens up new possibilities of using AqpZ-embedded biomimetic membranes for water purification with advantages that include high throughput with lesser energy consumption. A TFC aquaporin-based biomimetic membrane (ABM) can be prepared[323] by the IP method, where AqpZ-containing proteoliposomes is added to the MPD aqueous solution. Control membranes, either without aquaporins or with inactive (mutant) aquaporins, can also be similarly prepared. The separation performance of these membranes is evaluated by cross-flow RO tests. The active ABM give significantly higher water permeability (4 L/m^2 h bar) with comparable NaCl rejection (97%) at an applied pressure of 5 bar. Its permeability is about 40% higher compared to a commercial brackish water RO membrane (BW30) and an order of magnitude higher compared to a seawater RO membrane (SW30HR), which clearly demonstrates the great potential of the TFC ABM for desalination applications. In another study,[324] proteoliposome-containing aquaporin, a typical water channel protein, is fully encapsulated into the thin-film layer through cross-linking of PEI. This novel TFN membrane shows a significantly enhanced water permeability and typical NF rejection to $MgCl_2$.

4.6 Challenges

Progress in the development of polymer–matrix nanocomposite membranes for water treatment has been tremendous in recent years. Besides tuning the physicochemical properties of membranes (hydrophilicity, porosity, charge density, thermal and mechanical stability), the incorporation of nanomaterials can provide membranes with some unique properties of nanomaterials and also possibly induce new characteristics and functions based on their synergetic effects. It provides a new dimension to design the next generation of polymeric membranes with high-performance and antifouling properties. The potential applications of nanocomposite membranes could cover the whole filtration spectrum, including MF, UF, NF, RO, and FO.

Several challenges still need to be addressed to optimize the design of the nanocomposite membranes for industrial applications at a large scale:

- Fundamental understandings need to be developed to systematically reflect the effects of nanomaterials on membrane structures and correlate them to the membrane performance changes.
- Approaches for better dispersion of nanomaterials need to be further explored.
- Compatibility of nanofillers with polymers should be ensured to avoid leaching of nanomaterials into the environment.
- Large-scale production and industrial application with cost-effectiveness.

References

1. The United Nations World Water Development Report 4, 2012. Facts and figures: Managing Water under Uncertainty and Risk. http://unesdoc.unesco.org/images/0021/002154/215492e.pdf.
2. World Health Organization (WHO), 2000. Water Sanitation and Health (WSH)-Millennium Development Goals. http://www.who.int/water_sanitation_health/mdg1/en/.
3. Yin, J. and Deng, B. 2015. Polymer-matrix nanocomposite membranes for water treatment. *J. Membr. Sci.* 479: 256–275.
4. Robeson, L.M. 1991. Correlation of separation factor versus permeability for polymeric membranes. *J. Membr. Sci.* 62: 165–185.
5. Chung, T.S., Jiang, L.Y., Li, Y., and Kulprathipanja, S. 2007. Mixed matrix membranes (MMMs) comprising organic polymers with dispersed inorganic fillers for gas separation. *Prog. Polym. Sci. (Oxford)* 32: 483–507.

6. Jia, M., Peinemann, K.V., and Behling, R.D. 1991. Molecular sieving effect of the zeolite-filled silicone rubber membranes in gas permeation. *J. Membr. Sci.* 57: 289–296.
7. Duval, J.M., Folkers, B., Mulder, M.H.V., Desgrandchamps, G., and Smolders, C.A. 1993. Adsorbent filled membranes for gas separation. Part 1. Improvement of the gas separation properties of polymeric membranes by incorporation of microporous adsorbents. *J. Membr. Sci.* 80: 189–198.
8. Wu, H., Tang, B., and Wu, P. 2010. Novel ultrafiltration membranes prepared from a multi-walled carbon nanotubes/polymer composite. *J. Membr. Sci.* 362: 374–383.
9. Goh, P.S., Ng, B.C., Ismail, A.F., Aziz, M., and Sanip, S.M. 2010. Surfactant dispersed multi-walled carbon nanotube/polyetherimide nanocomposite membrane. *Solid State Sci.* 12: 2155–2162.
10. Shawky, H.A., Chae, S.R., Lin, S., and Wiesner, M.R. 2011. Synthesis and characterization of a carbon nanotube/polymer nanocomposite membrane for water treatment. *Desalination* 272: 46–50.
11. Mansourpanah, Y., Madaeni, S.S., Rahimpour, A., Adeli, M., Hashemi, M.Y., and Moradian, M.R. 2011. Fabrication new PES-based mixed matrix nanocomposite membranes using polycaprolactone modified carbon nanotubes as the additive: Property changes and morphological studies. *Desalination* 277: 171–177.
12. Vatanpour, V., Madaeni, S.S., Moradian, R., Zinadini, S., and Astinchap, B. 2011. Fabrication and characterization of novel antifouling nanofiltration membrane prepared from oxidized multiwalled carbon nanotube/polyethersulfone nanocomposite. *J. Membr. Sci.* 375: 284–294.
13. Majeed, S., Fierro, D., Buhr, K., Wind, J., Du, B., Boschetti-de-Fierro, A., and Abetz, V. 2012. Multi-walled carbon nanotubes (MWCNTs) mixed polyacrylonitrile (PAN) ultrafiltration membranes. *J. Membr. Sci.* 403–404: 101–109.
14. Zhao, X., Ma, J., Wang, Z., Wen, G., Jiang, J., Shi, F., and Sheng, L. 2012. Hyperbranched-polymer functionalized multi-walled carbon nanotubes for poly (vinylidene fluoride) membranes: From dispersion to blended fouling-control membrane. *Desalination* 303: 29–38.
15. De Lannoy, C.F., Jassby, D., Davis, D.D., Wiesner, M.R. 2012. A highly electrically conductive polymermultiwalled carbon nanotube nanocomposite membrane. *J. Membr. Sci.* 415–416: 718–724.
16. Daraei, P., Madaeni, S.S., Ghaemi, N., Ahmadi Monfared, H., and Khadivi, M.A. 2013. Fabrication of PES nanofiltration membrane by simultaneous use of multiwalled carbon nanotube and surface graft polymerization method: Comparison of MWCNT and PAA modified MWCNT. *Sep. Purif. Technol.* 104: 32–44.
17. Daraei, P., Madaeni, S.S., Ghaemi, N., Khadivi, M.A., Astinchap, B., and Moradian, R. 2013. Enhancing antifouling capability of PES membrane via mixing with various types of polymer modified multi-walled carbon nanotube. *J. Membr. Sci.* 444: 184–191.
18. De Lannoy, C.F., Soyer, E., and Wiesner, M.R. 2013. Optimizing carbon nanotube-reinforced polysulfone ultrafiltration membranes through carboxylic acid functionalization. *J. Membr. Sci.* 447: 395–402.
19. Yin, J., Zhu, G., and Deng, B. 2013. Multi-walled carbon nanotubes (MWNTs)/polysulfone (PSU) mixed matrix hollow fiber membranes for enhanced water treatment. *J. Membr. Sci.* 437: 237–248.

20. Phao, N., Nxumalo, E.N., Mamba, B.B., and Mhlanga, S.D. 2013. A nitrogen-doped carbon nanotube enhanced polyethersulfone membrane system for water treatment. *Phys. Chem.* 66: 148–156.

    Kim, E.S., Liu, Y., and Gamal El-Din, M. An in-situ integrated system of carbon nanotubes nanocomposite membrane for oil sands process-affected water treatment. *J. Membr. Sci.* 429: 418–417.
21. Zirehpour, A., Rahimpour, A., Jahanshahi, M., and Peyravi, M. 2014. Mixed matrix membrane application for olive oil wastewater treatment: Process optimization based on Taguchi design method. *J. Environ. Manage* 132: 113–120.
22. Saranya, R., Arthanareeswaran, G., and Dionysiou, D.D. 2014. Treatment of paper mill effluent using polyethersulfone/functionalised multiwalled carbon nanotubes based nanocomposite membranes. *Chem. Eng. J.* 236: 369–377.
23. Badawi, N.E., Ramadan, A.R., Esawi, A.M.K., and Mohamed El-Morsi. 2014. Novel carbon nanotube–cellulose acetate nanocomposite membranes for water filtration applications. *Desalination* 344: 79–85.
24. Nezam El-Dein, L.A., El-Gendi, A., Ismail, N., Abed, K.A., and Ahmed, A.I. 2015. Evaluation of cellulose acetate membrane with carbon nanotubes additives. *J. Ind. Eng. Chem.* 26: 259–264.
25. Kim, E.-S., Liu, Y., and Gamal El-Din, M. 2013. An in-situ integrated system of carbon nanotubes nanocomposite membrane for oils and process-affected water treatment. *J. Membr. Sci.* 429: 418–427.
26. Mauter, M.S. and Elimelech, M. 2008. Environmental applications of carbon-based nanomaterials. *Environ. Sci. Technol.* 42: 5843–5859.
27. Wang, X., Chen, X., Yoon, K., Fang, D., Hsiao, B.S., and Chu, B. 2005. High flux filtration medium based on nanofibrous substrate with hydrophilic nanocomposite coating. *Environ. Sci. Technol.* 39: 7684–7691.
28. Goh, K., Setiawan, L., Wei, L., Jiang, W., Wang, R., and Chen, Y. 2013. Fabrication of novel functionalized multi-walled carbon nanotube immobilized hollow fiber membranes for enhanced performance in forward osmosis process. *J. Membr. Sci.* 446: 244–254.
29. Choi, J.H., Jegal, J., and Kim, W.N. 2006. Fabrication and characterization of multi-walled carbon nanotubes/polymer blend membranes. *J. Membr. Sci.* 284: 406–415.
30. Tang, C., Zhang, Q., Wang, K., Fu, Q., and Zhang, C. 2009. Water transport behavior of chitosan porous membranes containing multi-walled carbon nanotubes (MWNTs). *J. Membr. Sci.* 337: 240–247.
31. Majeed, S., Fierro, D., Buhr, K., Wind, J., Du, B., Boschetti-de-Fierro, A., and Abetz, V. 2012. Multi-walled carbon nanotubes (MWCNTs) mixed polyacrylonitrile (PAN) ultrafiltration membranes. *J. Membr. Sci.* 403–404: 101–109.
32. Ebert, K., Fritsch, D., Koll, J., and Tjahjawiguna, C. 2004. Influence of inorganic fillers on the compaction behaviour of porous polymer based membranes. *J. Membr. Sci.* 233: 71–78.
33. Bae, T.H. and Tak, T.M. 2005. Effect of TiO_2 nanoparticles on fouling mitigation of ultrafiltration membranes for activated sludge filtration. *J. Membr. Sci.* 249: 1–8.
34. Arsuaga, J.M., Sotto, A., Rosario, G.D., Martínez, A., Molina, S., Teli, S.B., and de Abajo, J. 2013. Influence of the type, size, and distribution of metal oxide particles on the properties of nanocomposite ultrafiltration membranes. *J. Membr. Sci.* 428: 131–141.

35. Yang, Y., Wang, P., and Zheng, Q. 2006. Preparation and properties of polysulfone/TiO_2 composite ultrafiltration membranes. *J. Polym. Sci. Part B Polym. Phys.* 44: 879–887.
36. Cao, X., Ma, J., Shi, X., and Ren, Z. 2006. Effect of TiO_2 nanoparticle size on the performance of PVDF membrane. *Appl. Surf. Sci.* 253: 2003–2010.
37. Xiao, Y., Yu Wang, K., Chung, T.S., and Tan, J. 2006. Evolution of nano-particle distribution during the fabrication of mixed matrix TiO_2-polyimide hollow fiber membranes. *Chem. Eng. Sci.* 61: 6228–6233.
38. Li, J.B., Zhu, J.W., and Zheng, M.S. 2007. Morphologies and properties of poly(phthalazinone ether sulfone ketone) matrix ultrafiltration membranes with entrapped TiO_2 nanoparticles. *J. Appl. Polym. Sci.* 103: 3623–3629.
39. Yang, Y., Zhang, H., Wang, P., Zheng, Q., and Li, J. 2007. The influence of nano-sized TiO_2 fillers on the morphologies and properties of PSF UF membrane. *J. Membr. Sci.* 288: 231–238.
40. Fu, X., Matsuyama, H., and Nagai, H. 2008. Structure control of asymmetric poly(vinyl butyral)-TiO_2 composite membrane prepared by nonsolvent induced phase separation. *J. Appl. Polym. Sci.*, 108: 713–723.
41. Wu, G., Gan, S., Cui, L., and Xu, Y. 2008. Preparation and characterization of PES/TiO_2 composite membranes. *Appl. Surf. Sci.* 254: 7080–7086.
42. Yu, L.Y., Shen, H.M., and Xu, Z.L. 2009. PVDF-TiO_2 composite hollow fiber ultrafiltration membranes prepared by TiO_2 sol-gel method and blending method. *J. Appl. Polym. Sci.* 113: 1763–1772.
43. Li, J.F., Xu, Z.L., Yang, H., Yu, L.Y., and Liu, M. 2009. Effect of TiO_2 nanoparticles on the surface morphology and performance of microporous PES membrane. *Appl. Surf. Sci.* 255: 4725–4732.
44. Oh, S.J., Kim, N., and Lee, Y.T. 2009. Preparation and characterization of PVDF/TiO_2 organic-inorganic composite membranes for fouling resistance improvement. *J. Membr. Sci.* 345: 13–20.
45. Razmjou, A., Mansouri, J., and Chen, V. 2011. The effects of mechanical and chemical modification of TiO_2 nanoparticles on the surface chemistry, structure and fouling performance of PES ultrafiltration membranes. *J. Membr. Sci.* 378: 73–84.
46. Madaeni, S.S., Zinadini, S., and Vatanpour, V. 2011. A new approach to improve antifouling property of PVDF membrane using in situ polymerization of PAA functionalized TiO_2 nanoparticles. *J. Membr. Sci.* 380:155–162.
47. Sotto, A., Boromand, A., Balta, S., Kim, J., and Van Der Bruggen, B. 2011. Doping of polyethersulfone nanofiltration membranes: Antifouling effect observed at ultralow concentrations of TiO_2 nanoparticles. *J. Mater. Chem.*, 21: 10311–10320.
48. Hamid, N.A.A., Ismail, A.F., Matsuura, T., Zularisam, A.W., Lau, W.J., Yuliwati, E., and Abdullah, M.S. 2011. Morphological and separation performance study of polysulfone/titanium dioxide (PSF/TiO_2) ultrafiltration membranes for humic acid removal. *Desalination* 273: 85–92.
49. Abedini, R., Mousavi, S.M., and Aminzadeh, R. 2011. A novel cellulose acetate (CA) membrane using TiO_2 nanoparticles: Preparation, characterization and permeation study. *Desalination* 277: 40–45.
50. Yuliwati, E. and Ismail, A.F. 2011. Effect of additives concentration on the surface properties and performance of PVDF ultrafiltration membranes for refinery produced wastewater treatment. *Desalination* 273: 226–234.

51. Rahimpour, A., Jahanshahi, M., Rajaeian, B., and Rahimnejad, M. 2011. TiO_2 entrapped nano-composite PVDF/SPES membranes: Preparation, characterization, antifouling and antibacterial properties. *Desalination* 278: 343–353.
52. Sotto, A., Boromand, A., Zhang, R., Luis, P., Arsuaga, J.M., Kim, J., and Van der Bruggen, B. 2011. Effect of nanoparticle aggregation at low concentrations of TiO_2 on the hydrophilicity, morphology, and fouling resistance of PES-TiO_2 membranes. *J. Colloid Interface Sci.* 363: 540–550.
53. Vatanpour, V., Madaeni, S.S., Khataee, A.R., Salehi, E., Zinadini, S., and Monfared, H.A. 2012. TiO_2 embedded mixed matrix PES nanocomposite membranes: Influence of different sizes and types of nanoparticles on antifouling and performance. *Desalination* 292: 19–29.
54. Razmjou, A., Resosudarmo, A., Holmes, R.L., Li, H., Mansouri, J., and V. Chen. 2012. The effect of modified TiO_2 nanoparticles on the polyethersulfone ultrafiltration hollow fiber membranes. *Desalination* 287: 271–280.
55. Homaeigohar, S.S., Mahdavi, H., and Elbahri, M. 2012. Extraordinarily water permeable sol-gel formed nanocomposite nanofibrous membranes. *J. Colloid Interface Sci.* 366: 51–56.
56. Zhao, S., Wang, P., Wang, C., Sun, X., and Zhang, L. 2012. Thermostable PPESK/TiO_2 nanocomposite ultrafiltration membrane for high temperature condensed water treatment. *Desalination* 299: 35–43.
57. Abedini, R., Mousavi, S.M., and Aminzadeh, R. 2012. Effect of sonochemical synthesized TiO_2 nanoparticles and coagulation bath temperature on morphology, thermal stability and pure water flux of asymmetric cellulose acetate nanocomposite membranes prepared via phase inversion method. *Chem. Ind. Chem. Eng. Q.* 18: 385–398.
58. Vatanpour, V., Madaeni, S.S., Moradian, R., Zinadini, S., and Astinchap, B. 2012. Novel antibifouling nanofiltration polyethersulfone membrane fabricated from embedding TiO_2 coated multiwalled carbon nanotubes. *Sep. Purif. Technol.* 90: 69–82.
59. Ngang, H.P., Ahmad, A.L., Low, S.C., and Ooi, B.S. 2012. Preparation of mixed-matrix membranes for micellar enhanced ultrafiltration based on response surface methodology. *Desalination* 293: 7–20.
60. Zhang, G., Lu, S., Zhang, L., Meng, Q., Shen, C., and Zhang, J. 2013. Novel polysulfone hybrid ultrafiltration membrane prepared with TiO_2-g-HEMA and its antifouling characteristics. *J. Membr. Sci.* 436: 163–173.
61. Zhang, F., Zhang, W., Yu, Y., Deng, B., Li, J., and Jin, J. 2013. Sol-gel preparation of PAA-g-PVDF/TiO_2 nanocomposite hollow fiber membranes with extremely high water flux and improved antifouling property. *J. Membr. Sci.* 432: 25–32.
62. Teli, S.B., Molina, S., Sotto, A., Calvo, E.G., and Abajo, J.D. 2013. Fouling resistant polysulfone-PANI/TiO_2 ultrafiltration nanocomposite membranes. *Ind. Eng. Chem. Res.* 52: 9470–9479.
63. Teow, Y.H., Ahmad, A.L., Lim, J.K., and Ooi, B.S. 2013. Studies on the surface properties of mixed-matrix membrane and its antifouling properties for humic acid removal. *J. Appl. Polym. Sci.* 128: 3184–3192.
64. Joo Kim, H., Pant, H.R., Hee Kim, J., Jung Choi, N., and Kim, C.S. 2014. Fabrication of multifunctional TiO_2-fly ash/polyurethane nanocomposite membrane via electrospinning. *Ceram. Int.* 40: 3023–3029.
65. Rahimpour, A., Madaeni, S.S., Taheri, A.H., and Mansourpanah, Y. 2008. Coupling TiO_2 nanoparticles with UV irradiation for modification of polyethersulfone ultrafiltration membranes. *J. Membr. Sci.* 313: 158–169.

66. Damodar, R.A., You, S.J., and Chou, H.H. 2009. Study the self cleaning, antibacterial and photocatalytic properties of TiO_2 entrapped PVDF membranes. *J. Hazard. Mater.* 172: 1321–1328.
67. Ngang, H.P., Ooi, B.S., Ahmad, A.L., and Lai, S.O. 2012. Preparation of PVDF-TiO_2 mixed-matrix membrane and its evaluation on dye adsorption and UV-cleaning properties. *Chem. Eng. J.* 197: 359–367.
68. Sun, M., Su, Y., Mu, C., and Jiang, Z. 2010. Improved antifouling property of PES ultrafiltration membranes using additive of silica-PVP nanocomposite. *Ind. Eng. Chem. Res.* 49: 790–796.
69. Hoek, E.M.V., Ghosh, A.K., Huang, X., Liong, M., and Zink, J.I. 2011. Physical-chemical properties, separation performance, and fouling resistance of mixed-matrix ultrafiltration membranes. *Desalination* 283: 89–99.
70. Jomekian, A., Mansoori, S.A.A., and Monirimanesh, N. 2011. Synthesis and characterization of novel PEOMCM-41/PVDC nanocomposite membrane. *Desalination* 276: 239–245.
71. Klaysom, C., Moon, S.H., Ladewig, B.P., Lu, G.Q.M., and Wang, L. 2011. The influence of inorganic filler particle size on composite ion-exchange membranes for desalination. *J. Phys. Chem. C* 115: 15124–15132.
72. Shen, J.N., Ruan, H.M., Wu, L.G., and Gao, C.J. 2011. Preparation and characterization of PES-SiO_2 organic inorganic composite ultrafiltration membrane for raw water pretreatment. *Chem. Eng. J.* 168: 1272–1278.
73. Muriithi, B. and Loy, D.A. 2012. Processing, morphology, and water uptake of Nafion/ex situ Stöber silica nanocomposite membranes as a function of particle size. *ACS Appl. Mater. Interf.* 4: 6766–6773.
74. Huang, J., Zhang, K., Wang, K., Xie, Z., Ladewig, B., and Wang, H. 2012. Fabrication of polyethersulfone mesoporous silica nanocomposite ultrafiltration membranes with antifouling properties. *J. Membr. Sci.* 423–424: 362–370.
75. Pakizeh, M., Moghadam, A.N., Omidkhah, M.R., and Namvar-Mahboub, M. 2013. Preparation and characterization of dimethyldichlorosilane modified SiO_2/PSf nanocomposite membrane. *Korean J. Chem. Eng.* 30: 751–760.
76. Wu, H., Tang, B., and Wu, P. 2014. Development of novel SiO_2-GO nanohybrid/polysulfone membrane with enhanced performance. *J. Membr. Sci.* 451: 94–102.
77. Yan, L., Hong, S., Li, M.L., and Li, Y.S. 2009. Application of the Al_2O_3-PVDF nanocomposite tubular ultrafiltration (UF) membrane for oily wastewater treatment and its antifouling research. *Sep. Purif. Technol.* 66: 347–352.
78. Maximous, N., Nakhla, G.,Wan, W., and Wong, K. 2009. Preparation, characterization and performance of Al_2O_3/PES membrane for wastewater filtration, *J. Membr. Sci.* 341: 67–75.
79. Maximous, N., Nakhla, G., Wong, K., and Wan, W. 2010. Optimization of Al_2O_3/PES membranes for wastewater filtration. *Sep. Purif. Technol.* 73: 294–301.
80. Maximous, N., Nakhla, G., Wan, W., and Wong, K. 2010. Effect of the metal oxide particle distributions on modified PES membranes characteristics and performance. *J. Membr. Sci.* 361: 213–222.
81. Csetneki, I., Filipcsei, G., and Zrínyi, M. 2006. Smart nanocomposite polymer membranes with on/off switching control. *Macromolecules* 39: 1939–1942.
82. Daraei, P., Madaeni, S.S., Ghaemi, N., Salehi, E., Khadivi, M.A., Moradian, R., and Astinchap, B. 2012. Novel polyethersulfone nanocomposite membrane prepared by PANI/Fe_3O_4 nanoparticles with enhanced performance for Cu(II) removal from water. *J. Membr. Sci.* 415–416: 250–259.

83. Gholami, A., Moghadassi, A.R., Hosseini, S.M., Shabani, S., and Gholami, F. 2013. Preparation and characterization of polyvinyl chloride based nanocomposite nanofiltration-membrane modified by iron oxide nanoparticles for lead removal from water. *J. Ind. Eng. Chem.* 20: 1517–1522.
84. Alam, J., Dass, L.A., Ghasemi, M., and Alhoshan, M. 2013. Synthesis and optimization of PES-Fe_3O_4 mixed matrix nanocomposite membrane: Application studies in water purification. *Polym. Compos.* 34: 1870–1877.
85. Daraei, P., Madaeni, S.S., Ghaemi, N., Khadivi, M.A., Astinchap, B., and Moradian, R. 2013. Fouling resistant mixed matrix polyethersulfone membranes blended with magnetic nanoparticles: Study of magnetic field induced casting. *Sep. Purif. Technol.* 109: 111–121.
86. Gohari, R.J., Lau, W.J., Matsuura, T., Halakoo, E., and Ismail, A.F. 2013. Adsorptive removal of Pb(II) from aqueous solution by novel PES/HMO ultrafiltration mixed matrix membrane. *Sep. Purif. Technol.* 120: 59–68.
87. Jamshidi Gohari, R., Halakoo, E., Nazri, N.A.M., Lau, W.J., Matsuura, T., and Ismail, A.F. 2014. Improving performance and antifouling capability of PES UF membranes via blending with highly hydrophilic hydrous manganese dioxide nanoparticles. *Desalination* 335: 87–95.
88. Alhoshan, M., Alam, J., Dass, L.A., and Al-Homaidi, N. 2013. Fabrication of polysulfone/ZnO membrane: Influence of ZnO nanoparticles on membrane characteristics. *Adv. Polym. Technol.* 32: 21369.
89. Huang, J., Zhang, K., Wang, K., Xie, Z., Ladewig, B., and Wang, H. 2012. Fabrication of polyethersulfone-mesoporous silica nanocomposite ultrafiltration membranes with antifouling properties. *J. Membr. Sci.* 423–424: 362–370.
90. Sen, D., Ghosh, A.K., Mazumder, S., Bindal, R.C., and Tewari, P.K. 2014. Novel polysulfone–spray-dried silica composite membrane for water purification: Preparation, characterization and performance evaluation. *Sep. Purif. Technol.* 123: 79–86.
91. Razmjou, A., Resosudarmo, A., Holmes, R.L., Li, H., Mansouri, J., and Chen, V. 2012. The effect of modified TiO_2 nanoparticles on the polyethersulfone ultrafiltration hollow fiber membranes. *Desalination* 287: 271–280.
92. Rahimpour, A., Jahanshahi, M., Rajaeian, B., and Rahimnejad, M. 2011. TiO_2 entrapped nano-composite PVDF/SPES membranes: Preparation, characterization, antifouling and antibacterial properties. *Desalination* 278: 343–353.
93. Madaeni, S.S., Khataee, A.R., Salehi, E., Zinadini, S., Monfared, H.A., and Vatanpour, V. 2012. TiO_2 embedded mixed matrix PES nanocomposite membranes: Influence of different sizes and types of nanoparticles on antifouling and performance. *Desalination* 292: 19–29.
94. Zhao, S., Wang, P., Wang, C., Sun, X., and L. Zhang. 2012. Thermostable PPESK/TiO_2 nanocomposite ultrafiltration membrane for high temperature condensed water treatment. *Desalination* 299: 35–43.
95. Hwang, H.-Y., Kim, D.-J., Kim, H.-J, Y.-T. Hong, and Nam, S.-Y. 2011. Effect of nanoclay on properties of porous PVDF membranes. *Trans. Nonferrous Met. Soc.* 21: 141–147.
96. Lai, C.Y., Groth, A., Gray, S., and Duke, M. 2011. Investigation of the dispersion of nanoclays into PVDF for enhancement of physical membrane properties. *Desalin. Water Treat.* 34: 251–256.
97. Peng, Q.-Y., Cong, P.-H., Liu, X.-J., Liu, T.-X., Huang, S., and Li, T.-S. 2009. The preparation of PVDF/clay nanocomposites and the investigation of their tribiological properties. *Wear* 266: 713–720.

98. Dayma, N., Satapathy, B.K., and Patnaik, A. 2011. Structural correlations to sliding wear performance ofPA-6/PP-g-MA/nanoclay ternary nanocomposites. *Wear* 271: 827–836.
99. Pan, B., Xing, Y., Zhang, C., and Zhang, Y. 2010. Study on erosion wear behavior of PDCPD/MMT nanocomposite. *Adv. Mater. Res.* 123–125: 231–234.
100. Lai, C.Y., Groth, A., Gray, S., and Duke, M. 2014. Enhanced abrasion resistant PVDF/nanoclay hollow fibre composite membranes for water treatment. *J. Membr. Sci.* 449: 146–157
101. Monticelli, O., Bottino, A., Scandale, I., Capannelli, G., and Russo, S. 2007. Preparation and properties of polysulfone-clay composite membranes. *J. Appl. Polym. Sci.* 103: 3637–3644.
102. Leite, A.M.D., Maia, A.M.D., Araujo, E.M., and Lira, E.M. 2009. Nylon 6/ Brazilian clay membranes prepared by phase inversion. *J. Appl. Polym. Sci.* 113: 1488–1493.
103. Anadão, P., Sato, L.F., Wiebeck, H., and Valenzuela-Díaz, F.R. 2010. Montmorillonite as a component of polysulfone nanocomposite membranes. *Appl. Clay Sci.*, 48: 127–132.
104. Lai, C.Y., Groth, A., Gray, S., and Duke, M. 2011. Investigation of the dispersion of nanoclays into PVDF for enhancement of physical membrane properties. *Desalination Water Treat.* 34: 251–256.
105. Ghaemi, N., Madaeni, S.S., Alizadeh, A., Rajabi, H., and Daraei, P. 2011. Preparation, characterization and performance of polyethersulfone/organically modified montmorillonite nanocomposite membranes in removal of pesticides. *J. Membr. Sci.* 382: 135–147.
106. Ma, Y., Shi, F., Zhao, W., Wu, M., Zhang, J., Ma, J., and Gao, C. 2012. Preparation and characterization of PSf/clay nanocomposite membranes with LiCl as a pore forming additive. *Desalination* 303: 39–47.
107. Wang, P., Ma, J., Wang, Z., Shi, F., and Liu, Q. 2012. Enhanced separation performance of PVDF/PVP-g-MMT nanocomposite ultrafiltration membrane based on the NVP-grafted polymerization modification of montmorillonite (MMT). *Langmuir* 28: 4776–4786.
108. Anadão, P., Montes, R.R., Larocca, N.M., and Pessan, L.A. 2013. Influence of the clay content and the polysulfone molar mass on nanocomposite membrane properties. *Appl. Surf. Sci.* 275: 110–120.
109. Adams, F.V., Nxumalo, E.N., Krause, R.W.M., Hoek, E.M.V., and Mamba, B.B. 2012. Preparation and characterization of polysulfone/β-cyclodextrin polyurethane composite nanofiltration membranes. *J. Membr. Sci.* 405–406: 291–299.
110. Adams, F.V., Dlamini, D.S., Nxumalo, E.N., Krause, R.W.M., Hoek, E.M.V., and Mamba, B.B. 2013. Solute transport and structural properties of polysulfone/ β-cyclodextrin polyurethane mixed-matrix membranes. *J. Membr. Sci.* 429: 58–65.
111. Adams, F.V., Nxumalo, E.N., Krause, R.W.M., Hoek, E.M.V., and Mamba, B.B. 2013. The influence of solvent properties on the performance of polysulfone/ β-cyclodextrin polyurethane mixed-matrix membranes. *J. Appl. Polym. Sci.* 130: 2005–2014.
112. Adams, F.V., Nxumalo, E.N., Krause, R.W.M., Hoek, E.M.V., and Mamba, B.B. 2013. Application of polysulfone/cyclodextrin mixed-matrix membranes in the removal of natural organic matter from water. *Phys. Chem. Earth. 67: 71.*

113. Fan, Z., Wang, Z., Sun, N., Wang, J., and Wang, S. 2008. Performance improvement of polysulfone ultrafiltration membrane by blending with polyaniline nanofibers. *J. Membr. Sci.* 320: 363–371.
114. Sankir, M., Bozkir, S., and Aran, B. 2010. Preparation and performance analysis of novel nanocomposite copolymer membranes for Cr(VI) removal from aqueous solutions. *Desalination* 251: 131.
115. Zhao, S., Wang, Z., Wei, X., Tian, X., Wang, J., Yang, S., and Wang, S. 2011. Comparison study of the effect of PVP and PANI nanofibers additives on membrane formation mechanism, structure and performance. *J. Membr. Sci.* 385–386: 110–122.
116. Zhao, S., Wang, Z., Wang, J., Yang, S., and Wang, S. 2011. PSf/PANI nanocomposite membrane prepared by in situ blending of PSf and PANI/NMP. *J. Membr. Sci.* 376: 83–95.
117. Teli, S.B., Molina, S., Calvo, E.G., Lozano, A.E., and de Abajo, J. 2012. Preparation, characterization and antifouling property of polyethersulfone-PANI/PMA ultrafiltration membranes. *Desalination* 299: 113–122.
118. S. Zhao, Z. Wang, X. Wei, B. Zhao, J. Wang, S. Yang, and S. Wang. 2012. Performance improvement of polysulfone ultrafiltration membrane using well-dispersed polyaniline-poly(vinylpyrrolidone) nanocomposite as the additive. *Ind. Eng. Chem. Res.*, 51: 4661–4672.
119. Liao, Y., Wang, X., Qian, W., Li, Y., Li, X., and Yu, D.G. 2012. Bulk synthesis, optimization, and characterization of highly dispersible polypyrrole nanoparticles toward protein separation using nanocomposite membranes. *J. Colloid Interface Sci.* 386: 148–157.
120. Tetala, K.K.R. and Stamatialis, D.F. 2013. Mixed matrix membranes for efficient adsorption of copper ions from aqueous solutions. *Sep. Purif. Technol.* 104: 214–220.
121. Worthley, C.H., Constantopoulos, K.T., Ginic-Markovic, M., Markovic, E., and Clarke, S. 2013. A study into the effect of POSS nanoparticles on cellulose acetate membranes. *J. Membr. Sci.* 431: 62–71.
122. Zhao, W., Huang, J., Fang, B., Nie, S., Yi, N., Su, B., Li, H., and Zhao, C. 2011. Modification of polyethersulfone membrane by blending semi-interpenetrating network polymeric nanoparticles. *J. Membr. Sci.* 369: 258–266.
123. Tomalia, D.A., Naylor, A.N., and Goddard III, W.A. 1990. Starburst dendrimers: Molecular level control of size, shape, surface, chemistry and topology. *Angew. Chem. Int. Ed. Engl.* 29: 138–175.
124. Fréchet, J.M.J. 1994. Functional polymers and dendrimers: Reactivity, molecular architecture, and interfacial energy. *Science* 263: 1710–1715.
125. Fréchet, J.M.J. and Tomalia, D.A. 2001. *Dendrimers and Other Dendritic Polymers*, Wiley-VCH, Weinheim, Germany.
126. Cheng, Y. 2012. *Dendrimer Based in Drug Delivery Systems*, John Wiley & Sons, Hoboken, NJ.
127. Moorefield, C.N, Perera, S., and Newkome, G.R. 2012. Dendrimer chemistry: Supramolecular perspectives and applications, *Dendrimer-Based Drug Delivery Systems: From Theory to Practice*, Y. Cheng (ed.), John Wiley & Sons, Hoboken, NJ, pp. 1–54.
128. Boas, U. and Heegaard, P.M. 2004. Dendrimers in drug research. *Chem. Soc. Rev.* 33: 43–63.
129. Algarra, M., Campos, B.B., Miranda, M.S., and Esteves da Silva, J.C.G. 2011. CdSe quantum dots capped PAMAM dendrimer nanocomposites for sensing nitroaromatic compounds. *Talanta* 83: 1335–1340.

130. Esteves da Silva, J.C.G., Algarra, M., and Campos, B.B. 2011. Synthesis and analytical applications of quantum dots coated with different generations of DAB dendrimers, *Advances in Nanocomposites—Synthesis, Characterization and Industrial Applications*, B. Reddy (ed.), IN-TECH, Croatia, 23–38.
131. Algarra, M., Campos, B.B., Alonso, B., Miranda, M.S., Martínez, A.M., Casado, C.M., and Esteves da Silva, J.C.G. 2012. Thiolated DAB dendrimers and CdSe quantum dots nanocomposites for Cd(II) or Pb(II) sensing. *Talanta* 88: 403–407.
132. Jie, G. and Yuan, J. 2012. Quantum dots-based multifunctional dendritic superstructure for amplified electrochemiluminescence detection of ATP. *Biosens. Bioelectron.* 31: 69–76.
133. Oliveira, J.M., Salgado, A.J., Sousa, N., Mano, J.F., and Reisa, R.L. 2010. Dendrimers and derivatives as a potential therapeutic tool in regenerative medicine strategies—A review. *Prog. Polym. Sci.* 35: 1163–1194.
134. Vogtle, F., Richardt, G., and Werner, N. 2009. *Dendrimer Chemistry: Concepts, Syntheses, Properties, Applications*, Wiley-VCH, Weinheim, Germany.
135. Rether, A. and Schuster, M. 2003. Selective separation and recovery of heavy metal ions using water-soluble N-benzoylthiourea modified PAMAM polymers. *React. Funct. Polym.* 57: 13–21.
136. Diallo, M.S., Christie, S., Swaminathan, P., Balogh, L., Shi, X., Um, W., Papelis, C., Goddard 3rd, W.A., and Johnson Jr., J.H. 2004. Dendritic chelating agents. 1. Cu(II) binding to ethylene diamine core poly(amidoamine) dendrimers in aqueous solutions. *Langmuir* 20: 2640–2651.
137. Diallo, M.S., Christie, S., Swaminathan, P., Johnson Jr., J.H., and Goddard, W. 2005. Dendrimer enhanced ultrafiltration. 1. Recovery of Cu(II) from aqueous solutions using PAMAM dendrimers with ethylene diamine core and terminal NH_2 groups. *Environ. Sci. Technol.* 39: 1366–1377.
138. Diallo, M.S., Arasho, W., Johnson Jr., J.H., and Goddard, W.A. 2008. Dendritic chelating agents U(VI) binding to poly(amidoamine) and poly(propyleneimine) dendrimers in aqueous solutions. *Environ. Sci. Technol.* 42: 1572–1579.
139. Xu, Y. and Zhao, D. 2005. Removal of Cu(II) from contaminated soil using PAMAM dendrimers. *Environ. Sci. Technol.* 39: 2369–2377.
140. Barakat, M.A., Ramadan, M.H., Alghamadi, M.A., Algarny, S.S., Woodcock, H.L., and Kuhn, J. 2013. Remediation of Cu(II), Ni(II), and Cr(III) ions from simulated wastewater by dendrimer/titania composites. *J. Environ. Manage.* 117: 50–57.
141. Duan, S., Kouketsu, T., Kazama, S., and Yamada, K.J. 2006. Development of PAMAM dendrimer composite membranes for CO_2 separation. *J. Membr. Sci.* 283: 2–6.
142. Kouketsu, T., Duan, S., Kai, T., Kazama, S., and Yamada, K. 2007. PAMAM dendrimer composite membrane for CO_2 separation: Formation of a chitosan gutter layer. *J. Membr. Sci.* 287: 51–59.
143. Tomita, S., Sato, K., and Anzai, J. 2008. Layer-by-layer assembled thin films composed of carboxyl-terminated poly(amidoamine) dendrimer as a pH-sensitive nanodevice. *J. Colloid Interface Sci.* 326: 35–40.
144. Duan, S., Chowdhury, F.A., Kai, T., Kazama, S., and Fujioka, Y. 2008. PAMAM dendrimer composite membrane for CO_2 separation: Addition of hyaluronic acid in gutter layer and application of novel hydroxyl PAMAM dendrimer. *Desalination* 234: 278–285.
145. Sarkar, A., Carver, P.I., Zhang, T., Merrington, A., Bruza, K.J., Rousseau, J.L., Keinath, S.E., and Dvornic, P.R.J. Dendrimer-based coatings for surface modification of polyamide reverse osmosis membranes. *J. Membr. Sci.* 349: 421–428.

146. Zhang, Q., Wang, N., Zhao, L., Xu, T., and Cheng, Y. 2013. Polyamidoamine dendronized hollow fiber membranes in the recovery of heavy metal ions. *ACS Appl. Mater. Interfaces* 349: 1907–1912.
147. Vázquez, A.M.I., Alonso, B., Casado, C.M., Casado, J., and Benavente, J. 2014. Characterization of an engineered cellulose based membrane by thiol dendrimer for heavy metals removal Manuel Algarra. *Chem. Eng. J.* 253: 472–477.
148. Husain, S. and Koros, W.J. 2009. Macrovoids in hybrid organic/inorganic hollow fiber membranes. *Ind. Eng. Chem. Res.* 48: 2372–2379.
149. Junaidi, M.U.M., Leo, C.P., Kamal, S.N.M., and Ahmad, A.L. 2013. Fouling mitigation in humic acid ultrafiltration using polysulfone/SAPO-34 mixed matrix membrane. *Water Sci. Technol.* 67: 2102–2109.
150. Leo, C.P., Ahmad Kamil, N.H., Junaidi, M.U.M., Kamal, S.N.M., and Ahmad, A.L. 2013. The potential of SAPO-44 zeolite filler in fouling mitigation of polysulfone ultrafiltration membrane. *Sep. Purif. Technol.* 103: 84–91.
151. Liu, F., Ma, B.R., Zhou, D., Xiang, Y.H., and Xue, L.X. 2014. Breaking through tradeoff of Polysulfone ultrafiltration membranes by zeolite 4A. *Micropor. Mesopor. Mater.* 186: 113–120.
152. Ciobanu, G., Carja, G., and Ciobanu, O. 2007. Preparation and characterization of polymer-zeolite nanocomposite membranes. *Mater. Sci. Eng. C* 27: 1138–1140.
153. Liao, C., Yu, P., Zhao, J., Wang, L., and Luo, Y. 2011. Preparation and characterization of NaY/PVDF hybrid ultrafiltration membranes containing silver ions as antibacterial materials. *Desalination* 272: 59–65.
154. Sotiriou, G.A. and Pratsinis, S.E. 2010. Antibacterial activity of nanosil versions and particles. *Environ. Sci. Technol.* 44: 5649–5654.
155. Levard, C., Hotze, E.M., Lowry, G.V., and Brown, G.E. 2012. Environmental transformations of silver nano particles: Impact on stability and toxicity. *Environ. Sci. Technol.* 466: 900–6914.
156. Koseoglu-Imer, D.Y., Kose, B., Altinbas, M., and Koyuncu, I. 2013. The production of polysulfone (PS) membrane with silver nanoparticles (AgNP): Physical properties, filtration performances, and biofouling resistances of membranes. *J. Membr.Sci.* 428: 620–628.
157. Zodrow, K., Brunet, L., Mahendra, S., Li, D., Zhang, A., Li, Q., and Alvarez, P.J.J. 2009. Polysulfone ultrafiltration membranes impregnated with silver nanoparticles show improved biofouling resistance and virus removal. *WaterRes.*43: 715–723.
158. Sawada, I., Fachrul, R., Ito, T., Ohmukai, Y., Maruyama, T., and Matsuyama, H. 2012. Development of a hydrophilic polymer membrane containing silver nanoparticles with both organic antifouling and antibacterial properties. *J. Membr. Sci.* 387–388: 1–6.
159. Basri, H., Ismail, A.F., and Aziz, M. 2011. Polyethersulfone (PES)-silvercomposite UF membrane: Effect of silver loading and PVP molecular weight on membrane morphology and antibacterial activity. *Desalination* 273: 72–80.
160. Basri, H., Ismail, A.F., Aziz, M., Nagai, K., Matsuura, T., Abdullah, M.S., and Ng, B.C. 2010. Silver-filled polyethersulfone membranes for antibacterial applications—Effect of PVP and TAP addition on silver dispersion. *Desalination* 261: 264–271.
161. Huang, J., Arthanareeswaran, G., and Zhang, K. 2012. Effect of silver loaded sodium zirconium phosphate (nano AgZ) nanoparticles incorporation on PES membrane performance. *Desalination* 285: 100–107.

162. Li, J.H., Shao, X.S., Zhou, Q., Li, M.Z., and Zhang, Q.Q. 2013. The double effects of silver nanoparticles on the PVDF membrane: Surface hydrophilicity and antifouling performance. *Appl. Surf. Sci.* 265: 663–670.
163. De Gusseme, B., Hennebel, T., Christiaens, E., Saveyn, H., Verbeken, K., Fitts, J.P., Boon, N., and Verstraete, W. 2011. Virus disinfection in water by biogenic silver immobilized in polyvinylidene fluoride membranes. *Water Res.*45: 1856–1864.
164. Yin, J., Yang, Y., Hu, Z., and Deng, B. 2013. Attachment of silver nano particles (AgNPs) onto thin-film composite (TFC) membranes through covalent bonding to reduce membrane biofouling. *J. Membr. Sci.* 441: 73–82.
165. Lee, S.Y., Kim, H.J., Patel, R., Im, S.J., Kim, J.H., and Min, B.R. 2007. Silver nanoparticles immobilized on thin film composite polyamide membrane: Characterization, nanofiltration, antifouling properties. *Polym. Adv. Technol.* 18: 562–568.
166. Regiel, A., Irusta, S., Kyziol, A., Arruebo, M., and Santamaria, J. 2013. Preparation and characterization of chitosan-silver nanocomposite films and the antibacterial activity against *Staphylococcus aureus*. *Nanotechnology* 24: 015101.
167. Kim, E.-S., Hwang, G., Gamal El-Din, M., and Liu, Y. 2012. Development of nanosilver and multi-walled carbon nanotubes thin-film nanocomposite membrane for enhanced water treatment. *J. Membr. Sci.* 394–395: 37–48.
168. Zhu, X.Y., Bai, R.B., Wee, K.H., Liu, C.K., and Tang, S.L. 2010. Membrane surfaces immobilized with ionic or reduced silver and their anti-biofouling performances. *J. Membr.Sci.* 363: 278–286.
169. Yuksel, M.S., Tas, B., Koseoglu-Imer, D.Y., and Koyuncub, I. 2014. Effect of silver nanoparticle (AgNP) location in nanocomposite membrane matrix fabricated with different polymer type on antibacterial mechanism. *Desalination* 347: 120–130.
170. Mollahosseini, A., Rahimpour, A., Jahamshahi, M., Peyravi, M., and Khavarpour, M. 2012. The effect of silver nanoparticle size on performance and antibacteriality of polysulfone ultrafiltration membrane. *Desalination* 306: 41–50.
171. Zhang, M., Field, R.W., and Zhang, K. 2014. Biogenic silver nanocomposite polyether sulfone UF membranes with antifouling properties. *J. Membr. Sci.* 471: 274–284.
172. Sawada, I., Fachrul, R., Ito, T., Ohmukai, Y., Maruyama, T., and Matsuyama, H. 2012. Development of a hydrophilic polymer membrane containing silver nanoparticles with both organic antifouling and antibacterial properties. *J. Membr. Sci.* 387–388: 1–6.
173. Alpatova, A., Sikkim, E., Sun, X., Hwang, G., Liu, Y., and Gamal El-Din, M. 2013. Fabrication of porous polymeric nanocomposite membranes with enhanced anti-fouling properties: Effect of casting composition. *J. Membr. Sci.* 444: 449–460.
174. Kar, S., Subramanian, M., Ghosh, A.K., Bindal, R.C., Prabhakar, S., Nuwad, J., Pillaic, S., Chattopadhyay, C.G.S., and Tewari, P.K. 2011. Potential of nanoparticles for water purification: A case-study on anti-biofouling behaviour of metal based polymeric nanocomposite membrane. *Desalination Water Treat.* 27: 224–230.
175. Dasari, A., Quirós, J., Herrero, B., Boltes, K., García-Calvo, E., and Rosal, R. 2012. Antifouling membranes prepared by electrospinning polylactic acid containing biocidal nanoparticles. *J. Membr. Sci.* 405–406: 134–140.

176. Akar, N., Asar, B., Dizge, N., and Koyuncu, I. 2013. Investigation of characterization and biofouling properties of PES membrane containing selenium and copper nanoparticles. *J. Membr. Sci.* 437: 216–226.
177. Szabó, T., Tombácz, E., Illés, E., and Dékány, I. 2006. Enhanced acidity and pH-dependent surface charge characterization of successively oxidized graphite oxides. *Carbon* 44: 537–545.
178. Szabo, T., Berkesi, O., Forgo, P., Josepovits, K., Sanakis, Y., Petridis, D., and Dekany, I. 2006. Evolution of surface functional groups in a series of progressively oxidized graphite oxides. *Chem. Mater.* 18: 2740–2749.
179. Ganesh, B.M., Isloor, A.M., and Ismail, A.F. 2013. Enhanced hydrophilicity and salt rejection study of graphene oxide-polysulfone mixed matrix membrane. *Desalination* 313: 199–207.
180. Zhao, C., Xu, X., Chen, J., and Yang, F. 2013. Effect of graphene oxide concentration on the morphologies and antifouling properties of PVDF ultrafiltration membranes. *J. Environ. Chem. Eng.* 1: 349–354.
181. Jin, F., Lv, W., Zhang, C., Li, Z., Su, R., Qi, W., Yang, Q-H., and He, Z. 2013. High-performance ultrafiltration membranes based on polyethersulfone–graphene oxide composites. *RSC Adv.* 3: 21394–21397.
182. Wang, Z., Yu, H., Xia, J., Zhang, F., Li, F., Xia, Y., and Li, Y. 2012. Novel GO-blended PVDF ultrafiltration membranes. *Desalination* 299: 50–54.
183. Le-Clech, P., Chen, V., and Fane, T.A.G. 2006. Fouling in membrane bioreactors used in wastewater treatment. *J. Membr. Sci.* 284: 17–53.
184. Meng, F., Chae, S.-R., Drews, A., Kraume, M., Shin, H.-S., and Yang, F. 2009. Recent advances in membrane bioreactors (MBRs): Membrane fouling and membrane material. *Water Res.* 43: 1489–1512.
185. Zhang, H., Gao, J., Jiang, T., Gao, D., Zhang, S., Li, H., and Yang, F. 2011. A novel approach to evaluate the permeability of cake layer during cross-flow filtration in the flocculants added membrane bioreactors. *Bioresour. Technol.* 102: 11121–11131.
186. Wu, B., Kitade, T., Chong, T.H., Uemura, T., and Fane, A.G. 2012. Role of initially formed cake layers on limiting membrane fouling in membrane bioreactors. *Bioresour. Technol.* 118: 589–593.
187. Meng, F., Zhang, H., Yang, F., and Liu, L. 2007. Characterization of cake layer in submerged membrane bioreactor. *Environ. Sci. Technol.* 41: 4065–4070.
188. Zhao, C., Xu, X., Chena, J., Wang, G., and Yang, F. 2014. Highly effective antifouling performance of PVDF/graphene oxide composite membrane in membrane bioreactor (MBR) system. *Desalination* 340: 59–66.
189. Vatanpour, V., Madaeni, S.S., Moradian, R., Zinadini, S., and Astinchap, B. 2012. Novel antibiofouling nanofiltration polyethersulfone membrane fabricated from embedding TiO_2 coated multiwalled carbon nanotubes. *Sep. Purif. Technol.* 90: 69–82.
190. Crock, C.A., Rogensues, A.R., Shan, W., and Tarabara, V.V. 2103. Polymer nanocomposites with grapheme based hierarchical fillers as materials for multifunctional water treatment membranes. *Water Res.* 47: 3984–3996.
191. Wu, H., Tang, B., and Wu, P. 2014. Development of novel SiO_2-GO nanohybrid/polysulfone membrane with enhanced performance. *J. Membr. Sci.* 451: 94–102.
192. Filice, S., D'Angelo, D., Libertino, S., Nicotera, I., Kosma, V., Privitera, V., and Scalese, S. 2015. Graphene oxide and titania hybrid Nafion membranes for efficient removal of methyl orange dye from water. *Carbon* 8: 489–499.

193. Safarpour, M., Khataee, A., and Vatanpour, V. 2015. Effect of reduced graphene oxide/TiO_2 nanocomposite with different molar ratios on the performance of PVDF ultrafiltration membranes. *Sep. Purif. Technol.* 140: 32–42.
194. Li, G., Shen, L., Luo, Y., and Zhang, S. 2014. The effect of silver-PAMAM dendrimer nanocomposites on the performance of PVDF membranes. *Desalination* 338: 115–120.
195. Cadotte, J.E. 1981. Interfacially synthesized reverse osmosis membrane. US Patent No. US4277344.
196. Jeong, B.-H., Hoek, E.M.V., Yan, Y., Subramani, A., Huang, X., Hurwitz, G., Ghosh, A.K., and Jawor, A. 2007. Interfacial polymerization of thin film nanocomposites: A new concept for reverse osmosis membranes. *J. Membr. Sci.* 294: 1–7.
197. Glater, J., Hong, S.K., and Elimelech, M. 1994. The search for a chlorine-resistant reverse osmosis membrane. *Desalination* 95: 325–345.
198. Kwon, Y.N., Hong, S., Choi, H., and Tak, T. 2012. Surface modification of a polyamide reverse osmosis membrane for chlorine resistance improvement. *J. Membr. Sci.* 415–416: 192–198.
199. Shintani, T., Matsuyama, H., and Kurata, N. 2007. Development of a chlorine-resistant polyamide reverse osmosis membrane. *Desalination* 207: 340–348.
200. Park, J., Choi, W., Kim, S.H., Chun, B.H., Bang, J., and Lee, K.B. Enhancement of chlorine resistance in carbon nanotube-based nanocomposite reverse osmosis membranes. *Desalination Water Treat.* 15: 198–204.
201. Zhao, H., Qiu, S., Wu, L., Zhang, L., Chen, H., and Gao, C. 2014. Improving the performance of polyamide reverse osmosis membrane by incorporation of modified multi-walled carbon nanotubes. *J. Membr. Sci.* 450: 249–256.
202. Kim, S.G., Hyeon, D.H., Chun, J.H., Chun, B.H., and Kim, S.H. 2013. Nanocomposite poly(arylene ether sulfone) reverse osmosis membrane containing functional zeolite nanoparticles for seawater desalination. *J. Membr. Sci.* 443: 10–18.
203. Kim, S.G., Chun, J.H., Chun, B.H., and Kim, S.H. 2013. Preparation, characterization and performance of poly(aylene ether sulfone)/modified silica nanocomposite reverse osmosis membrane for seawater desalination. *Desalination* 325: 76–83.
204. Jadav, G.L. and Singh, P.S. 2009. Synthesis of novel silica-polyamide nanocomposite membrane with enhanced properties. *J. Membr. Sci.* 328: 257–267.
205. Jadav, G.L., Aswal, V.K., and Singh, P.S. 2010. SANS study to probe nanoparticle dispersion in nanocomposite membranes of aromatic polyamide and functionalized silica nanoparticles. *J. Colloid Interface Sci.* 351: 304–314.
206. Fathizadeh, M., Aroujalian, A., and Raisi, A. 2011. Effect of added NaX nanozeolite into polyamide as a top thin layer of membrane on water flux and salt rejection in a reverse osmosis process. *J. Membr. Sci.*, 375: 88–95.
207. Rajaeian, B., Rahimpour, A., Tade, M.O., and Liu, S. 2013. Fabrication and characterization of polyamide thin film nanocomposite (TFN) nanofiltration membrane impregnated with TiO_2 nanoparticles. *Desalination* 313: 176–188.
208. Kim, E.-S. and Deng, B. 2011. Fabrication of polyamide thin-film nanocomposite (PA-TFN) membrane with hydrophilized ordered mesoporous carbon (H-OMC) for water purifications. *J. Membr. Sci.* 375: 46–54.
209. Rana, D., Kim, Y., Matsuura, T., and Arafat, H.A. 2011. Development of antifouling thin-film-composite membranes for seawater desalination. *J. Membr. Sci.* 367: 110–118.

210. Lee, S.Y., Kim, H.J., Patel, R., Im, S.J., Kim, J.H., and Min, B.R. 2007. Silver nanoparticles immobilized on thin film composite polyamide membrane: Characterization, nanofiltration, antifouling properties. *Polym. Adv. Technol.* 18: 562–568.
211. Kim, E.S., Hwang, G., Gamal El-Din, M., and Liu, Y. 2012. Development of nanosilver and multi-walled carbon nanotubes thin-film nanocomposite membrane for enhanced water treatment. *J. Membr. Sci.* 394–395: 37–48.
212. Lind, M.L., Jeong, B.H., Subramani, A., Huang, X., and Hoek, E.M.V. 2009. Effect of mobile cation on zeolitepolyamide thin film nanocomposite membranes. *J. Mater. Res.* 24: 1624–1631.
213. Li, D. and Wang, H. 2010. Recent developments in reverse osmosis desalination membranes. *J. Mater. Chem.* 20: 4551–4566.
214. Bellona, C., Drewes, J.E., Xu, P., and Amy, G. 2004. Factors affecting the rejection of organic solutes during NF/RO treatment—A literature review. *Water Res.* 38: 2795–2809.
215. Roh, I.J., Greenberg, A.R. and Khare, V.P. 2006. Synthesis and characterization of interfacially polymerized polyamide thin films. *Desalination* 191: 279–290.
216. Zhang, L., Shi, G.Z., Qiu, S., Cheng, L.H., and Chen, H.L. 2011. Preparation of high-flux thin film nanocomposite reverse osmosis membranes by incorporating functionalized multi-walled carbon nanotubes. *Desalination Water Treat.* 34: 19–24.
217. Yin, J., Kim, E.-S., Yang, J., and Deng, B. 2012. Fabrication of a novel thin-film nanocomposite (TFN) membrane containing MCM-41 silica nanoparticles (NPs) for water purification. *J. Membr. Sci.* 423–424: 238–246.
218. Lind, M.L., Ghosh, A.K., Jawor, A., Huang, X., Hou, W., Yang, Y., and Hoek, E.M.V. 2009. Influence of zeolite crystal size on zeolite-polyamide thin film nanocomposite membranes. *Langmuir* 25: 10139–10145.
219. Kim, C.K., Kim, J.H., Roh, I.J., and Kim, J.J. 2000. The changes of membrane performance with polyamide molecular structure in the reverse osmosis process. *J. Membr. Sci.* 165: 189–199.
220. Ghosh, A.K., Jeong, B.H., Huang, X., and Hoek, E.M.V. 2008. Impacts of reaction and curing conditions on polyamide composite reverse osmosis membrane properties. *J. Membr. Sci.* 311: 34–45.
221. Lind, M.L., Suk, D.E., Nguyen, T.V., and Hoek, E.M.V. 2010. Tailoring the structure of thin film nanocomposite membranes to achieve seawater RO membrane performance. *Environ. Sci. Technol.* 44: 8230–8235.
222. Roy, S., Ntim, S.A., Mitra, S., and Sirkar, K.K. 2011. Facile fabrication of superior nanofiltration membranes from interfacially polymerized CNT-polymer composites. *J. Membr. Sci.* 375: 81–87.
223. Yin, J., Kim, E.S., Yang, J., and Deng, B. 2012. Fabrication of a novel thin-film nanocomposite (TFN) membrane containing MCM-41 silica nanoparticles (NPs) for water purification. *J. Membr. Sci.* 423–424: 238–246.
224. Saitua, H., Gil, R., and Padilla, A.P. 2011. Experimental investigation on arsenic removal with a nanofiltration pilot plant from naturally contaminated groundwater. *Desalination* 274: 1–6.
225. Pendergast, M.T.M., Nygaard, J.M., Ghosh, A.K., and Hoek, E.M.V. 2010. Using nanocomposite materials technology to understand and control reverse osmosis membrane compaction. *Desalination* 261: 255–263.
226. Pendergast, M.M., Ghosh, A.K., and Hoek, E.M.V. 2013. Separation performance and interfacial properties of nanocomposite reverse osmosis membranes. *Desalination* 308: 180–185.

227. McCutcheon, J.R. and Elimelech, M. 2006. Influence of concentrative and dilutive internal concentration polarization on flux behavior in forward osmosis. *J. Membr. Sci.* 284: 237–247.
228. Ma, N., Wei, J., Qi, S., Zhao, Y., Gao, Y., and Tang, C.Y. 2013. Nanocomposite substrates for controlling internal concentration polarization in forward osmosis membranes. *J Membr. Sci.* 441: 54–62.
229. Wang, Y., Ou, R., Ge, Q., Wang, H., and Xu, T. 2013. Preparation of polyethersulfone/carbon nanotube substrate for high-performance forward osmosis membrane. *Desalination* 330: 70–78.
230. Kwak, S.Y., Kim, S.H., and Kim, S.S. 2001. Hybrid organic/inorganic reverse osmosis (RO) membrane for bactericidal anti-fouling. 1. Preparation and characterization of TiO_2 nanoparticle self-assembled aromatic polyamide thin-film-composite (TFC) membrane. *Environ. Sci. Technol.* 35: 2388–2394.
231. Kim, S.H., Kwak, S.Y., Sohn, B.H., and Park, T.H. 2003. Design of TiO_2 nanoparticle self-assembled aromatic polyamide thin-film-composite (TFC) membrane as an approach to solve biofouling problem. *J. Membr. Sci.* 211: 157–165.
232. Bae, T.H. and Tak, T.M. 2005. Preparation of TiO_2 self-assembled polymeric nanocomposite membranes and examination of their fouling mitigation effects in a membrane bioreactor system. *J. Membr. Sci.* 266: 1–5.
233. Pendergast, M.M. and Hoek, E.M.V. 2011. A review of water treatment membrane nanotechnologies. *Energy Environ. Sci.* 4: 1946–1971.
234. Iijima, S. 1991. Helical microtubes of graphite carbon. *Nature (London)* 354: 56–58.
235. Li, Y.-H., Wang, S., Wei, J., Zhang, X., Xu, C., Luan, Z., Wu, D., and Wei, B. 2002. Lead adsorption on carbon nanotubes. *Chem. Phys. Lett.* 357: 263–266.
236. Li, Y.-H., Wang, S., Luan, Z., Ding, J., Xu, C., and Wu, D. 2003. Adsorption of cadmium(II) from aqueous solution by surface oxidized carbon nanotubes. *Carbon* 41: 1057–1062.
237. Li, Y.-H., Wang, S., Zhang, X., Wei, J., Xu, C., Luan, Z., and Wu, D. 2003. Adsorption of fluoride from water by aligned carbon nanotubes. *Mater. Res. Bull.* 38: 469–476.
238. Peng, X., Luan, Z., Ding, J., Di, Z., Li, Y., and Tian, B. 2005. Ceria nanoparticles supported nanotubes for the removal of arsenate from water. *Mater. Lett.* 59: 399–403.
239. Srivastava, A., Srivastava, O.N., Talapatra, S., Vajtai, R., and Ajayan, P.M. 2004. Carbon nanotube filters. *Nat. Mater.* 3: 610–614.
240. Kar, S., Bindal, R.C., Prabhakar, S., Tewari, P.K., Dasgupta, K., and Sathiyamoorthy, D. 2008. Potential of carbon nanotubes in water purification: An approach towards the development of an integrated membrane system. *Int. J. Nucl. Desalination* 3: 143–150.
241. Rosca, I.D., Watari, F., Uo, M., and Akasaka, T. 2005. Oxidation of multiwalled carbon nanotubes by nitric acid. *Carbon* 15: 3124–3131.
242. Sano, M., Kamino, A., Okamura, J., and Shinkai, S. 2001. Ring closure of carbon nanotubes. *Science* 293: 1299–1301.
243. Hamwi, A., Alvergnat, H., Bonnamy, S., and Beguin, F. 1997. Fluorination of carbon nanotubes. *Carbon* 35: 723–728.
244. Mickelson, E.T., Huffman, C.B., Rinzler, A.G, Smalley, R.E., Hauge, R.H., and Margrave, J.L. 1998. Fluorination of single-wall carbon nanotubes. *Chem. Phys. Lett.* 296: 188–194.
245. Touhara, H. and Okino, F. 2000. Property control of carbon material by fluorination. *Carbon* 38: 241–267.

246. Khare, B.N., Meyyappan, M., Cassell, A.M., Nguyen, C.V., Han, J., and Nano, J. 2002. Functionalisation of carbon nanotubes using atomic hydrogen from a glow discharge. *Nano Lett.* 2: 73–77.
247. Khare, B.N., Meyyappan, M., Kralj, J., Wilhite, P., Sisay, M., Imanaka, H., Koehne, J., and Baushchlicher, C.W. 2002. A glow-discharge approach for functionalisation of carbon nanotubes. *Appl. Phys. Lett.* 81: 5237–5239.
248. Kim, K.S., Bae, D.J., Kim, J.R., Park, K.A., Lim, S.C., Kim, J.J., Choi, W.B., Park, C.Y., and Lee, Y.H. 2002. Modification of electronic structures of a carbon nanotube by hydrogen functionalisation. *Adv. Mater.* 14: 1818–1821.
249. Chen, Y., Haddon, R.C., Fang, S., Rao, A.M., Eklund, P.C., Lee, W.H., Dickey, E.C. et al. 1998. Chemical attachment of organic functional groups to single-walled carbon nanotube material. *J. Mater. Res.* 13: 2423–2431.
250. Chen, J., Hamon, M.A., Hu, H., Chen, Y., Rao, A.M., Eklund, P.C., and Haddon, R.C. 1998. Solution properties of single-walled carbon nanotubes. *Science* 282: 95–98.
251. Lee, W.H, Kim, S.J., Lee, W.J., Lee, J.G., Haddon, R.C., and Reucroft, P.J. 2001. X-ray photoelectron spectroscopic studies of surface modified single-walled carbon nanotube material. *Appl. Surf. Sci.* 181: 121–127.
252. Kamaras, K., Itkis, M.E., Hu, H., Zhao, B., and Haddon, R.C. 2003. Covalent bond formation to a carbon nanotube metal. *Science* 301: 1501.
253. Hu, H., Zhao, B., Hamon, M.A., Kamaras, K., Itkis, M.E., and Haddon, R.C. 2003. Side wall functionalisation of single walled carbon nanotubes by addition of dichloro carbine. *J. Am. Chem. Soc.* 125: 14893–14900.
254. Khabashesku, V.N., Billups, W.E., and Margrave, J.L. 2002. Fluorination of single-wall carbon nanotubes and subsequent derivatization reactions. *Acc. Chem. Res.* 35: 1087–1095.
255. Stevens, J.L., Huang, A.Y., Peng, H.Q., Chiang, L.W., Khabashesku, V.N., and Margrave, J.L. 2003. Side wall amino-functionalization of single-walled carbon nanotubes through fluorination and subsequent reactions with terminal diamines. *Nano Lett.* 3: 331–336.
256. Dujardin, E., Ebbesen, T.W., Hiura, H., and Tanigaki, K. 1994. Capillarity and wetting of carbon nanotubes. *Science* 265: 1850–1852.
257. Ajayan, P.M. and Iijima, S. 1993. Capillarity induced filling of carbon nanotubes. *Nature* 361: 333–334.
258. Ajayan, P.M., Ebbesen, T.W., Ichihasi, T., Iijima, S., Tanigaki, K., and Hiura, H. 1993. Opening carbon nanotubes with oxygen and implications for filling. *Nature* 362: 522–525.
259. Ajayan, P.M., Stephan, O., Redlich, Ph., and Coliex, C. 1995. Carbon nanotubes as removable templates for metal-oxide nanocomposites. *Nature* 375: 564–567.
260. Yarin, A.L., Yazicioglu, A.G., Megaridis, C.M., Rossi, M.P., and Gogotsi, Y. 2005. Theoretical and experimental investigation of aqueous liquids contained in carbon nanotubes. *J. Appl. Phys.* 97: 124309/1–124309/13.
261. Sholl, D.S. and Fichthorn, K.A. 1997. Concerted diffusion of molecular clusters in a molecular sieve. *Phys. Rev. Lett.* 79: 3569–3572.
262. Sholl, D.S. 1999. Characterization of molecular cluster diffusion in AlPO4-5 using molecular dynamics. *Chem. Phys. Lett.* 305: 269–275.
263. Sholl, D.S. and Lee, C.K. 2000. Influences of concerted cluster diffusion on single-file diffusion of CF4 in AlPO4-5 and Xe in AlPO4-31. *J. Chem. Phys.* 112: 817–824.

264. Mao, Z. and Sinnott, S.B. 2001. Separation of organic molecular mixtures in carbon nanotubes and bundles: Molecular dynamics simulations. *J. Phys. Chem. B* 105: 6916–6924.
265. Skoulidas, A.I., Ackerman, D.M., Johnson, J.K., and Sholl, D.S. 2002. Rapid transport of gases in carbon nanotubes. *Phys. Rev. Lett.* 89: 185901/1–185901/4.
266. Bathia, S.K., Chen, H., and Sholl, D.S. 2005. Comparison of diffusive and viscous contributions to transport coefficients of light gases in single-walled carbon nanotubes. *Mol. Simul.* 31: 643–649.
267. Brovchenko, I., Geiger, A., and Oleinikova, A. 2001. Phase equilibria of water in cylindrical nanopores. *Phys. Chem. Chem. Phys.* 3: 1567–1569.
268. Allen, T.W., Kuyucak, S., and Chung, S.H. 1999. The effect of hydrophobic and hydrophilic channel walls on the structure and diffusion of water and ions. *J. Chem. Phys.* 111: 7985–7999.
269. Marti, J. and Gordillo, M.C. 2002. Microscopic dynamics of confined supercritical water. *Chem. Phys. Lett.* 354: 227–232.
270. Hummer, G., Rasaiah, J.C., and Noworyta, J.P. 2001. Water conduction through the hydrophobic channel of a carbon nanotube. *Nature* 414: 188–190.
271. Bear, J. 1972. *Dynamics of Fluids in Porous Media*. American Elsevier, New York.
272. Thomas, J.A. and McGaughey, A.J.H. 2008. Reassessing fast water transport through carbon nanotubes. *Nano Lett.* 8: 2788–2793.
273. Verwei, H. and van den Boom, H. 2007. Fast mass transport through carbon nanotube membranes. *Small* 3: 1996–2004.
274. Joseph, S. and Aluru, N.R. 2008. Pumping of confined water in carbon nanotubes by rotation translation coupling. *Phys. Rev. Lett.* 101: 064502/1–064502/4.
275. Striolo, A. 2006. The mechanism of water diffusion in narrow carbon nanotubes. *Nano Lett.* 6: 633–639.
276. Kang, S., Herzberg, M., Rodrigues, D.F., and Elimelech, M. 2008. Antibacterial effects of carbon nanotubes: Size does matter!. *Langmuir* 24: 6409–6413.
277. Narayan, R.J., Berry, C.J., and Brigmon, R.L. 2005. Structural and biological properties of carbon nanotube composite films. *Mater. Sci. Eng. B* 123: 123–129.
278. Wick, P., Manser, P., Limbach, L.K., Dettlaff-Weglikowska, U., Krumeich, F., Roth, S., Stark, S.W.J., and Bruinink, A. 2007. The degree and kind of agglomeration affect carbon nanotube cytotoxicity. *Toxicol. Lett.* 168: 121–131.
279. Brunet, L., Lyon, D.Y., Zodrow, K., Rouch, J.C., Caussat, B., Serp, P., Remigy, J.C., Wiesner, M.R., and Alvarez, P.J.J. 2008. Properties of membranes containing semi-dispersed carbon nanotubes. *Environ. Eng. Sci.* 25: 565–576.
280. Majumder, M. and Ajayan, P.M. 2010. Carbon nanotube membranes: A new frontier in membrane science, *Comprehensive Membrane Science and Engineering*, E. Drioli and L. Giorno (eds.), Vol. 1, Academic Press, Oxford, 291–310.
281. Miller, S.A., Young, V.Y., and Martin, C.R.J. Electro-osmotic flow in template-prepared carbon nanotube membranes. *Am. Chem. Soc.* 123, 12335–12342.
282. Hinds, B.J., Chopra, N., Rantell, T., Andrews, R., Gavalas, V., and Bachas, L.G. 2004. Aligned multiwalled carbon nanotube membranes. *Science* 303: 62–65.
283. Holt, J.K., Park, H.G., Wang, Y., Stadermann, M., Artyukhin, A.B., Grigoropoulosm, C.P., Noy, A., and Bakajin, O. 2006. Fast mass transport through sub-2-nanometer carbon nanotubes. *Science* 312: 1034–1037.
284. Majumder, M., Chopra, N., Andrews, R., and Hinds, B.J. 2005. Nanoscale hydrodynamics: Enhanced flow in carbon nanotubes. *Nature* 438: 44.

285. Majumder, M., Chopra, N., and Hinds, B.J. 2005. Effect of tip functionalization on transport through vertically oriented carbon nanotube membranes. *J. Am. Chem. Soc.* 127: 9062–9070.
286. Kang, S., Pinault, M., Pfefferle, L.D., and Elimelech, M. 2007. Single-walled carbon nanotubes exhibit strong antimicrobial activity. *Langmuir* 23: 8670–8673.
287. Brady-Estévez, A.S., Kang, S., and Elimelech, M. 2008. A single-walled-carbon-nanotube filter for removal of viral and bacterial pathogens. *Small* 4: 481–484.
288. Kar, S., Bindal, R.C., and Tewari, P.K. 2012. Carbon nanotube membranes for desalination and water purification: Challenges and opportunities. *Nano Today* 7: 385–389.
289. Goh, P.S., Ismail, A.F., and Ng, B.C. 2013. Carbon nanotubes for desalination: Performance evaluation and current hurdles. *Desalination* 308: 2–14.
290. Goh, P.S. and Ismail, A.F. 2015. Graphene-based nanomaterial: The state-of-the-art material for cutting edge desalination technology. *Desalination* 356: 115–128.
291. Mahmoud, K.A., Mansoor, B., Mansour, A., and Khraisheh, M. 2015. Functional graphene nanosheets: The next generation membranes for water desalination. *Desalination* 356: 208–225.
292. Ruan, M., Hu, Y., Guo, Z., Dong, R., Palmer, J., Hankinson, J., Berger, C., and de Heer, W.A. 2012. Epitaxial graphene on silicon carbide: Introduction to structured grapheme. *MRS Bull.* 37: 1138–1147.
293. Leenaerts, O., Partoens, B., and Peeters, F.M. 2008. Graphene: A perfect nanoballoon. *Appl. Phys. Lett.* 93: 193107.
294. Lu, Q. and Huang, R. 2009. Nonlinear mechanics of single-atomic layer graphene sheets. *Int. J. Appl. Mech.* 1: 443–467.
295. Xu, Z. and Gao, C. 2011. Graphene chiral liquid crystals and macroscopic assembled fibres. *Nat. Commun.* 2: 571.
296. Zaib, Q. and Fath, H. 2012. Application of carbon nano-materials in desalination processes. *Desalination Water Treat.* 51: 627–636.
297. Geim, A.K. 2009. Graphene: Status and prospects. *Science* 324: 1530–1534.
298. Lee, C., Wei, X., Kysar, J.W., and Hone, J. 2008. Measurement of the elastic properties and intrinsic strength of monolayer graphene. *Science* 321: 385–388.
299. Bunch, J.S., Verbridge, S.S., Alden, J.S., van der Zande, A.M., Parpia, J.M., Craighead, P.L., and McEuen, H.G. 2008. Impermeable atomic membranes from graphene sheets. *Nano Lett.* 8: 2458–2462.
300. Cohen-Tanugi, D. and Grossman, J.C. 2012. Water desalination across nanoporous graphene. *Nano Lett.* 12: 3602–3608.
301. Konatham, D., Yu, J., Ho, T.A., and Striolo, A. 2013. Simulation insights for graphene-based water desalination membranes. *Langmuir* 29: 11884–11897.
302. Wang, E.N. and Karnik, R. 2012. Graphene cleans up water. *Nat. Nanotechnol.* 7: 552–554.
303. Bae, S., Kim, H., Lee, Y., Xu, X., Park, J., Zheng, Y., Balakrishnan, J. et al. 2010. Roll-to-roll production of 30-inch graphene films for transparent electrodes. *Nat. Nanotechnol.* 5: 574–578.
304. Cohen-Tanugi, D. and Grossman, J.C. 2012. Water desalination across nanoporous graphene. *Nano Lett.* 12: 3602–3608.
305. Suk, M.E. and Aluru, N.R. 2010. Water transport through ultrathin grapheme. *J. Phys. Chem. Lett.* 1: 1590–1594.
306. Kannam, S.K., Todd, B.D., Hansen, J.S., and Daivis, P.J. 2012. Slip length of water on graphene: Limitations of non-equilibrium molecular dynamics simulations. *J. Chem. Phys.* 136: 024705.

307. Xiong, W., Liu, J.Z., Ma, M., Xu, Z.P., Sheridan, J., and Zheng, Q.S. 2011. Strain engineering water transport in graphene nanochannels. *Phys. Rev. E.* 84: 05632.
308. Dhiman, P., Yavari, F., Mi, X., Gullapalli, H., Shi, Y.F., Ajayan, P.M., and Koratkar, N. 2011. Harvesting energy from water flow over grapheme. *Nano Lett.* 11: 3123–3127.
309. Gordillo, M.C. and Marti, J. 2010. Water on graphene surfaces. *J. Phys. Condens. Matter* 22: 284111.
310. Hu, M. and Mi, B. 2013. Enabling graphene oxide nanosheets as water separation membranes. *Environ. Sci. Technol.* 47: 3715–3723.
311. Tang, C.Y., Zhao, Y., Wang, R., Hélix-Nielsen, C., and Fane, A.G. 2013. Desalination by biomimetic aquaporin membranes: Review of status and prospects. *Desalination* 308: 34–40.
312. Zhaoa, X., Jianga, Z., Lia, Z., Fana, X., Zhua,J., Wua, H., Yanlei, S., Yanga, D., Pana, F., and Shi, J. 2014. Biomimetic and bio-inspired membranes: Preparation and application. *Prog. Polym. Sci.* 39: 1668–1720.
313. Shen, Y.-X., Saboe, P.O., Sines, I.T., Erbakan, M., and Kumar, M. 2014. Biomimetic membranes: A review. *J. Membr. Sci.* 454: 359–381.
314. Agre, P., Sasaki, S., and Chrispeels, M.J. 1993. Aquaporins: A family of water channel proteins. *Am. J. Physiol. Ren. Physiol.* 265: F461.
315. Agre, P. 2003. *Aquaporin Water Channels,* Nobel Lecture, Stockholm, Sweden.
316. Meinild, A., Klaerke, D., and Zeuthen, T. 1998. Bidirectional water fluxes and specificity for small hydrophilic molecules in aquaporins 0–5. *J. Biol. Chem.* 273: 32446.
317. Agre, P., Preston, G.M., Smith, B.L., Jung, J.S., Raina, S., Moon, C., Guggino, W.B., and Nielsen, S. 1993. Aquaporin CHIP: The archetypal molecular water channel. *Am. J. Physiol. Ren. Physiol.* 265: F463–F476.
318. Zhu, F.Q., Tajkhorshid, E., and Schulten, K. 2004. Theory and simulation of water permeation in aquaporin-1. *Biophys. J.* 86: 50–57.
319. Kaufman, Y., Berman, A., and Freger, V. 2010. Supported lipid bilayer membranes for water purification by reverse osmosis. *Langmuir* 26: 7388–7395.
320. Kumar, M., Grzelakowski, M., Zilles, J., Clark, M., and W. Meier. 2007. Highly permeable polymeric membranes based on the incorporation of the functional water channel protein Aquaporin Z. *Proc. Natl. Acad. Sci. USA* 104: 20719–20724.
321. Wang, H., Chung, T.-S., and Tong, Y.W. 2013. Study on water transport through a mechanically robust AquaporinZ biomimetic membrane. *J. Membr. Sci.* 445: 47–52.
322. Zhonga, P.S., Chunga,T.-S., Jeyaseelanb, K., and Armugamb, A. 2012. Aquaporin-embedded biomimetic membranes for nanofiltration. *J. Membr. Sci.* 407–408: 27–33.
323. Zhao, Y., Qiu, C., Li, X., Vararattanavech, A., Shen, W., Torres, J., Hélix-Nielsen, C. et al. 2012. Synthesis of robust and high-performance aquaporin-based biomimetic membranes by interfacial polymerization-membrane preparation and RO performance characterization. *J. Membr. Sci.* 423–424: 422–428.
324. Li, X., Wang, R., Wicaksana, F., Tang, C., Torres, J., and Fane, A.G. 2014. Preparation of high performance nanofiltration (NF) membranes incorporated with aquaporin Z. *J. Membr. Sci.* 450: 181–188.

5

Nanocomposite Membranes in Gas Separation

5.1 Introduction

The development of gas separation membranes dates back to the early nineteenth century. It has got momentum after the discovery of high-flux asymmetric membranes by Loeb and Sourirajan in 1964. Mixed matrix membrane (MMM), comprising rigid permeable or impermeable particles, such as zeolites, carbon molecular sieves (CMSs), silica, titania, and carbon nanotubes (CNTs), dispersed in a continuous polymeric matrix presents an interesting approach for improving the separation properties of polymeric membranes. The particles are referred to as nanofillers if they are in nano dimensions, and correspondingly the MMMs are referred to as nanocomposite membranes. In this approach, using properties of both the organic and inorganic phases, a membrane with good permeability, selectivity, mechanical strength, and thermal as well as chemical stability and processing can be prepared. Superior gas separation properties of molecular sieve materials can be combined with desirable mechanical properties and convenient processing of polymers. Inorganic materials that are used as the dispersed phase in MMMs have unique structure, surface chemistry, and mechanical strength. When they are added to the polymer matrix, it is expected that the resulting membrane properties become better than those conventional polymer membranes. Very few studies have been carried out to increase the performance of gas separation membranes with MMMs.[1–4] Preparation of MMMs usually involves difficulties such as weak contact of particles in the polymer matrix and poor distribution of the dispersed phase in the continuous polymer matrix phase. In addition, particle size, particle pore size, dispersed phase load, and polymer type and properties can affect the mixed matrix properties. Preparation of defect-free membranes with desired performance characteristics is quite challenging.

5.2 Mixed Matrix Membrane

The separation of gases by membrane is a rapidly growing field.[5,6] In the membrane-based gas separation process, constituents are separated from their mixture due to differential permeation through membrane. Membrane-based separation offers several advantages such as low energy requirement and ease of operation.[7–12] As a result, gas separation by membrane process has acquired great significance in the industries, as gases occupy an important place in the chemical feedstock industry. As the name implies, membrane is the heart of the membrane-based process application. To fully utilize the growing opportunities in the field of gas separation, strong interest exists in the identification of new membrane materials which can comply with current requirements.[11] Criteria for selecting membrane materials for a given separation are complex and unique to particular application. Generally, durability, mechanical integrity, productivity, and separation efficiency are important considerations in selection of membrane material.[10] Of these requirements, selectivity and permeation rate are very important as high selectivity and permeability can lead to the requirement of lower driving force and smaller membrane area to achieve a given separation. In turn, a more efficient separation process results.

In the area of membrane-based gas separation, nonporous polymeric membranes working on solution–diffusion mechanism have been exclusively employed in current commercial devices.[13–15] Typically, polymers have the advantages of desirable mechanical properties and economical processing capabilities. The ability of a membrane to separate two molecules depends on the membrane selectivity, which is defined as the ratio of the permeabilities of the individual gas molecules:

$$\alpha_{AB} = \left(\frac{P_A}{P_B}\right) \tag{5.1}$$

where P is the permeability which is the product of D (diffusivity) and S (solubility).

Efforts on the development of polymers with improved gas separation properties focus on the rate of diffusion by systematic modification of polymer chemical structure or superstructure and chemical/thermal post-treatment of polymeric membranes.[16–22] Solubility selectivity may also be increased by modifying polymer structure to increase the solubility of one component in a mixture or adding special agents which can complex with the desired constituent in a mixture.[23] Inorganic membranes are usually formed from metals, ceramics, or pyrolyzed carbon.[24] These membranes are increasingly being explored to separate gas mixtures due to the well-known thermal and chemical stabilities as well as higher gas flux or selectivity compared to polymeric membranes. Inorganic materials such as zeolites and carbon are excellent materials with diffusivity and selectivity significantly higher than polymeric materials. The accurate size and shape and the narrow range of pore distribution ensure superior selectivity.[25] Microporous inorganic materials have also been modified to achieve solubility-based separation. In these materials, surface flow and capillary condensation play an important role in increasing the flow of larger species.[26] Early inorganic membranes were developed several decades back.[27] In the 1940s, membranes were developed by the Manhattan Project to enrich uranium by separating uranium isotopes as UF6. This was the first large-scale gas separation process using inorganic membranes. Membranes of various zeolites with large pores (Y-type,[28] X-type,[29] b-type[30]), medium pores (ZSM-5,[31] FER[32]), and small pores (A-type,[33] SAPO-34[34]) are used for gas separation. Some of the membranes have good selectivity. Excellent separation properties of CMS membranes (CMSMs) have been reported for the separation of gas mixtures such as natural gas, hydrocarbons, and air.[35–37]

Polymeric membranes tend to be more cost-effective than other membranes because of their ability to be easily spun into hollow fibers or spiral wound modules due to their flexibility and solution processability.[38] Despite these advantages, polymeric membranes are still restricted by the trade-off trend between gas permeability and selectivity.[39] The performance of various membrane materials available for the separation of O_2/N_2 is depicted in Figure 5.1. For the polymeric materials, a rather general trade-off exists between permeability and selectivity, with an *upper bound* limit. When materials with separation properties near this limit were modified based on the traditional structure–property relation, the resultant polymers have permeability and selectivity tracking along this line instead of exceeding it. However, the inorganic materials have properties lying far beyond the upper bound limit for the organic polymers.[40–44] Though tremendous improvements had been achieved in tailoring the polymer structure to enhance separation properties during the past two decades, further progress exceeding the trade-off line seems to present a severe challenge in the near future. Similarly, the immediate application of inorganic membranes is still seriously hindered by the lack of technology to form continuous and defect-free membranes, the extremely high cost for the membrane production, and handling issues (e.g., inherent brittleness).[45,46] In view of this situation, a new approach is needed to provide an alternate and cost-effective membrane with separation properties well above the upper-bound limit between permeability and selectivity.

The membrane morphology emerging with the potential for future applications involves MMM, consisting of organic polymer and inorganic particle phases, as shown in Figure 5.2. The bulk phase (phase A) is typically a polymer; the dispersed phase (phase B) represents the inorganic particles, which may be zeolites, CMSs, or nanosized particles. MMMs have the potential to achieve higher selectivity, permeability, or both relative to the existing polymeric membranes, resulting from the addition of the inorganic particles with their inherent superior separation characteristics. At the same time, the

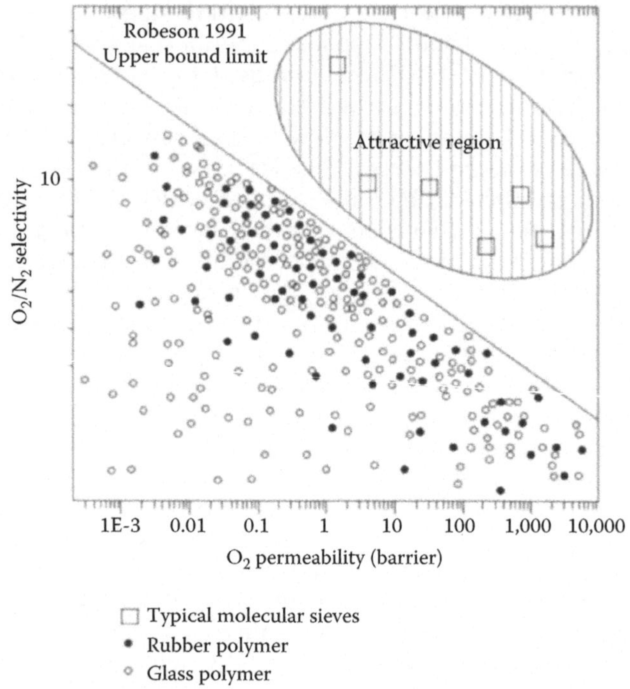

FIGURE 5.1
Relationship between the O_2/N_2 selectivity and the O_2 permeability for polymeric membranes and inorganic membranes. The dots indicate the performance of polymeric materials.

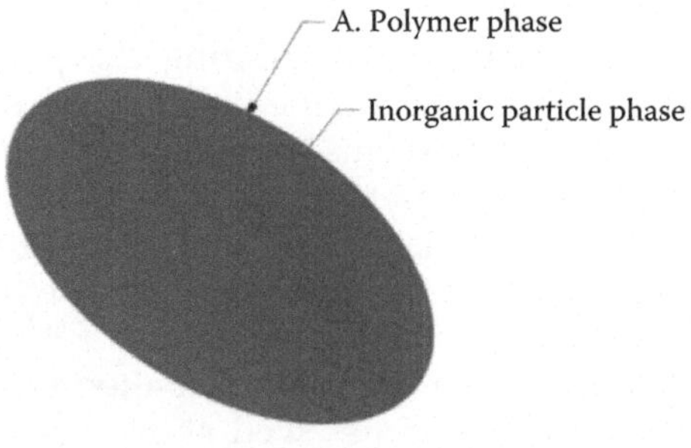

FIGURE 5.2
Schematic of an MMM.

fragility inherent in the inorganic membranes may be avoided by using a flexible polymer as the continuous matrix.

The investigation of MMMs for gas separation has been reported in the 1970s with the discovery of a delayed diffusion time lag effect for CO_2 and CH_4 when adding 5A zeolite into rubbery polymer polydimethyl siloxane (PDMS).[47] The addition of 5A into the polymer matrix causes large increase in the diffusion time lag. However, it has only minor effect on the steady-state permeation. The mixed matrix system of polymer/adsorbent gives better separation performance than that of pure polymeric system[48] such as an enhanced O_2/N_2 selectivity from 3.0 to 4.3 when increasing silicalite content in the polymer cellulose acetate (CA) matrix. The concept of MMM has been demonstrated in the mid-1980s, which also used CA/silicalite MMMs for CO_2/H_2 separation, indicating that silicalite in the membrane phase has reversed the selectivity from H_2 to CO_2.[49]

5.3 Nanocomposite Membrane Preparation

Because of the huge difference between the polymer and inorganic materials in their properties and strong aggregation of the nanofillers, polymer–inorganic nanocomposite membranes cannot be prepared by common methods such as melt blending and roller mixing. The commonly used preparation technologies for the fabrication of nanocomposite membranes are (1) solution bending, (2) *in situ* polymerization, and (3) sol–gel method.[50]

5.3.1 Solution Blending

Solution blending is a simple way to fabricate polymer–inorganic nanocomposite membranes. A polymer is first dissolved in a solvent to form a solution, and then inorganic nanoparticles are added into the solution and dispersed by stirring. The nanocomposite membrane is cast by removing the solvent.[51–55] This method is easy to operate and suitable for all kinds of inorganic materials. The concentration of the polymer and inorganic components is easy to control. However, the inorganic ingredients are liable to aggregate in the membranes.

5.3.2 *In Situ* Polymerization

In this method, the nanoparticles are mixed well with organic monomers, and then the monomers are polymerized. There are often some functional groups, such as hydroxyl or carboxyl, on the surface of inorganic particles, which can generate initiating radicals, cations, or anions under high-energy radiation or plasma to initiate the polymerization of the monomers on their surface.[56–60] In this method, inorganic nanoparticles with functional groups can be connected with polymer chains by covalent bonds. However, it is still difficult to avoid the aggregation of inorganic nanoparticles in the resultant membrane.

5.3.3 Sol–Gel Method

This method is the most widely used preparation technology for nanocomposite membranes. In this method, organic monomers, oligomers, or polymers and inorganic nanoparticle precursors are mixed together in the solution.

The inorganic precursors then hydrolyze and condense into well-dispersed nanoparticles in the polymer matrix. The reaction conditions are moderate, usually room temperature and ambient pressure. The concentration of organic and inorganic components is easy to control in the solution. Additionally, the organic and inorganic ingredients are dispersed at the molecular or nanometer level in the membranes, and thus, the membranes are homogeneous.[61–67]

5.4 MMM Materials

Proper material selection for both the matrix and the inorganic phase is very important in the development of MMM. It has been found that the properties of polymer as well as inorganic phase can affect the morphology and separation performance of MMM. Usually highly selective polymers can result into MMMs with better separation performance.[68–71] Therefore, glassy polymers with superior gas selectivity are preferred to highly permeable but poorly selective rubbery polymers.[68–73] Although glassy polymers are better than rubbery polymers because of their rigid structure, adhesion between the polymer phase and the external surface of the particles is a major challenge when glassy polymers are used in the preparation of MMMs. In these cases, weak organic–inorganic interaction causes voids to form in the polymer–filler interface. Therefore, in the selection of the matrix phase, gas separation properties and adhesion between the two phases must be considered.

The dispersed inorganic phase as well as the continuous phase can affect the separation properties and morphology of MMM. As mentioned above, porous and nonporous fillers are the two major inorganic phase materials that are used for MMM fabrication. Among porous materials, CMSs[74,75] and zeolites[76–86] are the most commonly used inorganic fillers for MMM development. The type, structure, and properties of zeolites play an important role in the development of MMM.[87–89] CMSs are characterized by the pore size distribution of molecular dimensions and have potential as an important component for producing MMMs.[90] These materials have hydrophobic internal surfaces and are being used in industry to separate air by adsorption of oxygen and to remove carbon dioxide from landfill gases. Metal organic frameworks (MOFs)[91] and activated carbon and CNTs[92,93] are other types of porous materials that have been used as the dispersed phase in MMM fabrication.

When a porous material is used as a filler in the polymer matrix, its pore size distribution, surface chemistry, and functional groups must be consistent with the gas molecule pairs. For example, activated carbon is suitable for the separation of carbon dioxide (CO_2) and methane (CH_4) because it has a higher adsorption selectivity for CO_2 (polar gas) than for CH_4 (nonpolar compound), but this filler is not suitable for oxygen (O_2)/nitrogen (N_2) separation. Also, the molecular sieving phase must accurately correspond to the size and shape differences of the gas molecules.[94]

The effect of the nonporous material on separation potential of MMM is different from porous inorganic materials with sieving function; interaction between polymer–chain segments and nanofillers as well as functional groups on the surface of the inorganic phase must be considered when these materials are added to a polymer matrix. For instance, adding silica to a polyimide (PI) matrix can disrupt polymer chain packing and thus increase the oxygen and nitrogen permeation rates.[95] However, adding TiO_2 to the PI matrix can increase the CO_2/CH_4 and H_2/CH_4 selectivity because interactions of CO_2 and H_2 with TiO_2 are stronger than TiO_2–CH_4 interactions.[96] Silica,[97] TiO_2,[98] and fullerene (C60)[99] are the common impermeable inorganic particles used for nanocomposite MMM development.

5.5 Effect of the Inorganic Dispersed Phase

Several parameters affect the performance of the MMM. The major challenges are a suitable combination of polymers and particles, the physical properties of the inorganic fillers such as particle size and particle agglomeration, and the polymer/particle interface morphologies.

The effect of the inorganic dispersed phase on the MMM properties is related to its chemical structure, its surface chemistry, and the type of particles. The inorganic materials used for the MMM can be classified into porous and nonporous types. The effect of porous fillers on the MMM is different from nonporous inorganic fillers and can be related to their structure and pore size. Generally, porous fillers act as molecular sieving agents in the polymer matrix and separate gas molecules by their shape or size. Due to their concise apertures, porous inorganic particles have usually high permeability and selectivity, which is above the Robeson upper bound. When these highly selective porous fillers are added to the polymer matrix, they selectively allow the desired component to pass through the pores, and thus, an MMM, whose permselectivity is higher than that of the neat polymeric membrane, can be obtained. In other words, addition of porous inorganic fillers to a polymer matrix increases not only the permeability of the desired component but also the overall selectivity of the desired component. Addition of porous fillers to the polymer matrix is an ideal way to overcome the traditional permeability–selectivity trade-off of the polymeric membranes. It is important to note that the above-mentioned ideal effect is justified when the polymer chains wet the porous particles completely and there are no defects in the polymer–particle interface.

By contrast, nonporous material fillers can improve the separation properties of the resultant MMMs by increasing the matrix tortuous pattern and decreasing the diffusion of the larger molecules.[100] In addition, nanoscale inorganic materials may disrupt the polymer chain packing and increase the free volume between polymer chains and thus increase gas diffusion. For instance, it has been shown that with addition of 20 vol.% nonporous silica particles to the polysulfone (PSF) matrix, the void volume increases from 0.2% to 2.8%.[98] This small increase in void volume along with the insufficient polymer chain packing causes an increase in the total free volume. The increase in free volume increases the diffusion and solubility coefficients of the silica-filled polymer and causes an increase in the permeability of the gases such as H_2, He, O_2, CO_2, N_2, and CH_4. It has been reported that with addition of 20 vol.% silica to the PSF matrix, the CO_2 and CH_4 permeabilities increase by 212% and 400%, respectively. The hydroxyl and other functional groups on the surface of these nanomaterials may also interact with polar gases (CO_2 and SO_2) and thus improve the penetrant solubility in the resulting nanocomposite MMMs.

5.6 Porous Material Fillers

5.6.1 Carbon Molecular Sieve

CMS membranes exhibit attractive characteristics such as outstanding selectivity, high permeability, stability in corrosive environments, and applicability for operation at high temperatures.[101] CMS membranes can be made by carbonization of a suitable polymeric precursor under controlled conditions such as temperature, pressure, ramp rate, dwelling time, and atmosphere (vacuum or inert gas).[102] Selection of the polymer precursor, pyrolysis atmosphere, and temperature are considered to be the critical factors affecting the properties of the final membrane.[103,104] Several polymers have been used to produce CMS membranes, including polyacrylonitrile (PAN),[105] phenolic resin,[106] polyfurfuryl alcohol,[107] polyvinylidene-based polymers,[108] and various PIs.[109–115] PIs are preferred as the precursors normally due to their rigidity, high-melting point, high glass transition temperature (T_g), thermal stability, as well as attractive separation performance.[116,117] Many classes of PIs with different dianhydrides, including those containing the pyromellitic dianhydride group such as Kapton,[118–122] benzophenone-tetracarboxylic dianhydride such as Matrimid and P84,[123–125] 3,3′,4,4′-biphenyltetracarboxylic dianhydride such as polyimide resins in powder form and solutions of polyimide precursors,[126–129] and hexafluoroisopropylidene (6FDA) such as pyralin[130] have been synthesized and investigated for their suitability as membrane materials.

Besides the choice of membrane precursors, it is shown that the final pyrolysis temperature has a significant influence on the microstructure, separation performance, and gas transport mechanism of CMS membranes. CMS membrane prepared from 6FDA-durene with a decomposition temperature of 496°C shows that the permeability of the membrane improves by an increase in final pyrolysis temperature from 325°C to 475°C. However, for final pyrolysis temperature in the range of 475°C–525°C, the permeability of the smaller gas molecules remains constant, whereas the permeability of the larger gas declines. CMS membranes have been prepared by carbonization of Kapton dense films at temperatures in the range of 500°C–1000°C and under 10^{-5} torr vacuum. These membranes exhibited high gas permeability and selectivity that were influenced by the final pyrolysis temperature.

In addition to the parameters mentioned above, the carbonization atmosphere must be controlled in order to inhibit deterioration or undesired burn-off of the precursor during pyrolysis. The pyrolysis of precursors is typically performed under vacuum or in the presence of an inert gas. It has been observed that the pyrolysis of PIs under vacuum, compared to using an inert gas, can generally offer more selective but less permeable CMS membranes.[131,132] This is due to the role of the inert gas in controlling the heat transfer to the precursor and its subsequent effect on the formation of a more porous microstructure with an enhanced permeability.

Studies have demonstrated the interesting advantages by employing blending technology for gas separation membranes.[133,134] Polymer blending not only can provide the opportunity for altering the properties of the constituent polymers for obtaining synergistic properties, but also can offer new features that may not be found in any one of the constituents. Therefore, blending of suitably selected materials can enable reconciling classes of polymers with different separation properties and physicochemical characteristics through a simple yet reproducible procedure.

Due to the complexities involved in the preparation of CMS membranes, there is need to explore the effect of the precursor molecular structures as well as pyrolysis temperature and atmosphere along with the use of polymer blending to provide a guideline for designing and fabricating high-performance CMS membranes for gas separation applications. The role of key parameters in the fabrication and performance analysis of CMS membranes prepared through blending of polybenzimidazole (PBI) and three PIs containing different dianhydride moieties in their chemical structure has been investigated.[135] Results indicate that the chemical structure of the blend components, the microstructure of the precursor, the blend composition, and the pyrolysis conditions play important roles in the transport properties of the resulting membranes. Pyrolysis protocols with distinguished final pyrolysis temperature and degree of vacuum (Pf) used for the preparation of CMSMs are shown in Figure 5.3. It indicates that two pyrolysis temperatures, 580°C and 800°C, and two degree of vacuum, Pf1 (10^{-3} Torr) and Pf2 (10^{-7} Torr), have been used. The heating rates have been ramped by progressively increasing the pyrolysis temperature starting from 10°C/min, and also

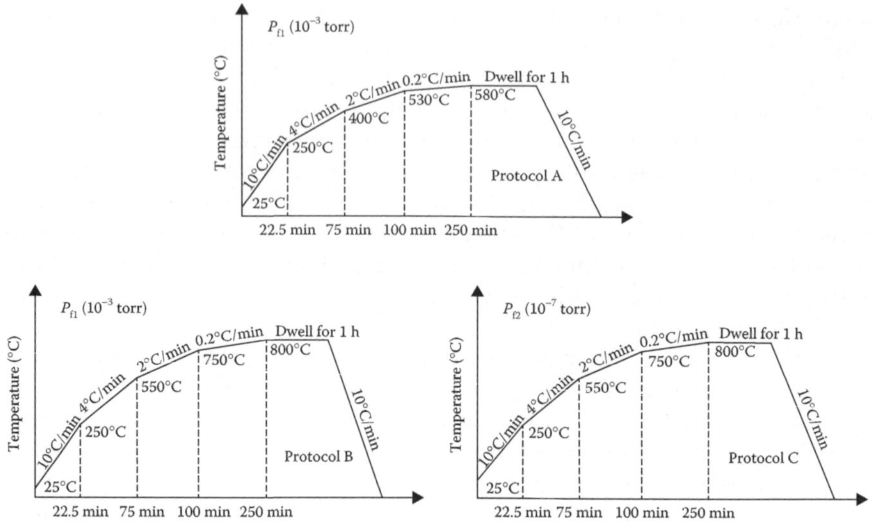

FIGURE 5.3
Pyrolysis protocols with distinguished final pyrolysis temperature and degree of vacuum (P_f) used for the preparation of CMSMs. (Reprinted from *Sep. Purif. Technol.*, 122, Hosseini, S.S., Omidkhah, M.R., Moghaddam, A.Z., Pirouzfar, V., Krantz, W.B., and Tan, N.R., Enhancing the properties and gas separation performance of PBI–polyimides blend carbon molecular sieve membranes via optimization of the pyrolysis process, 278–289, Copyright 2014, Figures 2, 5, and 6, with permission from Elsevier.)

at the lower rates of 4, 2, and 0.2°C/min. The dwell time at the final temperature was 1 h for all protocols. The resultant membranes are cooled gradually to room temperature at the end of each protocol.

The influence of the type of PI used in the blend on the permeability of the carbon membranes follows a trend. A higher pyrolysis temperature results into a membrane with lower permeability but higher selectivity. In addition, a higher degree of vacuum in the pyrolysis chamber increases the selectivity of the membranes by as much as 40% at the expense of permeability. The highest gas pair selectivity for O_2/N_2, CO_2/CH_4, and CO_2/N_2 can be obtained from PBI–Kapton carbonized at 10^{-7} torr and 800°C. The N_2/CH_4 and CO_2/CH_4 separation performances of PBI–Kapton blend precursors with various compositions and their corresponding CMS membranes together with the trade-off lines are shown in Figures 5.4 and 5.5, respectively. It can be noted that the separation performances of the developed membranes are very distinguished particularly for CO_2/CH_4 and selectivities as high as 150–200 can be obtained from PBI–Kapton-derived CMS membrane. As a result, CMS membranes developed from the blends of high-performance polymers according to the recommended procedures and having suitable blend composition can be considered as potential breakthroughs for nitrogen and carbon dioxide removal from natural gas.

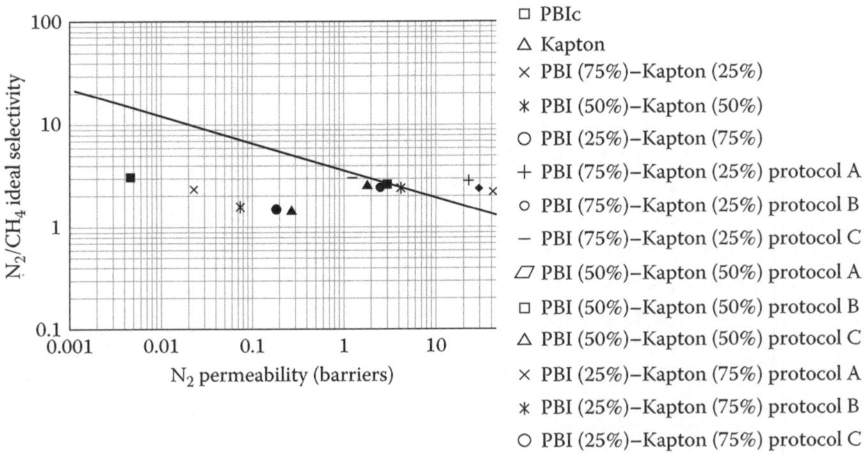

FIGURE 5.4
N_2/CH_4 separation performance of PBI–Kapton blend precursors and the corresponding CMS membranes with respect to trade-off line. (Reprinted from *Sep. Purif. Technol.*, 122, Hosseini, S.S., Omidkhah, M.R., Moghaddam, A.Z., Pirouzfar, V., Krantz, W.B., and Tan, N.R., Enhancing the properties and gas separation performance of PBI–polyimides blend carbon molecular sieve membranes via optimization of the pyrolysis process, 278–289, Copyright 2014, Figures 2, 5, 6, with permission from Elsevier.)

In order to evaluate the performance of supported carbon membranes, the transport mechanism of supported carbon membranes has been investigated in the range 32°C–150°C and 1–2.5 bar.[136] PI (Matrimid5218) material is coated and pyrolyzed under N_2 atmosphere on TiO_2—ZrO_2 macroporous tubes. The supported carbon membrane has been evaluated to determine its permeation for low-molecular-weight gases such as H_2, CH_4, CO, N_2, and CO_2. For these gases, the permeance of the composite-supported carbon membranes obtained after pyrolysis at 550°C increases with an inverse square root of molecular weight. The temperature dependence of the permeance has been described using an Arrhenius law with the negative activation energies for hydrogen, carbon dioxide, and nitrogen providing evidence of a nonactivated process. CMS membranes have been prepared by pyrolysis of PIs containing carboxylic acid groups, in order to investigate their permeation properties for He, CO_2, O_2, and N_2.[137] The PIs are synthesized with a varying number of carboxylic acid groups in the diamine. Wide-angle X-ray diffraction and nitrogen gas sorption isotherms are employed to characterize the carbon structures. From pure gas permeation experiments, it is found that the removal of carboxylic acid groups during the pyrolysis results in a large pore volume in the carbon matrix affecting the gas separation performance of the final CMS membranes.

Studies on CMSMs have shown that adding inorganic nanofillers to the carbon matrix improves the permeability and/or the selectivity.[138–148]

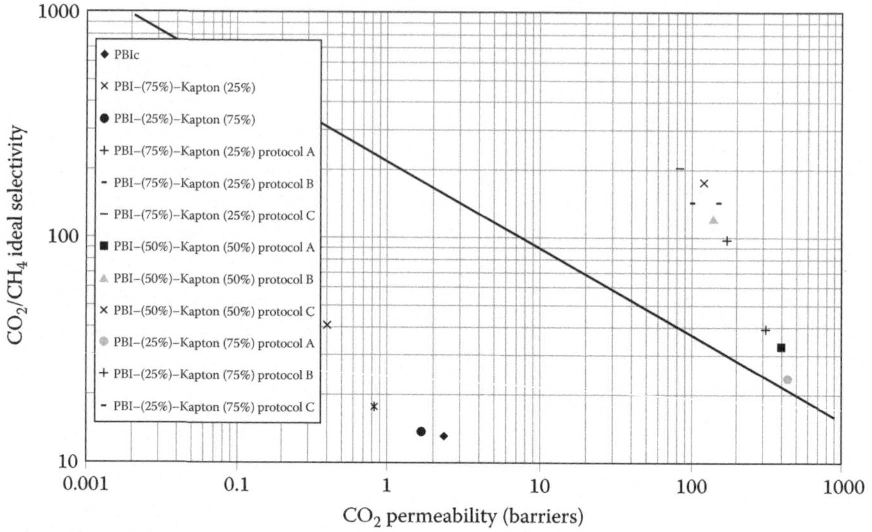

FIGURE 5.5
CO_2/CH_4 separation performance of PBI–Kapton blend precursors and the corresponding CMS membranes with respect to trade-off line. (Reprinted from *Sep. Purif. Technol.*, 122, Hosseini, S.S., Omidkhah, M.R., Moghaddam, A.Z., Pirouzfar, V., Krantz, W.B., and Tan, N.R., Enhancing the properties and gas separation performance of PBI–polyimides blend carbon molecular sieve membranes via optimization of the pyrolysis process, 278–289, Copyright 2014, Figures 2, 5, and 6; *Prog. Mater. Sci.*, 69, Gadipelli, S. and Guo, Z.X., Graphene-based materials: Synthesis and gas sorption, storage and separation, 1–60, Copyright 2015; *J. Membr. Sci.*, 62, Robeson, L.M., Correlation of separation factor versus permeability for polymeric membranes, 165–185, Copyright 1991, with permission from Elsevier.)

Permeability increases with the ionic radius of the metal for composite carbon membranes prepared from Li^+-, Na^+-, and K^+-substituted sulfonated PI due to the increase in interplanar *d*-spacing in the membranes. Studies have shown that by adding silica to polymer precursors, the resulting carbon composite membrane gives better performance due to the faster diffusion given by silica microporous domains. For CMSMs derived from PI containing Pd, the membranes exhibit higher H_2/N_2 selectivity than nondoped ones because the Pd particles act as an inhibitor to the permeation of N_2.[149] Adding Na^+, Mg^{2+}, Al^{3+}, Ag^+, Cu^{2+}, Fe^{3+}, and Ca^{2+} to the precursor results in c-CMSMs with different gas permeation performances, depending on the feature and valency of the metal cations.[150,151] By dispersing silver (Ag) nanoparticles into P84[152] and SPEEK precursors, higher O_2/N_2 selectivity is achieved due to the selective diffusion pathways which occur along the silver nanoparticle surface. Silver-doped carbon membranes exhibit higher CO_2/CH_4 selectivities due to the greater interconnection among the ultra-micropores formed by the migration of silver and better permeabilities due to the more disordered packing of the graphitic sheets of Ag–carbon membranes. There is need for more studies to be undertaken for establishing the cause for the improved separation properties by carbon composite membranes.

5.6.2 Carbon Nanotubes

CNTs are composed of sp^2 bonds. The CNTs consisting of one single graphite layer rolled up in to a hollow cylinder are called single-walled CNTs (SWCNTs) and those consisting of up to several tens of graphitic shells are known as multiwalled CNTs (MWCNTs).[153] CNTs, discovered in 1991, have received great attention since the finding of rapid transport of gases in CNTs, where gas transport is faster than in any other known materials as a result of the inherent smoothness of CNTs.[154–156] The high selectivity of CNTs for hydrogen isotopes has been reported.[157] The adsorption properties of CNTs for CO_2, CH_4, and N_2 suggest that CNTs are more selective for CO_2.[158–160] CNTs have been used to prepare MMMs with different polymer matrix. It is reported that the interfacial contact of polyimidesiloxane-CNT-based MMMs is enhanced by siloxane segment, and the gas permeabilities increase in proportion to the content of CNTs.[161] MWNTs are found relatively more effective than SWNTs in enhancing gas permeabilities of brominated poly(2,6-diphenyl-1,4-phenylene oxide) CNT-based MMMs.[162] Studies on the water/ethanol pervaporation separation properties of polyvinyl alcohol (PVA)/CNTs are also reported.[163] CNTs tend to form entangled agglomerates due to the presence of van der Waals force; thus, the homogeneous dispersion of CNTs in polymer matrix is a major challenge.[164,165] It is possible to obtain homogeneous and fine dispersion of CNTs in polymer matrix by introducing functional groups to CNTs.[166] Surfactant has also been used to facilitate the dispersion of CNTs in a polymer matrix.[167–169]

A CNT-reinforced polyvinyl amine/PVA blend nanocomposite membrane with facilitated transport fixed-site carriers for CO_2/CH_4 separation has been developed[170] with the focus on improving membrane separation performance at elevated pressures and the scaling up of the membrane for potential industrial applications. With the addition of a small amount of CNTs (1.0 wt.%), the membrane shows an enhanced water-swelling capacity as well as good durability against the compaction effect in operations at elevated pressures. The CO_2 permeance of the CNT-reinforced nanocomposite membranes improves significantly compared to the membrane without CNTs, whereas the CO_2/CH_4 selectivity remains the same.

PDMS composites synthesized with different amount of MWCNTs as membranes give better selectivity for the separation of H_2 from CH_4 gas species. Membranes with MWCNT concentrations of 1% increase the selectivity of H_2 gas by 94.8%.[171] It is noted that the permeation of CH_4 is almost totally blocked through membranes with MWCNT concentrations greater than 5%. With the incorporation of MWCNTs, a decrease in the number of available Si—CH_3 and Si—O bonds as well as an increase in the formation of Si—C bonds occur, which initiates the reduction in CH_4 permeation.

A series of MWCNTs/poly(bisphenol A-co-4-nitrophthalic anhydride-co-1,3-phenylenediamine) nanocomposite membrane with a nominal MWCNT content between 1 and 15 wt.% have been prepared by solution-casting method, in which the very fine MWCNTs are embedded into glassy polymer membranes[172] giving increased permeability and enhanced selectivity as well as useful ability to filter gases and organic vapors at the molecular level.

Novel nanocomposite membranes containing SWCNTs inside a PSF matrix have been prepared and characterized.[173] The CNTs are functionalized with long-chain alkyl amines to facilitate dispersion in the polymer. Both permeability and diffusivity of the membranes increase with increasing weight fraction of CNTs. Experimental sorption isotherms of H_2, O_2, CH_4, and CO_2 are consistent with isotherms predicted by atomistic simulations.

A comprehensive method for the fabrication and characterization of nanocomposite membranes containing vertically aligned SWCNTs has been demonstrated[174] by preparing hydrodynamic self-assembly of CNTs, followed by encapsulation of the oriented CNTs in an *in situ* polymerized polyacrylic matrix. The top surface of the polymer film is removed by plasma etching to expose the CNT tips, leading to fast gas transport rates.

Composite membranes with different contents of MWCNTs dispersed in polystyrene can be prepared by electrocasting,[175] in which an alternating electric field of 2000 V/cm is vertically exerted on the thin layer of pre-membrane solution until the solvent is evaporated completely. The electrocasting results in vertical alignment as well as uniform dispersion of MWCNTs in the membranes. The electrocast composite membranes exhibit higher oxygen and nitrogen permeability; however, the permeability of the oxygen is higher than that of the nitrogen, resulting in improved selectivities of the membrane.

Performance test of neat PI and PI/beta-cyclodextrin-functionalized MWCNT (bCfMWCNT) MMMs prepared[176] by immersion precipitation method, shows that permeation and CO_2/CH_4 selectivity is considerably higher in case of PI/bCfMWCNTs MMMs as compared to neat PI membrane.

5.6.3 Graphene

Graphene is a two-dimensional (2D) sp^2-bonded carbon sheet, arranged in a hexagonal honeycomb lattice.[177–180] It is basically a single layer of graphite, which is made up of stacked layers of graphene. The layers in graphite interact weakly through van der Waals forces. Several large-scale processing methods have been involved for different graphene-based materials through either graphitic top-down or molecular carbon precursor bottom-up approaches (Figure 5.6).[181] Flexibility in modification and functionalization of the graphene surface has opened up many possibilities for the development of tailored functional materials. Graphene is mostly obtained from graphite precursors through oxidation–exfoliation–reduction, that is, in the form of graphene oxide (GO). Generally, the oxidation and reduction create

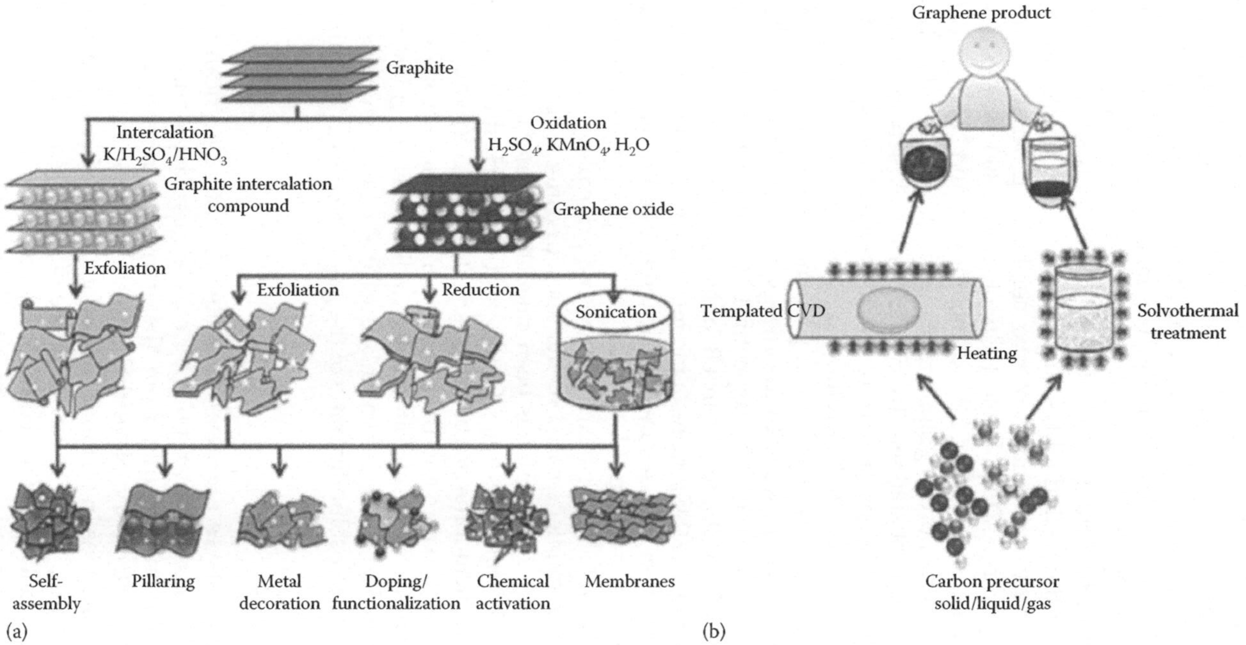

FIGURE 5.6
(a) Graphitic top-down approach and (b) molecular carbon precursor bottom-up approach for producing a wide variety of graphene-based materials in large quantities.

many sites on the graphene, which offers clear advantage in gas sorption, storage, and separation and further functionalization.

Pristine and porous 2D graphene membranes with and without functionalization have been investigated for gas purification.[182] The pristine graphene is impermeable to He and H_2.[183–185] Protons can readily pass through a perfect graphene sheet with a low tunneling barrier compared to a substantially higher barrier for H_2. First-principles density functional theory (DFT) calculations estimate that the H_2/CH_4 diffusion selectivities are on the order of 10^8 and 10^{23} for the N- and all H-functionalized pores, respectively.[186] The pores are created by the removal of two neighboring benzene rings on a single-layer graphene. A CO_2/N_2 selectivity of around 300 with a free energy barrier for permeation of 24.7 and 9.6 kJ/mol for CO_2 and N_2, respectively, is reported for the same porous graphene using classical molecular dynamics simulations.[187] A controllable selectivity and permeable separation of H_2/N_2 is demonstrated in a series of porous graphene membranes with different pore sizes and shapes created by removing 11–16 carbons.[188] First-principles DFT and MD simulations on 0.135 nm pore membrane show a high selectivity for H_2 over other gases with the energy barrier of 0.12, 0.26, 0.25, and 0.82 eV for H_2, N_2, CO, and CH_4, respectively.[189] High permeable selectivity for H_2 relative to CH_4, CO, and CO_2 in a B- or N-doped (pore rim of polyphenylene) porous graphene (PG, one hexagon missing) has been demonstrated.[190]

Electronic properties and separation capabilities of line defect-containing graphene (LD-G) as well as LD-G with missing octagons passivated by hydrogen (H-LD-G) have been explored by first principles.[191] The selectivity of the two porous graphene-based membranes for different gas species (He, H_2, N_2, O_2, CO, CO_2, and CH_4) has been calculated. Water solubility-driven separation of CO_2 from CO_2/O_2, CO_2/N_2, and CO_2/CH_4 mixtures using a porous graphene membrane has been investigated.[192] By introducing a water slab between a gas mixture and the graphene membrane, it has been shown that the gas mixture can be separated based on the water solubility of the gas molecule. By considering various gas mixtures, it is shown that the separation ratio follows the water solubility ratio of the gas molecules in the mixture.

5.6.4 Zeolites

Zeolite molecular sieves are excellent materials with significantly higher diffusivity and selectivity than polymeric materials. The accurate size and shape discrimination resulting from the narrow pore size distribution ensures superior selectivity. Nevertheless, zeolite membranes have expensive cost and difficulties in forming continuous and defect-free membranes of practical meaning.[193–196] The performance of MMM suffers from defects caused by poor contact at the molecular sieve/polymer interface. Several methods have been proposed to improve the polymer–zeolite interaction, and hence to avoid nonselective voids. Polymer chain flexibility is maintained during membrane preparation either by annealing the membranes above glass transition temperature of polymer[197–200] or by adding a plasticizer into the membrane formulation. Alternatively, the external surface of zeolites is modified by silane coupling agents or surface-initiated polymerization with preformed particles to promote the adhesion between zeolite and glassy polymer.[201,202] Also, the low-molecular-weight additives (LMWAs) are employed to prepare glassy polymer/LMWA blend membranes.[203–206]

Diffusion rates of different sized molecules in zeolite pores can differ by orders of magnitude, especially when one molecule is approximately the same size as or larger than the pores. If a mixture contains some molecules that fit into zeolite pores and other molecules that cannot, the membrane separates the mixture by molecular sieving. The structure properties of the some commonly applied zeolites are summarized in Table 5.1.

Material selection for both matrix and sieve phases is a key aspect in the development of MMMs.[207,208] When designing a mixed matrix system for separating a certain gas pair, the molecular sieving phase must provide precise size and shape discrimination ability to distinguish the molecules. Moreover, zeolites with three-dimensional (3D) networks are generally preferred for gas separation because they offer less restricted diffusion paths. Matrix polymer selection fixes minimum membrane performance in the absence of defects. Although rubbery polymers conform more readily to zeolites than glassy polymers, they typically exhibit high permeabilities and low selectivities, and therefore push the overall performance of MMMs considerably below the upper bound trade-off curve. As a result, the attractive

TABLE 5.1
Properties of Major Zeolite Types

Zeolite	Structural Type	Structural Dimension	Pore Size (Å)
4A	LTA	3D	3.8
5A	LTA	3D	4.3
ITQ-29	LTA	3D	4
13X	Faujasite	3D	7.4
NaY	Faujasite	3D	7.4
ZSM-2	Faujasite	3D	7.4
L	LTL	2D	7.1
Beta	BEA	3D	(5.5 × 5.5) and (6.4 × 7.6)
Silicate-1	MFI	2D	(5.1 × 5.5) and (5.3 × 5.6)
ZSM-5	MFI	2D	(5.1 × 5.5) and (5.3 × 5.6)
SSZ-13	CHA	3D	3.8
SAPO-34	CHA	3D	3.8

Source: Reprinted from *J. Ind. Eng. Chem.*, 19, Bastani, D., Esmaeili, N., and Asadollahi, M., Polymeric mixed matrix membranes containing zeolites as a filler for gas separation applications: A review, 375–393, Copyright 2013, Tables 2 and 3, with permission from Elsevier.

Notes: LTA, Linde type A; LTL, Linde type L; MFI, ZSM-5 (five); BEA, Beta polymorphs; CHA, Chabazite.

polymer matrix materials are generally glassy with relatively lower permeability and much higher selectivities. Indeed, addition of zeolites or another highly selective media would only improve the already industrially acceptable properties, if defects can be eliminated.

Flat dense MMMs have been actively pursued in industry and academia for gas separation; however, they have thick dense selective layer and much lower gas permeation flux.[209] To further expand the application of the promising MMMs, a more effective membrane structure, the asymmetric hollow-fiber membranes, needs to be explored. Mixed matrix hollow fibers appear promising for gas and hydrocarbon separations. Compared with flat membranes, hollow fiber is more favored due to the following advantages: (1) a larger membrane area per volume and (2) good flexibility and easy handling in the module fabrication.[210–213] There are challenges in the synthesis of hollow-fiber membranes. The important issues in hollow-fiber MMMs are (1) how to make mixed matrix layer thickness as thin as possible, (2) how to reduce defects in the selective skin, and (3) how to really take advantage of the high-selective nature of zeolite molecular sieves. In Table 5.2,[214] different kinds of zeolite–polymer MMMs are given.

5.6.5 Metal Organic Framework

MOFs, a class of porous hybrid inorganic–organic materials, are made by 3D or 2D porous networks of transition metal complexes with proper organic ligands.[215–217] They are used as inorganic fillers due to their high porosity and improved interaction between particles and polymeric matrices owing to the organic linkers in MOF particles.[218,219] Zeolitic imidazolate frameworks (ZIFs) are a subclass of MOFs that exhibit exceptional thermal and chemical stability.[220,221] Due to their high stability, ZIFs are attractive for various separation applications, particularly for CO_2/CH_4 and C_3H_6/C_3H_8 separation.[222–225] Studies have been carried out for the room-temperature ionic liquid/ZIF-8 MMMs for natural gas sweetening, indicating that the incorporation of 25.8 wt.% ZIF-8 tripled the CO_2 permeability.[226] Incorporation of Cu_3BTC_2 particles into P84 for ethylene/ethane separation[227,228] gives the increase in separation factor by 73% with a 20 wt.% loading of Cu_3BTC_2. An increase of 250% in permeability and 150% in selectivity has been observed for propylene/propane separation by incorporation of 48% ZIF-8 into the 6FDA-diaminomesitylene polymer. So far, only a limited variety of MOFs and ZIFs have been studied as MMMs for gas separation such as MOF-5,[229] ZIF-7,[230] ZIF-8,[231–233] ZIF-90,[234,235] Cu_3BTC_2,[236] MIL-53, and FeBTC. ZIF-71 [Zn(4,5-dichloroimidazole)$_2$] has been recently used as an inorganic filler in fabrication of membranes for pervarporation.[237–239] ZIF-71 has a rhombic structure with zinc(II) as the metal center and 4,5-dichloroimidazole as the organic linker. It possesses a rhombic crystal structure with an aperture size of 4.2 Å and a pore cavity of 16.5 Å. The aperture size of ZIF-71 lies between the kinetic diameters of propylene and propane, 4.0 and 4.3 Å, respectively,[240]

TABLE 5.2

Different Kinds of Zeolite–Polymer MMMs

Polymer	Zeolite	Zeolite Particle Size	Zeolite Loading	Solvent	Polymer Concentration (wt.%)	Additive	Type of Effect of Additive
PC	4A	Average size: 3 μm	Z/Pb: 5%–30% (w/w)	DCM	P/S: 12% (w/v)	pNA	LMWA
PDMS	Silicalite-1, NaX, NaA, Graphite	1.7 μm, 2.3, 4, 1.5	Z/Z + P: 15, 30 and 50 wt.%	DCM	P/S: 1:10 wt.%	–	–
Cross-linked PDMS	SSZ-13	–	15 wt.%	–	–	APDMES	Silanation
PEBA	ZSM-5	1–5 μm	Z/P: 10 and 30 wt.%	*n*-Butanol and *n*-propanol (4:1 wt.%)	P/S: 15 wt.%	–	–
PEEK-WC	4A	3 μm	Z/Z + P: 30 wt.%	DCM, DMA	15 wt.%	APDEMS, DEA	Silanation
PES	4A	Nanosize 4A: 50–140 nm, commercial 4A: 1–5 μm	Z/P: 20 wt.%	NMP	–	–	–
PES	NaA, AgA	1–2 μm	20, 30, 40, and 50 wt.%	NMP	30 wt.%	–	–
PES	4A	Average size: 100 nm	20 wt.%	NMP	30 wt.%		
PES	3A, 4A, 5A	1–5 μm	20 wt.%	NMP	30 wt.%		
PES	SAPO-34	0.5–1 μm	20% (w/w)	Dimethyl sulfoxide	20% (w/v)	HMA	LMWM
PES/PI (=20/80 wt.%)	4A	Less than 2 μm	Z/(Z + P): 20 wt.%	NMP	P/S: 25/75 wt.%	–	–

(Continued)

TABLE 5.2 (*Continued*)

Different Kinds of Zeolite–Polymer MMMs

Polymer	Zeolite	Zeolite Particle Size	Zeolite Loading	Solvent	Polymer Concentration (wt.%)	Additive	Type of Effect of Additive
PI	Nu-6(2)	60 nm × 1000 nm × 1000 nm	4, 8, and 15 wt.%	DCM	P/S: 88 wt.%	–	–
PI	4A	–	Z/total materials: 15 vol.%	–	–	1. RDP fyroflex 2. Di-butyl phthalate 3. 4-Hydroxy	Benzophenone plasticizer
PI	13X, NaY	0.64–4.33 mm	43 wt.%	DMSO	P/S: 1/(5.71) wt.%	TAP	LMWM
PI	L	250–300 nm	Z/P: 20/80 wt.%	THF	P/S: 24 g/L	APTES	Silanation
PI	FAU/EMT	500–800 nm	25 wt.%	NMP	P/S: 10 wt.%	APDEMS	Silanation
Fluorinated PI	ZSM-2	–	20 wt.%	THF	–	APTES	Silanation
Polyimide siloxane	L	–	0–20 wt.%	THF	–	–	–
PI/PSF	Silicalite-1	Hollow zeolite sphere: 4 µm Crystals: 0.3 µm × 1 µm × 2.0 µm	4, 8, and 16 wt.%	Chloroform	90 wt.% of solvent, 10 wt.% polymer + zeolite	–	–
PI/PSF (0/100, 30/70, 50/50, 70/30, and 100/0 wt.%)	ZSM-5	–	0, 10, and 20 wt.% of total	DCM	15 wt.% polymer blend, 85 wt.% DC	–	–

(*Continued*)

TABLE 5.2 (*Continued*)
Different Kinds of Zeolite–Polymer MMMs

Polymer	Zeolite	Zeolite Particle Size	Zeolite Loading	Solvent	Polymer Concentration (wt.%)	Additive	Type of Effect of Additive
Cross-linkable PI-PDMC	(SM) SSZ-13	–	25 wt.%	–	–	–	–
Cross-linkable PI	FAU/EMT	700–900 nm	25 wt.%	Chloroform	P/S: 10–12 wt.%	APTES, APMDES, APDMES	Silanation
Cross-linkable PI	SSZ-13	500 nm	25 wt.%	THF	–	–	–
PSF	ITQ-29	2.5 µm	Z/P: 4, 8, and 12 wt.%	DCM, THF	–	–	–
PSF	Nu-6(2)	5–10 µm	Z/P + S + Z: 0.47, 0.87, and 1.92 wt.%	DCM			
PSF-Ac	3A, 5A		40 wt.%	TCM	Additive/P: 25 wt.%, P/S: 6.5 wt.%	APTMS	Silanation
Toluene	4A	2–2.5 µm	PVA Z/P: 15, 25, and 40 vol.%	DCM	P + Z: 20–25 wt.%	–	–
PVA	4A, SSZ-13	–	Z/P: 15 vol.%	4A: DCM SSZ-13: isopropanol	(Z + P)/solvent: 1:4 wt.%	–	–
PVA	4A	0.5–1.5 µm	50 vol.%	Toluene	25 wt.%	–	–
PVA, Ultem	4A	5 µm	15, 30, and 40 wt.%	DCM, toluene	P/S: 20 wt.%	–	–
SEBS-29S	BEA	Average size: 0.62 µm	10 wt.% of the dry polymer	Toluene	20 wt.% in toluene	Various organosilane	Silanation

(*Continued*)

TABLE 5.2 (***Continued***)

Different Kinds of Zeolite–Polymer MMMs

Polymer	Zeolite	Zeolite Particle Size	Zeolite Loading	Solvent	Polymer Concentration (wt.%)	Additive	Type of Effect of Additive
Teflon AF 1600	Silicalite-1	0.35 and 0.080 μm	350 nm: 29 wt.%, 80 nm: 30.0 and 40.2 wt.%	Galden HT 110	–	–	–
PES	4A	2 μm	20 wt.% of zeolite in total solid	NMP	P/S: 25/75 wt.%	Dynasylan Ameo	Silanation
Outer layer: PES Inner layer: PI	Outer layer: BEA Inner layer: –	0.3 μm	Z/(Z + P): 20 wt.%	NMP	Outer layer: PES/NMP/EtOH: 35/50/15 wt.% Inner layer: PI/NMP/EtOH: 20/67/13 wt.%	–	–
Outer layer: PSF Inner layer: PI	Outer layer: BEA	BEA: 0.4 μm	Z/(Z + P): 10, 20, and 30 wt.%	NMP, EtOH as a solvent	Outer layer: 30 wt.% Inner layer: 23 wt.%	*p*-xylenediamine/methanol: 2.5/100 (w/v)	Hydrogen bonding between Z and P
Outer layer: PSF Inner layer: PI	Outer layer: BEA Inner layer: –	Average size: 0.4 mm	Z/P: 20 wt.%	NMP and EtOH as the solvent and the nonsolvent	Outer layer: P/S:30 wt.% Inner layer: P/S: 23 wt.%	–	–
Ultem, Matrimid	MFI	2,5 mm; 100 and 300 nm	20,30, and 35 wt.%	–	–	–	–

Source: Reprinted from *J. Ind. Eng. Chem.*, 19, Bastani, D., Esmaeili, N., and Asadollahi, M., Polymeric mixed matrix membranes containing zeolites as a filler for gas separation applications: A review, 375–393, Copyright 2013, Tables 2 and 3, with permission from Elsevier.

Notes: BEA, Beta polymorphs; PC, Polycarbonate; PDMS, polydimethyl siloxane; PEBA, poly(ether-block-amide); PEEK-WC, modified polyetheretherketone; PES, polyethersulfone; PI, polyimide; DCM, dichloromethane; DMA, dimethylacetamide; MFI, ZSM-5 (five); NMP, N-Methyl-2-pyrrolidone; PNA, p-nitroaniline; APDMES, 3-aminopropyldimethylethoxysilane; APDEMS, (3-aminopropyl)-diethoxymethyl silane; APMDES, 3-aminopropylmethyldiethoxysilane; DEA, diethanolamine; HMA, 2-hydroxy 5-methyl aniline; LMWA, low molecular weight additive; LMWM, low molecular weight material; NaY, zeolite Y in sodium form; FAU/EMT, faujasite; ZSM-2, zeolite socony mobil-2; DMSO, dimethylsulfoxide; THF, tetrahydrofuran; TAP, 2,4,6-triaminopyrimidine; APTES, aAminopropyltriethoxysilane; SEBS-29S, elastometric block copolymer of styrene-b-styrene(Kraton G1652, Kraton Polymers).

which has a desirable molecular sieving effect. Thus, the resultant MMM may also be effective for propylene/propane separation. Studies have been attempted to investigate the potential of utilizing ZIF-71[241] nanoparticles as inorganic fillers in preparing MMMs for gas separation. The incorporation of ZIF-71 into 6FDA-Durene is expected to produce MMMs with improved gas permeability and possibly C_3H_6/C_3H_8 selectivity. The 6FDA-Durene is selected as the polymer matrix due to its high intrinsic gas separation performance, in particular, for its high gas permeability and reasonable CO_2/CH_4 and C_3H_6/C_3H_8 selectivity. This is the first study to synthesize ZIF-71 nanoparticles with sizes less than 100 nm and employ them as inorganic fillers to fabricate MMMs for gas separation.

5.6.6 Polyhedral Oligomeric Silsesquioxane

Polyhedral oligomeric silsesquioxane (POSS) is an organic/inorganic hybrid material with a cage structure that comprises 6–12 silicon atoms together with oxygen atoms. Different kinds of functional or nonfunctional organic groups are attached to the corner silicon molecules. Depending on the objective and application, the corner organic groups can vary from one to another. The silica core of the POSS is inert and rigid, whereas the organic groups at the vertices provide the excellent compatibility with the polymeric matrix and promote the membrane processability. POSS molecules can be incorporated into the polymer matrix by copolymerization[242,243] or physical blending,[244,245] thus improving the properties of materials in thermal and mechanical stability as well as its rheological and viscoelastic behavior.[245–247]

Hybrid POSS Octa Amic Acid–Matrimid nanocomposite membranes have been successfully made.[248] The excellent dispersion of POSS with carboxylic functional groups, provides a high-density ionic binding platform for the introduction of Zn^{2+}, which is the engine of the facilitated transport to specific gases. The best gas separation performance is observed for the hybrid POSS–Matrimid–Zn^{2+} nanocomposite membrane (20 wt.% POSS–Matrimid–0.3M $ZnCl_2$), in which, the selectivity of CO_2/CH_4 and O_2/N_2 increases by 70% and 30%, respectively, compared with untreated nanocomposite membrane (20 wt.% POSS–Matrimid). Nanocomposite membranes are prepared[249] by incorporation of commercial polyethylene glycol (PEG)-functionalized POSS in two grades of poly-ether-*block*-amide: PEBAX MH 1657 and PEBAX 2533. CO_2 permeability increases twofold after incorporation of 30 wt.% PEG-POSS in PEBAX MH 1657. Simultaneous enhancement in permeability and selectivity is observed up to 30 wt.% loading of PEG-POSS in PEBAX 2533 at 30°C. The effect of temperature upon CO_2 permeability and CO_2 selectivity over N_2, O_2, CH_4, and H_2 has been evaluated between 30°C and 70°C. Substantial influence upon the thermal transition of the polyether domain of both polymers has been observed due to incorporation of PEG-POSS. Influence of temperature upon properties of tailor-made PEBAX MH 1657 nanocomposite membranes for postcombustion CO_2 capture has been investigated.[250] Synthesis of methoxy PEG-functionalized POSS nanoparticles by epoxy ring-opening reaction in three different solvents has been carried out.[251] Although ideal selectivity characteristics for the matrix polymer are not significantly affected by the presence of fillers, the single gas permeability is remarkably increased in both cases.

5.6.7 Layered Silicates

Advances in MMMs have shown growing interest in utilizing layered structure inorganic fillers such as Nu-6(2), zeolite,[252,253] AMH-3,[254–256] JDF-L1,[257–259] and layered aluminophosphate.[260,261] These types of fillers offer marked changes in membrane characteristics and the separation performance by the incorporation of a small amount of loading (10 wt.%) due to their very high aspect ratio.

Clay mineral is a natural form of layered silicate, abundantly available, thus economically attractive compared to synthetic or laboratory-synthesized layered silicates. Such mineral is composed of extremely thin flakes approximately 1 nm in thickness and can range from 30 to 1000 nm in lateral dimensions and hence possess very high aspect ratio.[262] Upon dispersion in organic solvents, the clay tactoids break into smaller units, whereas the inter-gallery space swells, which later facilitates the intercalation by polymer chain. Further intercalation by polymer chain results in exfoliation of the silicate layers to their individual thin fragments.

Several attempts are directed to investigate the potential of different types of clay fillers on the membrane gas separation behavior. Such attempts involve the incorporation of montmorillonite,[263–265] sepiolite,[266] halloysite,[267] and laponite[268] in different polymer membrane hosts. There are four important parameters[269] associated with the clay addition on the permeability of a nanocomposite: the volume fraction of the nano-platelets, their aspect ratio, their dispersion structure, and their orientation in the matrix relative to the diffusion direction. The effect of the above-mentioned parameters on polymeric membrane selectivity needs detailed studies.

5.7 Nonporous Material Fillers

5.7.1 Metal Oxide Nanoparticles

Metal oxide nanoparticles such as MgO and TiO_2 are emerging materials due to their potential applications for membrane-based separation. The primary particles with diameter in nanoscale and high specific area of these metal oxides allow improvement in particle distribution and prevent nonselective void formation in nanoparticles/polymer matrix interface. These nanoparticles are not inherently fused together and have potential to be dispersed individually or in nanoscale aggregates. The incorporation of metal oxides causes the alteration of gas transport behavior in such MMM, which is the result of chain packing disruption and nanoscale agglomeration of nanoparticle in polymer matrix.

Nanocomposite materials based on TiO_2 and PI derived from 1,4-bis(3,4-dicarboxyphenoxy) benzene dianhydride and 2′,2′-dimethyl-4,4′-methylene dianiline prepared[270] by blending of TiO_2 sol and PI solution show that the high TiO_2 content in TiO_2/PI nanocomposite membranes results in the great enhancement of gas separation performance of the TiO_2/PI nanocomposite materials compared to PI. The permeability of H_2 and O_2 in TiO_2/PI nanocomposite membrane with TiO_2 content of 25 wt.% has been observed as 14.1 and 0.718 barrier, respectively, which is 3.7 times and 4.3 times higher than that of pure PI membrane. The effect of TiO_2 nanoparticle addition on MMMs based on Matrimid 5218 prepared by using solution-casting method has demonstrated that the inclusion of TiO_2 increases the gas permeability of MMMs due to chain packing disruption, void formation at polymer–nanoparticle interface as well as within nanoparticle aggregates.[271] For example, in MMMs containing 15 vol.% TiO_2, the permeability of N_2, CH_4, and CO_2 increases up to 2.76, 3.3, and 1.86 times higher than that of pure Matrimid, respectively. It is noted that the use of TiO_2 nanoparticles improves membrane performance in CO_2/CH_4 separation and presents a trade-off line with similar slope compared to Robeson upper insertion bound. Organic–inorganic nanocomposite membranes prepared from polyvinyl acetate (PVAc) and TiO_2 using the solution-casting method and compared with pure PVAc membranes[272] show that the crystallinity of PVAc is decreased by the incorporation of TiO_2. The addition of TiO_2 nanoparticles to PVAc improved the thermal stability of the resulting membranes through an increase in the glass transition temperature (T_g). It is found that the addition of TiO_2 up to 10 wt.% improves both the permeability and the selectivity of the nanocomposite membranes. The permeability of O_2, CO_2, and H_2 is increased by 95%, 79%, and 62%, respectively. Selectivity of the gas pairs O_2/N_2, H_2/N_2, and CO_2/N_2 is also increased by 38%, 26.5%, and 14%, respectively. Cardo PI/TiO_2 MMMs prepared[273] by blending of TiO_2 sol and the PI solution show that MMMs containing 24 wt.% of TiO_2 sol are nanocomposite membranes with strong interactions between inorganic TiO_2 phase and organic PI phase. Compared with PI membrane, the addition of TiO_2 greatly improves the gas separation properties of cardo PI membranes and MMMs. The gas permeation (P_{O2}) and ideal separation factor ($\alpha_{O2/N2}$) of the cardo PI/TiO_2 MMMs prepared in this work are beyond Robeson's trade-off upper bound and appear promising for gas separation application. MMMs composed of sulfonated polyether ether ketone (SPEEK) and amine-functionalized titania submicrospheres have been prepared.[274] Grafting PEI with abundant amine groups onto the titania fillers remarkably increases the content of facilitated transport sites in the membranes, leading to an increment in both gas permeability and selectivity. High humidity also contributes to the facilitated transport of CO_2 by the generation of HCO_3^-. Mechanical and thermal stabilities of the membranes are also enhanced compared to pure SPEEK membrane.

Studies have been conducted to investigate the influence of TiO_2 particle surface chemistry on the gas transport properties of the MMM based on both glassy[275] and rubbery polymer.[276] The MMMs are found to be defect free at low TiO_2 particle loading in the polymer matrix in which the nanoparticles are dispersed individually and in nanoscale aggregates. Whereas when high loading of TiO_2 is filled into the polymer, some nanoparticles form micron-sized aggregates, indicating the presence of transmembrane defects at such high filler concentration. Besides demonstrating great influence on the dispersion properties, the loading of TiO_2 is also found to manipulate the gas transport properties of the MMMs.

Several studies related to MgO embedded in MMMs have been reported.[277–278] Nanocomposite membranes prepared by incorporating nanoscale magnesium oxide particles with different loadings into the Matrimid matrix lead to an increase in gas permeability of membranes. The highest permeability is observed for the membranes containing 40 wt.% MgO loading. However, the selectivity of nanocomposite membranes is less than that of the neat Matrimid. These changes can be mainly ascribed to the effect of pore dimensions of MgO nanoparticles, which are larger than the size range of gas molecules. Also nanocomposite membranes with 20 wt.% MgO have been modified by silver treatment for stipulated periods of time. It is found that the major driving force for penetration of silver ions is offered by MgO nanoparticles and polymer played the role of a carrier. The modified membranes exhibit enhanced gas separation performance for selected gas pairs compared to the neat Matrimid membrane. The best performance is observed for nanocomposite membranes (with 20 wt.% MgO) after a 10-day silver treatment in which the CO_2/CH_4 and H_2/N_2 selectivity increased by 50% and 35%, respectively. In principle, the presence of affinity and interaction between MgO surface and some gas species, for example, CO_2, has also prompted the enhancement in the transport of the molecules. For instance, it has been reported that CO_2 permeability increases from 52 barrier in the neat polybutadiene to 650 barrier in MMM containing 27 vol.% (nominal) MgO. It is ascribed to the surface properties of MgO nanoparticles showing high affinity of physiosorption toward acidic CO_2 molecules hence adsorbed large concentration of CO_2 even at low pressure and consequently enhanced the gas permeability.[279]

5.7.2 Silica

Silica is a conventional class of inorganic fillers that have received significant attention throughout the development of MMM. Silica nanoparticles that can be further categorized into nonporous silica and ordered mesoporous silica. These fillers are in general introduced in the polymeric matrix to form a heterogenous membrane through sol–gel reaction to develop nanoscale particles of the inorganic oxide in a polymer matrix. In this reaction, the silica precursors are hydrolyzed and condensed into the dispersed nanoparticles in the polymer matrix. Because the dispersion is at the molecular or nanoscale level, the interactions between the silica and the organic part can be tailored in order to manipulate the morphological structure at the interface of the two phases.[280,281]

Ordered mesoporous silica with a variety of particle sizes, shapes, and pore diameters is used for the development of new generation of MMM. Mesoporous molecular sieves possess pores large enough (2–50 nm) to readily allow the penetration of polymer chains, resulting in better wetting and dispersion of particles.[282] Ordered mesoporous silica materials have properties such as high mechanic and thermal stability, facility of chemical functionalization, and high specific surface areas (>500 m^2/g).[283] The most common ordered mesoporous silica fillers are MCM-41, MCM-48, and SBA-15.[284–288] In spite of the good adhesion to the polymer matrix, ordered mesoporous materials would offer some limitations concerning gas separation performance due to the gas transport through the inorganic mesoporous membranes, which normally follows the Knudsen diffusion model where the permeance is inversely proportional to the square root of the molecular weight of the penetrants. Also, the pores of mesoporous materials are too large to achieve size selectivity. Therefore, the pores need to be chemically modified to facilitate selective adsorption.

The addition of nonporous nanosized silica is of great potential to affect polymer chain packing in glassy and high free-volume polymers. Due to the nonpermeability of the nonporous silica particles, the addition of this filler into the polymer matrix does not directly contribute to the change of transport property, but it alters the molecular packing of the polymer chains, resulting in an improvement of the permeation as well as the selectivity.[289] The nature of the interface morphology, corresponding to changes in the free volume concentration, free volume cage size as well as the total frictional free volume of the MMM, is strongly influenced by the type of silica, that is, methyl phenyl or silanol groups that may interact with the polymer matrix.[290] In addition, the silica particle loading also significantly affects the permeability of the gases because high loading of the nanoparticles and increased aggregate size may also give rise to the nonlinear expansion of the free volume.[291] Studies have shown that the weak interaction between the silica particle and the polymer matrix may induce void formation during film fabrication, which has a significant effect on the physical properties as well as the gas transport performance of the hybrid membrane.[292–295]

Enhancement in gas permeability as much as 20 times increase in CO_2/N_2 and CO_2/CH_4 gas selectivity in MMM prepared by incorporating high loading of silica nanoparticles within ethylene-vinyl acetate[296] and PBI[297] by sol–gel method has been noted. The performance evaluation of MMM prepared by incorporating fumed silica into cross-linked PVA–polysiloxane[298] reveals that the addition of silica particles has resulted in significant enhancement in the separation performance in which the silica content is the most influential parameter for giving simultaneously high CO_2 permeability and CO_2/H_2 selectivity, which can be attributed to the decrease of H_2 transport within the MMM as the packing density of nonporous silica increased.

Studies on gas permeation properties of polyurethane–silica nanocomposite membranes[299] with silica content of 2.5, 5, 10, and 20 wt.% for pure CO_2, CH_4, N_2, and O_2 gases, respectively, indicate reduction in permeability of all gases, but enhancement in CO_2/N_2, CO_2/CH_4, and O_2/N_2 selectivity by increasing the content of nonpermeable silica nanoparticles in polyurethane–silica membranes. In the case of polyurethane–silica (20 wt.%) nanocomposite membrane, the obtained CO_2/N_2 selectivity is 1.65 times of pure polyurethane, whereas the CO_2 permeability reduction of polyurethane–silica membranes is 35.6% in comparison with pure polyurethane. MMMs have been developed by incorporating inorganic silica nanoparticles into blends of PSF/PI asymmetric membranes[300] for gas separation using phase inversion technique. Permeation results show that CO_2 permeance increases with the introduction of 5.2 wt.% silica contents (73.7 ± 0.2 GPU) in PSF/PI-20% blend, and it increases with the increase in silica contents. With 15.2 wt.% silica content, the highest permselectivity of $\alpha_{CO2/CH4}$ (i.e., 61.0 ± 0.3 to 60.2 ± 0.4) is observed for the treated membrane at 2–10 bar. Nanocomposite membranes of PDMS with different amounts of fumed silica have been synthesized over a porous support of PAN.[301] Thermal analysis indicates that increasing the silica content enhances the thermal stability of membranes. The permeation results show that the nanocomposite membrane containing 11 wt.% of fumed silica exhibits 38% increase in the permeability of *n*-butane and simultaneously 30% increase in the selectivity of *n*-butane over methane. Fumed nano-SiO_2 particles have been modified with a silane coupling agent (3-aminopropyl)triethoxysilane and subsequently incorporated into the P84 polymer matrix to form nanocomposite membranes.[302] The composite membranes based on the modified SiO_2 show better bonding morphology at the polymer/particle interfaces, higher stress modulus and tensile strength, and significantly higher glass transition temperature compared with those of the composite membranes based on the pristine SiO_2 particles. The permeability of gases (such as CO_2, O_2, N_2, and He) increases with increasing volume fraction of the inorganic filler. The composite membranes with less than 0.14 vol.% of the modified SiO_2 have comparable selectivity as that of the pure P84 membrane and significantly higher tensile modulus and stress than the latter. Polyurethane and polyurethane–silica nanocomposite membranes have been prepared by solution blending and casting–evaporation methods.[303] The results show reduction in gas permeability but enhancement in CO_2/N_2, CO_2/CH_4, and O_2/N_2 ideal selectivity.

5.8 Permeation Models for MMMs

MMM permeation models can be classified into two models: one that is used to predict the permeation of an MMM comprising permeable particles and polymer matrix and the other that is applicable for an MMM comprising impermeable particles and polymer matrix.

5.8.1 Porous Particles

Permeation models for MMMs with porous particles are used to predict the effective permeability of a gaseous penetrant in an MMM as a function of continuous phase (polymer matrix) permeability, dispersed phase (porous particles) permeability and volume fraction of dispersed phase. Bouma et al. used the Maxwell–Wagner–Sillar model to calculate the effective permeability of an MMM with a dilute dispersion of ellipsoids[304–306]:

$$P_M = \frac{P_c\left[nP_d + (1-n)P_c - (1-n)\varphi_d(P_c - P_d)\right]}{\left[nP_d + (1-n)P_c + n\varphi_d(P_c - P_d)\right]} \tag{5.2}$$

where:

P_M is the effective permeability of a gaseous penetrant in an MMM
P_c is the continuous phase permeability
P_d is the dispersed phase permeability
φ_d is the volume fraction of the dispersed phase
n is the shape factor of the particle

In the above equation, $n = 1/3$ corresponds to the dilute suspension of spherical particles and leads to the following equation known as Maxwell's equation:

$$P_M = \frac{P_c\left[P_d + 2P_c - 2\varphi_d(P_c - P_d)\right]}{\left[P_d + 2P_c + \varphi_d(P_c - P_d)\right]} \tag{5.3}$$

Maxwell's model predicts the MMM permeability. The model is applicable to a dilute suspension of spheres and is applicable for low loadings, when the volume fraction of filler particles is less than about 20%. Maxwell's model has limitations in predicting the permeability of MMMs at the maximum packing volume fraction of filler particles. The model does not account for particle size distribution, particle shape, and aggregation of particles. To calculate the permeability of an MMM with a high filler volume fraction, the Bruggeman model[307] can be used. Although the Bruggeman model is applicable for high loadings, it has limitations with respect to the other issues as highlighted for the Maxwell model. The Lewis–Nielsen[308,309] and Pal models[310] in such case can be used to calculate the effective permeability of MMMs with vmaximum packing volume fraction of filler particles and may include the effects of morphology on permeability.

All models presented above are applicable to predict the MMM behavior when there are no interfacial defects between two phases. Interfacial defects affect the membrane separation performance and should be taken into account in permeability models. In a modified two-phase Maxwell model,[311] the polymer matrix dispersed phase and interphase has been taken as an idealized two-phase system with the matrix polymer being one phase and the combined particle and interphase constituting the other. A similar three-phase model can be applied to describe the MMMs with poor organic–inorganic contact.[312] The two-phase modified Maxwell model to predict the permeability of the pseudo-dispersed phase is as follows:

$$P_{ps} = \frac{P_c\left[P_d + 2P_I - 2\varphi_s\left(P_I - P_d\right)\right]}{\left[P_d + 2P_I + \varphi_s\left(P_I - P_d\right)\right]} \tag{5.4}$$

where:

P_{ps} is the effective permeability of the pseudo-dispersed phase (i.e., dispersed particles and rigidified layer or dispersed particles and voids or dispersed particles and pore blockage)

P_d is the permeability of the particles

P_I is the permeability of the interface (rigidified layer or voids or particle pore blockage)

φ_s is the volume fraction of the filler dispersed phase within the pseudo-dispersed phase[311,312]

5.8.2 Impermeable Particles

Several models have been developed to predict the permeability of nanocomposite membranes with nonporous impermeable fillers. The Maxwell model for composite membranes containing nonporous impermeable fillers dispersed into the polymer matrix can be written as follows[313,314]:

$$P_M = \frac{P_c(1 - \varphi_d)}{\left(1 + 0.5\,\varphi_d\right)} \tag{5.5}$$

where:

P_M is the permeability of the composite

P_c is the permeability of the polymer matrix (continuous phase)

φ_d is the volume fraction of impermeable fillers (dispersed phase)

Based on the Maxwell equation, the permeability of composite membranes should decrease with increasing particle volume fraction, but several researchers observed an increase in gas permeability upon adding impermeable particles to polymer matrix.[315–317] This phenomenon can be attributed to the fact that in the Maxwell model, the interactions between the nanofillers and the polymer chains, and the nanofillers and the penetrants have been neglected.

If nanofillers disrupt the polymer-chain packing and increase the free volume between the polymer chains, penetrant gas diffusion and gas permeability of the resulting mixed matrix nanocomposite membrane increase based on the following Cohen and Turnbull equation[318]:

$$D_M = A\exp\left(\frac{-\gamma V^*}{V_f}\right) \tag{5.6}$$

where:

A is a constant depending on the temperature

γ is an overlap parameter for avoiding double-counting of the free volume elements

V^* is the minimum free volume element size nearly equal to the penetrant size

V_f is the accessible average free volume in the media for transport of the penetrants

According to this equation, an increase in the free volume of the polymer continuous phase (V_f) causes an increase in the penetrant diffusivity and results in an enhancement in the permeability.

References

1. Chung, T.S., Jiang, L.Y., Li, Y., and Kulprathipanja, S. 2007. Mixed matrix membranes (MMMs) comprising organic polymers with dispersed inorganic fillers for gas separation. *Prog. Polym. Sci. (Oxford)* 32: 483–507.
2. Cong, H., Radosz, M., Towler, B.F., and Shen, Y. 2007. Polymer-inorganic nanocomposite membranes for gas separation. *Sep. Purif. Technol.* 55: 281–291.
3. Aroona, M.A., Ismailb, A.F., Matsuurab, T., and Montazer-Rahmatia, M.M. 2010. Performance studies of mixed matrix membranes for gas separation: A review. *Sep. Purif. Technol.* 75: 229–242.
4. Rousseau, R.W. 1987. *Handbook of Separation Process Technology*, Wiley, New York.
5. Nunes, S.P. and Peinemann, K.V. 2001. *Membrane Technology: In the Chemical Industry*, Wiley-VCH, Weinheim, Germany.
6. Berry, R.I. 1981. Membrane separate gas. *Chem. Eng.* 88: 63–67.
7. Strathman, H. 1981. Membrane separation process. *J. Membr. Sci.* 9: 121–189.
8. Spillman, R.W. and Sherwin, M.B. 1990. Gas separation membranes: The first decade. *Chem. Technol.* 20: 378–408.
9. Paul, D.R. and Yampol'skii, Y.P. 1994. *Polymeric Gas Separation Membranes*, CRC Press, Boca Raton, FL.

10. Stern, S.A. 1994. Polymers for gas separations: The next decade. *J. Membr. Sci.* 94: 1–65.
11. Mulder, M. 1996. *Basic Principles of Membrane Technology*, Kluwer Academic, Dordrecht, the Netherlands.
12. Spillman, R.W. 1989. Economics of gas separation membranes. *Chem. Eng. Prog.* 85: 41–62.
13. Meindersma, G.W. and Kuczynskyi, M. 1996. Implementing membrane technology in the process industry: Problems and opportunities. *J. Membr. Sci.* 113: 285–292.
14. Freeman, B.D. and Pinnau, I. 1999. *Polymer Membrane for Gas and Vapor Separations: Chemistry and Material Science*, American Chemical Society, Washington, DC.
15. Pinnau, I. and Freeman, B.D. 1990. *Membrane Formation and Modification: Overview*, American Chemical Society, Washington, DC.
16. Koros, W.J. and Flemming, G.K. 1993. Membrane-based gas separation. *J. Membr. Sci.* 83: 1–80.
17. Bos, A., Punt, I.G., Wessling, M., and Strathmann, H. 1998. Plasticization-resistant glassy polyimide membrane for CO_2/CH_4 separations. *Sep. Purif. Technol.* 14: 27–39.
18. McCaig, M.S. and Paul, D.R. 1999. Effect of UV cross-linking and physical aging on the gas permeability of the thin glassy polyacrylate films. *Polymer* 40: 7209–7225.
19. Bos, A., Punt, I., Strathmann, H., and Wessling, M. 2001. Suppression of gas separation membrane plasticization by homogeneous polymer blending. *AIChE J.* 47:1088–1093.
20. Tin, P.S., Chung, T.S., Liu, Y., Wang, R., Liu, S.L., and Pramoda, K.P. 2003. Effect of cross-linking on gas separation performance of Matrimid membranes. *J. Membr. Sci.* 225: 77–90.
21. Shao, L., Chung, T.S., Hong, G.H., and Pramoda, K.P. 2005. Polyimide modification by a linear aliphatic diamine to enhanced transport performance and plasticization resistance. *J. Membr. Sci.* 256: 46–56.
22. Chung, T.S., Shao, L., and Tin, P.S. 2006. Surface modification of polyimide membranes by diamines for H_2 and CO_2 separation. *Macromol. Rapid Commun.* 27: 998–1003.
23. Morisato, A., He, Z., Pinnau, I., and Merkel, T.C. 2002. Transport properties of PA 12-PTMO/AgBF4 solid polymer electrolyte membranes for olefin/paraffin separation. *Desalination* 145: 347–351.
24. Morooka, S. and Kusakabe, K. 1999. Microporous inorganic membranes for gas separation. *MRS Bull.* 24: 25–29.
25. Koresh, J.E. and Soffer, A. 1980. Study of molecular sieve carbons. Part 2. Estimation of cross-sectional diameters of nonspherical molecules. *J. Chem. Soc. Faraday I* 76: 2472–2480.
26. Rao, M.B. and Sircar, S. 1996. Performance and pore characterization of nanoporous carbon membranes for gas separation. *J. Membr. Sci.* 110: 109–118.
27. Pandy, P. and Chauhan, R.S. 2001. Membranes for gas separation. *Prog. Polym. Sci.* 26: 853–893.
28. Kusakabe, K., Kuroda, T., and Morooka, S. 1998. Separation of carbon dioxide from nitrogen using ion-exchanged faujasite-type zeolite membranes formed on porous support tubes. *J. Membr. Sci.* 48: 13–23.

29. Kita, H., Fuchida, K., Horita, T., Asamura, H., and Okamoto, K. 2001. Preparation of faujasite membranes and their permeation properties. *Sep. Purif. Technol.* 25: 261–268.
30. Tuan, V.A., Li, S., Falconer, J.L., and Noble, R.D. 2002. In situ crystallization of beta zeolite membranes and their permeation and separation properties. *Chem. Mater.* 14: 489–492.
31. Yan, Y., Davis, M.E., and Gavalas, G.R. 1995. Preparation of zeolite ZSM-5 membranes by in situ crystallization on porous Al_2O_3. *Ind. Eng. Chem. Res.* 34: 1652–1661.
32. Nishiyama, N., Ueyama, K., and Matsukata, M. 1997. Synthesis of FER membrane on an alumina support and its separation properties. *Stud. Surf. Sci. Catal.* 105: 2195–2205.
33. Aoki, K., Kusakabe, K., and Morooka, S. 1998. Gas permeation properties of A-type zeolite membranes formed on porous substrate by hydrothermal synthesis. *J. Membr. Sci.* 141: 197–209.
34. Li, S., Falconer, J.L., and Noble, R.D. 2004. SAPO-34 membranes for CO_2/CH_4 separation. *J. Membr. Sci.* 241: 121–135.
35. Vu, D.Q., Koros, W.J., and Miller, S.J. 2002. High pressure CO_2/CH_4 separation using carbon molecular sieve hollow fiber membranes. *Ind. Eng. Chem. Res.* 41: 367–380.
36. Okamoto, K., Kawamura, S., Yoshino, M., Kita, H., Hirayama, Y., and Tanihara, N. 1999. Olefin/paraffin separation through carbonized membranes derived from an asymmetric polyimide hollow fiber membrane. *Ind. Eng. Chem. Res.* 38: 4424–4432.
37. Xiao, Y.C., Dai, Y., Chung, T.S., and Guiver, M.D. 2005. Effects of brominating Matrimid polyimide on the physical and gas transport properties of derived carbon membranes. *Macromolecules* 38: 10042–10049.
38. Koros, W.J. 2002. Gas separation membranes: Needs for combined materials science and processing approaches. *Macromol. Symp.* 188: 13–22.
39. Robeson, L.M. 1991. Correlation of separation factor versus permeability for polymeric membranes. *J. Membr. Sci.* 62: 165–185.
40. Singh, A.S. and Koros, W.J. 2000 Air separation properties of flat sheet homogeneous pyrolytic carbon membranes. *J. Membr. Sci.* 174: 177–188.
41. Feuters, A.B. and Centeno, T.A. 1999. Preparation of supported carbon molecular membranes. *Carbon* 37: 679–684.
42. Kim, Y.K., Lee, J.M., Park, H.B., and Lee, Y.M. 2004. The gas separation properties of carbon molecular sieve membranes derived from polyimides having carboxylic acid groups. *J. Membr. Sci.* 235: 139–146.
43. Park, H.B., Kim, Y.K., Lee, J.M., Lee, S.Y., and Lee, Y.M. 2004. Relationship between chemical structure of aromatic polyimides and gas permeation properties of their carbon molecular sieve membranes. *J. Membr. Sci.* 229: 117–127.
44. Tin, P.S., Chung, T.S., Kawi, S., and Guiver, M.D. 2004. Novel approaches to fabricate carbon molecular sieve membranes based on chemical modified and solvent treated polyimides. *Micropor. Mesopor. Mater.* 73: 151–160.
45. Saracco, G., Neomagus, H.W.J.P., Versteeg, G.F., and Van Swaaij, W.P.M. 1995. High-temperature membrane reactor: Potential and problems. *Chem. Eng. Sci.* 54: 1997–2017.
46. Caro, J., Noack, M., Kolsch, P., and Schaefer, R. 2000. Zeolite membrane-state of their development and perspective. *Micropor. Mesopor. Mater.* 38: 3–24.

47. Paul, D.R. and Kemp, D.R. 1973. The diffusion time lag in polymer membranes containing adsorptive fillers. *J. Polym. Sci. Polym. Phys.* 41: 79–93.
48. Kulprathipanja, S., Neuzil, R.W., and Li, N.N. 1988. Separation of fluids by means of mixed matrix membranes, US patent 4,740,219.
49. Kulprathipanja, S., Neuzil, R.W., and Li, N.N. 1992. Separation of gases by means of mixed matrix membranes, US patent 5,127,925.
50. Li, C., Shao, H., and Zhong, S. 2003. Preparation technology of organic–inorganic hybrid membrane. *Huxue Jinzhan* 16: 83–89.
51. Genne, I., Kuypers, S., and Leysen, R. 1996. Effect of the addition of ZrO_2 to polysulfone based UF membranes. *J. Membr. Sci.* 113: 343–350.
52. Wara, N.M, Francis, L.F., and Velamakanni, B.V. 1995. Addition of alumina to cellulose acetate membranes. *J. Membr. Sci.* 104: 43–49.
53. Stephen, R., Ranganathaiah, C., Varghese, S., Joseph, K., and Thomas, S. 2006. Gas transport through nano and micro composites of natural rubber (NR) and their blends with carboxylated styrene butadiene rubber (XSBR) latex membranes. *Polymer* 47: 858–870.
54. Mascia, L., Zhang, Z., and Shaw, S.J. 1996. Carbon fibre composites based on polyimide/silica ceramers: Aspects of structure–properties relationship. *Composites A* 27: 1211–1221.
55. Shi, D., Kong, Y., Yang, J., and Du, H. 2000. Study on transitional metal organic complex-polyimide hybrid material for gas separation membranes. *Acta Polym. Sin.* 4: 457–461.
56. Liu, H. 1997. Synthesis of TiO_2 nanopowder enwrapped by organic membrane with microwave induced plasma method. *Huaxue Tongbao* 10: 44–46.
57. Doucoure, A., Guizard, C., Durand, J., Berjoan, R., and Cot, L. 1996. Plasma polymerization of fluorinated monomers on mesoporous silica membranes and application to gas permeation. *J. Membr. Sci.* 117: 143–150.
58. Patel, N.P., Miller, A.C., and Spontak, R.J. 2003. Highly CO_2-permeable and selective polymer nanocomposite membranes. *Adv. Mater.* 15: 729–733.
59. Patel, R.J., Aberg, C.M., Sanchez, A.M., Capracotta, M.D., Martin, J.D., and Spontak, R.J. 2004. Morphological, mechanical and gas-transport characteristics of crosslinked poly(propylene glycol): Homopolymers, nanocomposites and blends. *Polymer* 45: 5941–5950.
60. Nunes, S.P., Peinenmann, K.V., Ohlorogge, K., Alpers, A., Keller, M., and Pires, A.T.N. 1999. Membranes of poly(ether imide) and nanodispersed silica. *J. Membr. Sci.* 157: 219–226.
61. Brzesowsky, R.H., de With, G., van de Cruijsem, S., Snijkers-Hendrickx, I.J.M., Wolter, W.A.M., and van Lierop, J.G. 1998. Glass strengthening by silica particle reinforced organic–inorganic coatings. *J. Non-Cryst. Solids* 241: 27–37.
62. Livage, J. and Sanchez, C. 1992. Sol–gel chemistry. *J. Non-Cryst. Solids* 145: 11–19.
63. Kioul, A. and Mascia, L. 1994. Compatibility of polyimide-silicate ceramers induced by alkoxysilane silane coupling agents. *J. Non-Cryst. Solids* 175: 169–186.
64. Smaihi, M., Jermoumi, T., Marignan, J., and Noble, R.D. 1996. Organic–inorganic gas separation membranes: Preparation and characterization. *J. Membr. Sci.* 116: 211–220.
65. Iwata, M., Adachi, T., Tomidokoro, M., Ohta, M., and Kobayashi, T. 2003. Hybrid sol–gel membranes of polyacrylonitrile–tetraethoxysilane composites for gas permselectivity. *J. Appl. Polym. Sci.* 88: 1752–1759.

66. Gomes, D., Nunes, S.P., and Peinemann, K.V. 2005. Membranes for gas separation based on poly(1-trimetylsilyl-1-propyne)–silica nanocomposites. *J. Membr. Sci.* 246: 13–25.
67. Duval, J.M. 1995. Adsorbent filled polymeric membranes, PhD thesis, University of Twente, Enschede, the Netherlands.
68. Mahajan, R. and Koros, W.J. 2000. Factors controlling successful formation of mixed matrix gas separation materials. *Ind. Eng. Chem. Res.* 39: 2692–2696.
69. Mahajan, R. and Koros, W.J. 2002. Mixed matrix membrane materials with glassy polymers. Part 1. *Polym. Eng. Sci.* 42: 1420–1431.
70. Mahajan, R. and Koros, W.J. 2002. Mixed matrix membrane materials with glassy polymers. Part 2. *Polym. Eng. Sci.* 42: 1432–1441.
71. Paul, D.R. and Kemp, D.R. 1973. Diffusion time lag in polymer membranes containing adsorptive fillers. *J. Polym. Sci. Part C Polym. Symp.* 41: 79–93.
72. Jia, M., Peinemann, K.V., and Behling, R.D. 1991. Molecular sieving effect of zeolite filled silicone rubber membranes. *J. Membr. Sci.* 57: 289–296.
73. Anson, M., Marchese, J., Garis, E., Ochoa, N., and Pagliero, C. 2004. ABS copolymer-activated carbon mixed matrix membranes for CO_2/CH_4 separation. *J. Membr. Sci.* 243: 19–28.
74. Vu, D.Q., Koros, W.J., and Miller, S.J. 2003. Mixed matrix membranes using carbon molecular sieves. I. Preparation and experimental results. *J. Membr. Sci.* 211: 311–334.
75. Vu, D.Q., Koros, W.J., and Miller, S.J. 2003. Mixed matrix membranes using carbon molecular sieves. II. Modeling permeation behavior. *J. Membr. Sci.* 211: 335–348.
76. Jiang, L.Y., Chung, T.S., Cao, C., Huang, Z., and Kulprathipanja, S. 2005. Fundamental understanding of nano-sized zeolite distribution in the formation of the mixed matrix single and dual-layer asymmetric hollow fiber membranes. *J. Membr. Sci.* 252: 89–100.
77. Li, Y., Chung, T.S., Huang, Z., and Kulprathipanja, S. 2006. Dual-layer polyethersulfone (PES)/BTDA-TDI/MDI co-polyimide (P84) hollow fiber membranes with a submicron PES–zeolite beta mixed matrix dense-selective layer for gas separation. *J. Membr. Sci.* 277: 28–37.
78. Pechar, T.W., Kim, S., Vaughan, B., Marand, E., Tsapatsis, M., Jeong, H.K., and Cornelius, C.J. 2006. Fabrication and characterization of polyimide–zeolite L mixed matrix membranes for gas separations. *J. Membr. Sci.* 277: 195–202.
79. Jiang, L.Y. and Chung, T.S. 2006. Fabrication of mixed matrix hollow fibers with intimate polymer–zeolite interface for gas separation. *AIChE J.* 52: 2898–2908.
80. Gorgojo, P., Uriel, S., Tellez, C., and Coronas, J. 2008. Development of mixed matrix membranes based on zeolite Nu-6(2) for gas separation. *Micropor. Mesopor. Mater.* 115: 85–92.
81. Sen, D., Kalıpcilar, H., and Yilmaz, L. 2007. Development of polycarbonate based zeolite 4A filled mixed matrix gas separation membranes. *J. Membr. Sci.* 303: 194–203.
82. Ismail, A.F., Kusworo, T.D., and Mustafa, A. 2008. Enhanced gas permeation performance of polyethersulfone mixed matrix hollow fiber membranes using novel Dynasylan Ameo silane agent. *J. Membr. Sci.* 319: 306–312.
83. Ciobanu, G., Carja, G., and Ciobanu, O. 2008. Structure of mixed matrix membranes made with SAPO-5 zeolite in polyurethane matrix. *Micropor. Mesopor. Mater.* 115: 61–66.

84. Husain, S. and Koros, W.J. 2007. Mixed matrix hollow fiber membranes made with modified HSSZ-13 zeolite in polyetherimide polymer matrix for gas separation. *J. Membr. Sci.* 288: 195–207.
85. Rafizah, W.A.W. and Ismail, A.F. 2008. Effect of carbon molecular sieve sizing with poly (vinylpyrrolidone) K-15 on carbon molecular sieve–polysulfone mixed matrix membrane. *J. Membr. Sci.* 307: 53–61.
86. Funk, C.V. and Lloyd, D.R. 2008. Zeolite-filled microporous mixed matrix (ZeoTIPS) membranes: Prediction of gas separation performance. *J. Membr. Sci.* 313: 224–231.
87. Duval, J.M., Folkers, B., Mulder, M.H.V., Desgrandchamps, G., and Smolders, C.A. 1993. Adsorbent filled membranes for gas separation. Part 1. Improvement of the gas separation properties of polymeric membranes by incorparation of microporous adsorbents. *J. Membr. Sci.* 80: 189–198.
88. Breck, D.W. 1974. *Zeolite Molecular Sieves*, John Wiley & Sons, New York.
89. Yang, R.T. 1997. *Gas Separation by Adsorption Process*, Imperial College Press, London.
90. Ismail, A.F. and David, L.I.B. 2001. A review on the latest development of carbon membranes for gas separation. *J. Membr. Sci.* 193: 1–18.
91. Zhang, Y., Musselman, I.H., Ferraris, J.P., and Balkus, K.J. 2008. Gas permeability properties of Matrimid membranes containing the metal-organic framework Cu–BPY–HFS. *J. Membr. Sci.* 313: 170–181.
92. Kim, S., Pechar, T.W., and Marand, E. 2006. Poly(imide siloxane) and carbon nanotube mixed matrix membranes for gas separation. *Desalination* 192: 330–339.
93. Kim, S., Chen, L., Johnson, J.K., and Marand, E. 2007. Polysulfone and functionalized carbon nanotube mixed matrix membranes for gas separation: Theory and experiment. *J. Membr. Sci.* 294: 147–158.
94. Mahajan, R. 2000. Formation, characterization and modeling of mixed matrix membrane materials, PhD thesis, University of Texas, Austin, TX.
95. Moaddeb, M. and Koros, W.J. 1997. Gas transport properties of thin polymeric membranes in the presence of silicon dioxide particles. *J. Membr. Sci.* 125: 143–163.
96. Hu, Q., Marand, E., Dhingra, S., Fritsch, D., Wen, J., and Wilkes, G. 1997. Poly(amideimide)/TiO_2 nano-composite gas separation membranes: Fabrication and characterization. *J. Membr. Sci.* 135: 65–79.
97. Ahn, J., Chung, W.-J., Pinnau, I., and Guiver, M.D. 2008. Polysulfone/silica nanoparticle mixed-matrix membranes for gas separation. *J. Membr. Sci.* 314: 123–133.
98. Kong, Y., Du, H., Yang, J., Shi, D., Wang, Y., Zhang, Y., and Xin, W. 2002. Study on polyimide/TiO_2 nanocomposite membranes for gas separation. *Desalination* 146: 49–55.
99. Higuchi, A., Agatsuma, T., Uemiya, S., Kojima, T., Mizoguchi, K., Pinnau, I., Nagai, K., and Freeman, B.D. 2000. Preparation and gas permeation of immobilized fullerene membranes. *J. Appl. Polym. Sci.* 7: 529–537.
100. Bertelle, S., Gupta, T., Roizard, D., Vallieres, C., and Favre, E. 2006. Study of polymer–carbon mixed matrix membranes for CO_2 separation from flue gas. *Desalination* 199: 401–402.
101. Kyotani, T. 2000. Control pore in carbon. *Carbon* 38: 269–286.
102. Schindler, E. and Maier, F. 1990. Manufacture of porous carbon membranes, US patent 4,919,860.

103. Saufi, S.M. and Ismail, A.F. 2004. Fabrication of carbon membranes for gas separation—A review. *Carbon* 42: 241–259.
104. Hosseini, S.S. and Chung, A.F. 2009. Carbon membranes from blends of PBI and polyimides for N_2/CH_4 and CO_2/CH_4 separation and hydrogen purification. *J. Membr. Sci.* 328: 174–185.
105. David, L.I.B. and Ismail, A.F. 2003. Influence of the thermostabilization process and soak time during pyrolysis on the polyacrylonitrile carbon membranes for O_2/N_2 separation. *J. Membr. Sci.* 213: 285–291.
106. Fuertes, A.B. and Menendez, I. 2002. Separation of hydrocarbon gas mixtures using phenolic resin-based carbon membranes. *Sep. Purif. Technol.* 28: 29–41.
107. Shiflett, M.B. and Foley, H.C. 2001. Reproducible production of nanoporous carbon membranes. *Carbon* 39: 1421–1446.
108. Centeno, T.A. and Fuertes, A.B. 2000. Carbon molecular sieve gas separation membranes based on poly (vinylidene chloride-co-vinyl chloride). *Carbon* 38: 1067–1073.
109. Barsema, J.N., Klijnstra, S.D., Van der Vegt, N.F.A., Koops, G.H., and Wessling, M. 2004. Intermediate polymer to carbon gas separation membranes based on Matrimid PI. *J. Membr. Sci.* 328: 93–102.
110. Park, H.B., Kim, T.K., Lee, J.M., Lee, S.Y., and Lee, Y.M. 2004. Relationship between chemical structure of aromatic polyimides and gas permeation properties of their carbon molecular sieve membranes. *J. Membr. Sci.* 229: 117–127.
111. Shao, L., Chung, T.S., and Pramoda, K.P. 2005. The evolution of physiochemical and transport properties of 6FDA-durene toward carbon membranes; from polymer, intermediate to carbon. *Micropor. Mesopor. Mater.* 84: 59–68.
112. Shao, L., Chung, T.S., Wensley, G., Goh, S.H., and Pramoda, K.P. 2004. Casting solvent effects on morphologies, gas transport properties of a novel 6FDA/PMDA-TMMDA copolyimide membrane and its derived carbon membranes. *J. Membr. Sci.* 244: 77–87.
113. Steel, K.M. and Koros, W.J. 2003. Investigation of porosity of carbon materials and related effects on gas separation properties. *Carbon* 41: 253–266.
114. Tin, P.S., Chung, T.S., Liu, Y., and Wang, R. 2004. Separation of CO_2/CH_4 through carbon molecular sieve membranes derived from P84 polyimide. *Carbon* 42: 3123–3131.
115. Hosseini, S.S., Teoh, M.M., and Chung, T.S. 2008. Hydrogen separation and purification in membranes of miscible polymer blends with interpenetration networks. *Polymer* 49: 1594–1603.
116. Li, N.N., Fane, A.G., Winston, W.S., and Matsuura, T. 2008. *Advanced Membrane Technology and Applications*, John Wiley & Sons, Hoboken, NJ.
117. Brunetti, A., Scura, F., Barbieri, G., and Drioli, E. 2010. Membrane technologies for CO_2 separation. *J. Membr. Sci.* 359: 115–125.
118. Suda, H. and Haraya, K. 1997. Gas permeation through micropores of carbon molecular sieve membranes derived from Kapton polyimide. *J. Phys. Chem. B* 101: 3988–3994.
119. Wang, K., Suda, H., and Haraya, K. 2003. The characterization of CO_2 permeation in a CMSM derived from polyimide. *Sep. Purif. Technol.* 31: 61–69.
120. Petersen, J., Matsuda, M., and Haraya, K. 1997. Capillary carbon molecular sieve membranes derived from Kapton for high temperature gas separation. *J. Membr. Sci.* 131: 85–94.

121. Lua, A.C. and Su, J. 2006. Effects of carbonisation on pore evolution and gas permeation properties of carbon membranes from Kapton polyimide. *Carbon* 44: 2964–2972.
122. Fuertes, A.B., Nevskaia, D.M., and Centeno, T.A. 1999. Carbon composite membranes from Matrimid and Kapton polyimides for gas separation. *Micropor. Mesopor. Mater.* 33: 115–125.
123. Kim, Y.K., Park, H.B., and Lee, Y.M. 2005. Preparation and characterization of carbon molecular sieve membranes derived from BTDA–ODA polyimide and their gas separation properties. *J. Membr. Sci.* 255: 265–273.
124. Xiao, Y., Dai, Y., Chung, T.S., and Guiver, M.D. 2005. Effects of brominating matrimid polyimide on the physical and gas transport properties of derived carbon membranes. *Macromolecules* 38: 10042–10049.
125. Hayashi, J., Yamamoto, M., Kusakabe, K., and Morooka, S. 1995. Simultaneous improvement of permeance and permselectivity of 3,30,4,40-biphenyltetracarboxylicd ianhydride-4,40-oxydianiline polyimide membrane by carbonization. *Ind. Eng. Chem. Res.* 34: 4364–4370.
126. Hayashi, J., Mizuta, H., Yamamoto, M., Kusakabe, K., and Morooka, S. 1996. Pore size control of carbonized BPDA-ODA polyimide membrane by chemical vapor deposition of carbon. *J. Membr. Sci.* 124: 243–251.
127. Takeichi, T., Eguchi, Y., Kaburagi, Y., Hishiyama, Y., and Inagaki, M. 1999. Carbonization and graphitization of BPDA/PDA polyimide films: Effect of structure of polyimide precursor. *Carbon* 37: 569–575.
128. Pirouzfar, V., Moghaddam, A.Z., Omidkhah, M.R., and Hosseini, S.S. 2013. Investigating the effect of dianhydride type and pyrolysis condition on the gas separation performance of membranes derived from blended polyimides through statistical analysis. *J. Ind. Eng. Chem.* 20: 1061–1070.
129. Lin, W.H., Vora, R.H., and Chung, T.S. 2000. Gas transport properties of 6FDA-durene/1,4-phenylenediamine (PDA) copolyimides. *J. Polym. Sci. Polym. Phys.* 38: 2703–2713.
130. Vu, D.Q., Koros, W.J., and Miller, S.J. 2003. Mixed matrix membranes using carbon molecular sieves. I. Preparation and experimental results. *J. Membr. Sci.* 211: 311–334.
131. Geiszler, V.C. and Koros, W.J. 1996. Effect of polyimide pyrolysis conditions on carbon molecular sieve membrane properties. *Ind. Eng. Chem. Res.* 35: 2999–3003.
132. Kim, Y.K., Park, H.B., and Lee, Y.M. 2005. Gas separation properties of carbon molecular sieve membranes derived from polyimide/polyvinylpyrrolidone blends: Effect of the molecular weight of polyvinylpyrrolidone. *J. Membr. Sci.* 251: 159–167.
133. Lee, H.J., Suda, H., Haraya, K., and Moon, S.H. 2007. Gas permeation properties of carbon molecular sieving membranes derived from the polymer blend of polyphenylene oxide (PPO)/polyvinylpyrrolidone (PVP). *J. Membr. Sci.* 296: 139–146.
134. Hosseini, S.S. and Chung, T.S. 2011. Polymer blends and carbonized polymer blends, US Patent 2011/0192, 281.
135. Hosseini, S.S., Omidkhah, M.R., Moghaddam, A.Z., Pirouzfar, V., Krantz, W.B., and Tan, N.R. 2014. Enhancing the properties and gas separation performance of PBI–polyimides blend carbon molecular sieve membranes via optimization of the pyrolysis process. *Sep. Purif. Technol.* 122: 278–289.

136. Bricen, K., Iulianelli, A., Montane, D., Valls, R.G., and Basile, A. 2012. Carbon molecular sieve membranes supported on nonmodified ceramic tubes for hydrogen separation in membrane reactors. *Int. J. Hydrogen Energy*. 37: 13536–13544.
137. Kim, Y.K., Lee, J.M., Park, H.B., and Lee, Y.M. 2004. The gas separation properties of carbon molecular sieve membranes derived from polyimides having carboxylic acid groups. *J. Membr. Sci.* 235: 139–146.
138. Barsema, J.N., Balster, J., Van der Vegt, N.F.A., Koops, G.H., Jordan, V., and Wessling, M. 2003. Ag functionalized carbon molecular sieves membranes for separating O_2 and N_2. *Mater. Res. Soc. Symp. Proc.* 752: 207–212.
139. Barsema, J.N., Kapantaidakis, G.C., Van der Vegt, N.F.A., Koops, G.H., and Wessling, M. 2003. Preparation and characterization of highly selective dense and hollow fiber asymmetric membranes based on BTDA-TDI/MDI co-polyimide. *J. Membr. Sci.* 216: 195–205.
140. Chng, M.L., Xiao, Y.C., Chung, T.S., Toriida, M., and Tamai, S. 2009. Enhanced propylene/propane separation by carbonaceous membrane derived from poly(aryl-ether-ketone)/2,6-bis(4-azidobenzylidene)-4-methyl-cyclohexanone interpenetrating network. *Carbon* 47: 1857–1866.
141. Kim, Y.K., Park, H.B., and Lee, Y.M. 2003. Carbon molecular sieve membranes derived from metal-substituted sulfonated polyimide and their gas separation properties. *J. Membr. Sci.* 226: 145–158.
142. Kong, C.L., Wang, J.Q., Yang, J.H., Lu, J.M., Wang, A.F., and Zhao, Q.Y. 2007. Thin carbon-zeolite composite membrane prepared on ceramic tube filter by vacuum slip casting for oxygen/nitrogen separation. *Carbon* 45: 2848–2850.
143. Liu, Q.L., Wang, T.H., Qiu, J.S., and Cao, Y.M. 2006. A novel carbon/ZSM-5 nanocomposite membrane with high performance for oxygen/nitrogen separation. *Chem. Commun.* 11: 1230–1232.
144. Park, H.B., Jung, C.H., Kim, Y.K., Nam, S.Y., Lee, S.Y., and Lee, Y.M. 2004. Pyrolytic carbon membranes containing silica derived from poly(imide siloxane): The effect of siloxane chain length on gas transport behavior and a study on the separation of mixed gases. *J. Membr. Sci.* 235: 87–98.
145. Park, H.B., Lee, S.Y., and Lee, Y.M. 2005. Pyrolytic carbon membranes containing silica: Morphological approach on gas transport behavior. *J. Mol. Struct.* 739: 179–190.
146. Teixeira, M., Campo, M.C., Tanaka, D.A.P., Tanco, M.A.L., Magen, C., and Mendes, A. 2011. Composite phenolic resin-based carbon molecular sieve membranes for gas separation. *Carbon* 49: 4348–4358.
147. Xiao, Y.C., Chng, M.L., Chung, T.S., Toriida, M., Tamai, S., Chen, H.M., and Jean, Y.C.J. 2010. Asymmetric structure and enhanced gas separation performance induced by in situ growth of silver nanoparticles in carbon membranes. *Carbon* 48: 408–416.
148. Yin, X.Y., Wang, J.Q., Chu, N.B., Yang, J.H., Lu, J.M., Zhang, Y., and Yin, D.H. 2010. Zeolite L/carbon nanocomposite membranes on the porous alumina tubes and their gas separation properties. *J. Membr. Sci.* 348: 181–189.
149. Yoda, S., Hasegawa, A., Suda, H., Uchimaru, Y., Haraya, K., Tsuji, T., and Otake, K. 2004. Preparation of a platinum and palladium/polyimide nanocomposite film as a precursor of metal-doped carbon molecular sieve membrane via supercritical impregnation. *Chem. Mater.* 16: 2363–2368.
150. Yoshimune, M., Fujiwara, I., Suda, H., and Haraya, K. 2006. Gas transport properties of carbon molecular sieve membranes derived from metal containing sulfonated poly(phenylene oxide). *Desalination* 193: 66–72.

151. Zeng, C.F., Zhang, L.X., Cheng, X.H., Wang, H.T., and Xu, N.P. 2008. Preparation and gas permeation of nano-sized zeolite NaA-filled carbon membranes. *Sep. Purif. Technol.* 63: 628–633.
152. Barsema, J.N., Balster, J., Jordan, V., van der Vegt, N.F.A., and Wessling, M. 2003. Functionalized carbon molecular sieve membranes containing Ag-nanoclusters. *J. Membr. Sci.* 219: 47–57.
153. Popov, V.N. 2004. Carbon nanotubes: Properties and application. *Mater. Sci. Eng. R* 43: 61–102.
154. Iijima, S. 1991. Helical microtubules of graphitic carbon. *Nature* 354: 56–58.
155. Skoulidas, A.I., Ackerman, D.M., Johnson, J.K., and Sholl, D.S. 2002. Rapid transport of gases in carbon nanotubes. *Phys. Rev. Lett.* 89: 185901.
156. Chen, H.B., Johnson, J.K, and Sholl, D.S. 2006. Transport diffusion of gases is rapid in flexible carbon nanotubes. *J. Phys. Chem. B* 110: 1971–1975.
157. Wang, Q.Y., Challa, S.R., Sholl, D.S., and Johnson, J.K. 1999. Quantum sieving in carbon nanotubes and zeolites. *Phys. Rev. Lett.* 82: 956–959.
158. Razavi, S.S., Hashemianzadeh, S.M, and Karimi, H. 2011. Modeling the adsorptive selectivity of carbon nanotubes for effective separation of CO_2/N_2 mixtures. *J.Mol. Model.* 17: 1163–1172.
159. Kim, S., Chen, L., Johnson, J.K., and Marand, E. 2007. Polysulfone and functionalized carbon nanotube mixed matrix membranes for gas separation: Theory and experiment. *J. Membr. Sci.* 294: 147–158.
160. Huang, L.L., Zhang, L.Z, Shao, Q, Lu, L.H, Lu, X.H., Jiang, S.Y., and Shen, W.F. 2007. Simulations of binary mixture adsorption of carbon dioxide and methane in carbon nanotubes: Temperature, pressure, and pore size effects. *J. Phys. Chem. C* 111: 11912–11920.
161. Kim, S., Pechar, T.W., and Marand, E. 2006. Poly(imidesiloxane) and carbon nanotube mixed matrix membranes for gas separation. *Desalination* 192: 330–339.
162. Cong, H.L., Zhang, J.M., Radosz, M., and Shen, Y.Q. 2007. Carbon nanotube composite membranes of brominated poly(2,6-diphenyl-1,4-phenyleneoxide) for gas separation. *J. Membr. Sci.* 294: 178–185.
163. Choi, J.H., Jegal, J., and Kim, W.N. 2007. Modification of performances of various membranes using MWNTs as a modifier. *Macromol. Symp.* 249: 610–617.
164. Liu, L.Q., Barber, A.H., Nuriel, S., and Wagner, H.D. 2005. Mechanical properties of functionalized single walled carbon-nanotube/poly(vinylalcohol) nanocomposites. *Adv. Funct. Mater.* 15: 975–980.
165. Liu, T., Tong, Y., and Zhang, W.-D. 2007. Preparation and characterization of carbon nanotube/polyetherimide nanocomposite films. *Compos. Sci. Technol.* 67: 406–412.
166. Deng, J., Zhang, X., Wang, K., Zou, H., Zhang, Q., and Fu, Q. 2007. Synthesis and properties of poly(ether urethane)membranes filled with isophorone diisocyanate-grafted carbon nanotubes. *J. Membr. Sci.* 288: 261–267.
167. Rausch, J., Zhuang, R.C., and Mader, E. 2010. Surfactant assisted processing of carbon nanotube/polypropylene composites: Impact of surfactants on the matrix polymer. *J. Appl. Polym. Sci.* 117: 2583–2590.
168. Tang, Q.Y., Chan, Y.C., Wong, N.B., and Cheung, R. 2010. Surfactant-assisted processing of polyimide/multiwall carbon nanotube nanocomposites for microelectronics applications. *Polym. Int.* 59: 1240–1245.
169. Gong, X.Y., Liu, J., Baskaran, S., Voise, R.D., and Young, J.S. 2000. Surfactant-assisted processing of carbon nanotube/polymer composites. *Chem. Mater.* 12: 1049–1052.

170. Deng, L. and Hägg, M.-B. 2014. Carbon nanotube reinforced PVAm/PVA blend FSC nanocomposite membrane for CO_2/CH_4separation. *Int. J. Greenhouse Gas Control* 26: 127–134.
171. Nour, M., Berean, K., Balendhran, S., Oua, J.Z., Plessis, J.D., McSweeney, C., Bhaskaran, M., Sriram, S., and Kalantar-zadeh, K. 2013. CNT/PDMS composite membranes for H_2 and CH_4 gas separation. *Int. J. Hydrogen Energy* 38: 10494–10501.
172. Wenga, H., Tseng, H.-H., and Wey, M.-Y. 2009. Preparation and characterization of multi-walled carbon nanotube/PBNPI nanocomposite membrane for H_2/CH_4 separation. *Int. J. Hydrogen Energy* 34: 8707–8715.
173. Kima, S., Chenb, L., Johnsom, J.K., and Marand, E. 2007. Polysulfone and functionalized carbon nanotube mixed matrix membranes for gas separation: Theory and experiment. *J. Membr. Sci.* 294: 147–158.
174. Surapathi, A., Alonso, J.H., Rabie, F., Martin, S., and Marand, E. 2011. Fabrication and gas transport properties of SWNT/polyacrylic nanocomposite membranes. *J. Membr. Sci.* 375: 150–156.
175. Wu, B., Li, X., Ana, D., Zhao, S., and Wang, Y. 2014. Electro-casting aligned MWCNTs/polystyrene composite membranes for enhanced gas separation performance. *J. Membr. Sci.* 462: 62–68.
176. Aroon, M.A., Ismail, A.F., and Matsuura, T. 2013. Beta-cyclodextrin functionalized MWCNT: A potential nano-membrane material for mixed matrix gas separation membranes development. *Sep. Purif. Technol.* 115: 39–50.
177. Pierson, H.O. 1993. *Handbook of Carbon, Graphite, Diamond and Fullerenes*, Noyes Publication, Saddle River, NJ.
178. Dai, L., Chang, D.W., Baek, J.B., and Lu, W. 2012. Carbon nanomaterials for advanced energy conversion and storage. *Small* 8: 1130–1166.
179. Dreyer, D.R., Ruoff, R.S., and Bielawski, C.W. 2012. From conception to realization: A historial account of graphene and some perspectives for its future. *Angew. Chem. Int. Ed.* 51: 7640–7654.
180. Novoselov, K.S., Falko, V.I., Colombo, L., Gellert, P.R., Schwab, M.G., and Kim, K.A. 2012. Roadmap for graphene. *Nature* 490: 192–200.
181. Gadipelli, S. and Guo, Z.X. 2015. Graphene-based materials: Synthesis and gas sorption, storage and separation. *Prog. Mater. Sci.* 69: 1–60.
182. Katsnelson, M.I. and Fasolino, A. 2013. Graphene as a prototype crystalline membrane. *Acc. Chem. Res.* 46: 97–105.
183. Leenaerts, O., Partoens, B., and Peeters, F.M. 2008. Graphene: A perfect nanoballoon. *Appl. Phys. Lett.* 93: 193107.
184. Miao, M., Nardelli, M.B., Wang, Q., and Liu, Y. 2013. First principles study of the permeability of graphene to hydrogen atoms. *Phys. Chem. Chem. Phys.* 15: 16132–16137.
185. Tsetseris, L. and Pantelides, S.T. 2014. Graphene: An impermeable or selectively permeable membrane for atomic species? *Carbon* 67: 58–63.
186. Jiang, D., Cooper, V.R., and Dai, S. 2009. Porous graphene as the ultimate membrane for gas separation. *Nano Lett.* 9: 4019–4024.
187. Liu, H., Dai, S., and Jiang, D. 2013. Insights into CO_2/N_2 separation through nanoporous graphene from molecular dynamics. *Nanoscale* 5: 9984–9987.
188. Du, H., Li, J., Zhang, J., Su, G., Li, X., and Zhao, Y. 2011. Separation of hydrogen and nitrogen gases with porous graphene membrane. *J. Phys. Chem. C* 115: 23261–23266.

189. Tao, Y., Xue, Q., Liu, Z., Shan, M., Ling, C., Wu, T., and Li, X. 2014. Tunable hydrogen separation in porous graphene membrane: First principle and molecular dynamic simulation. *ACS Appl. Mater. Interf.* 6: 8048–8058.
190. Lu, R., Rao, D., Lu, Z., Qian, J., Li, F., Wu, H., Wang, Y. et al. 2012. Prominently improved hydrogen purification and dispersive metal binding for hydrogen storage by substitutional doping in porous graphene. *J. Phys. Chem. C* 116: 21291–21296.
191. Qin, X., Meng, Q., Feng, Y., and Gao, Y. 2013. Graphene with line defect as a membrane for gas separation: Design via a first-principles modeling. *Surf. Sci.* 607: 153–158.
192. Lee, J. and Aluru, N.R. 2013. Water-solubility-driven separation of gases using graphene membrane. *J. Membr. Sci.* 428: 546–553.
193. Pandey, P. and Chauhan, R.S. 2001. Membranes for gas separation. *Prog. Polym. Sci.* 26: 853–893.
194. Caro, J. and Noack, M. 2008. Zeolite membranes—Recent developments and progress. *Micropor. Mesopor. Mater.* 115: 215–233.
195. Hong, M. 2007. Zeolite membranes for hydrogen purification and carbon dioxide removal, PhD thesis, University of Colorado, Boulder, CO.
196. Xu, X., Bao, Y., Song, C., Yang, W., Liu, J., and Lin, L. 2005. Synthesis, characterization and single gas permeation properties of NaA zeolite membrane. *J. Membr. Sci.* 249: 51–64.
197. Mahajan, R., Burns, R., Schaeffer, M., and Koros, W.J. 2002. Challenges in forming successful mixed matrix membranes with rigid polymeric materials. *J. Appl. Polym. Sci.* 86: 881–890.
198. Li, Y., Chung, T.S., Cao, C., and Kulprathipanja, S. 2005. The effects of polymer chain rigidification, zeolite pore size and pore blockage on polyethersulfone (PES)-zeolite A mixed matrix membranes. *J. Membr. Sci.* 260: 45–55.
199. Ismail, A.F., Rahim, R.A., and Rahman, W.A.W.A. 2008. Characterization of polyethersulfone/Matrimid® 5218 miscible blend mixed matrix membranes for O_2/N_2 gas separation. *Sep. Purif.* Technol. 63: 200–206.
200. Clarizia, G., Algieri, C., Regina, A., and Drioli, E. 2008. Zeolite-based composite PEEK-WC membranes: Gas transport and surface properties. *Micropor. Mesopor. Mater.* 115: 67–74.
201. Ismail, A.F., Kusworo, T.D., and Mustafa, A. 2008. Enhanced gas permeation performance of polyethersulfone mixed matrix hollow fiber membranes using novel Dynasylan Ameo silane agent. *J. Membr. Sci.* 319: 306–312.
202. Hu, C.C., Liu, T.C., Lee, K.R., Ruaan, R.C., and Lai, J.Y. 2006. Zeolite-filled PMMA composite membranes: Influence of coupling agent addition on gas separation properties. *Desalination* 193: 14–24.
203. Yong, H.H., Park, H.C., Kang, Y.S., Won, J., and Kim, W.N. 2001. Zeolite-filled polyimide membrane containing 2,4,6-triaminopyrimidine. *J. Membr. Sci.* 188: 151–163.
204. Karatay, E., Kalipcilar, H., and Yilmaz, L. 2010. Preparation and performance assessment of binary and ternary PES-SAPO 34-HMA based gas separation membranes. *J. Membr. Sci.* 364: 75–81.
205. Sen, D., Kalipcilar, H., and Yilmaz, L. 2007. Development of polycarbonate based zeolite 4A filled mixed matrix gas separation membranes. *J. Membr. Sci.* 303: 194–203.

206. Baerlocher, C., McCusker, L.B., and Olson, D.H. 2007. *Atlas of Zeolite Framework Types*, 6th Edition, Elsevier, Amsterdam, the Netherlands.
207. Mahajan, R. and Koros, W.J. 2000. Factors controlling successful formation of mixed-matrix gas separation materials. *Ind. Eng. Chem. Res.* 39: 2692–2696.
208. Tantekin-Ersolmaz, S.B., Atalay-Oral, C., Tatlier, M., Senatalar, A.E., Schoeman, B., and Sterte, J. 2000. Effect of zeolite particle size on the performance of polymer–zeolite mixed matrix membranes. *J. Membr. Sci.* 175: 285–288.
209. Li, Y., Chung, T.S., Huang, Z., and Kulprathipanja, S. 2006. Dual-layer polyethersulfone (PES)/BTDA-TDI/MDI co-polyimide (P84) hollow fiber membranes with a submicron PES–zeolite beta mixed matrix dense-selective layer for gas separation. *J. Membr. Sci.* 277: 28–37.
210. Miller, S.J., Munson, C.L., Kulkarni, S.S., Hasse, J., and David, J. 2002. Purification of p-xylene using composite mixed matrix membranes, US Patent 6,500,233.
211. Ekiner, O.M. and Kulkarni, S.S. 2003. Process for making hollow fiber mixed matrix membranes, US Patent 6,663,805, Air Liquide.
212. Koros, W.J., Wallace, D., Wind, J.D., Miller, S.J., Bickel, C.S., and Vu, D.Q. 2003. Crosslinked and crosslinkable hollow fiber mixed matrix membrane and method of making same, US Patent 20030,140,789.
213. Husain, S. and Koros, W.J. 2007. Polymeric mixed matrix membranes containing zeolites as a filler for gas separation applications: A review. *J. Membr. Sci.* 288: 195.
214. Bastani, D., Esmaeili, N., and Asadollahi, M. 2013. Polymeric mixed matrix membranes containing zeolites as a filler for gas separation applications: A review. *J. Ind. Eng. Chem.* 19: 375–393.
215. Rowsell, J.L.C. and Yaghi, O.M. 2004. Metal–organic frame works: A new class of porous materials. *Micropor. Mesopor. Mater.* 73: 3–14.
216. Caro, J. 2011. Are MOF membranes better in gas separation than those made of zeolites? *Curr. Opin. Chem. Eng.* 1: 77–83.
217. Huang, A, Dou, W, and Caro, J. 2010. Steam-stable zeolitic imidazolate framework ZIF-90 membrane with hydrogen selectivity through covalent functionalization. *J. Am. Chem. Soc.* 132: 15562–15564.
218. Yang, T.X, Xiao, Y.C, and Chung, T.S. 2011. Poly-/metal-benzimidazole nanocomposite membranes for hydrogen purification. *Energy Environ. Sci.* 4: 4171–4180.
219. Dong, G., Li, H., and Chen, V. 2013. Challenges and opportunities for mixed-matrix membranes for gas separation. *J. Mater. Chem. A* 1: 4610–4630.
220. Park, K.S., Ni, Z., Côté, A.P., Choi, J.Y., Huang, R., Uribe-Romo, F.J., Chae, H.K., O'Keeffe, M., and Yaghi, O.M. 2006. Exceptional chemical and thermal stability of zeolitic imidazolate frameworks. *Proc. Natl. Acad. Sci. USA* 103: 10186–10191.
221. Fan, L., Xue, M., Kang, Z., Li, H., and Qiu, S. 2012. Electro spinning technology applied in zeolitic imidazolate framework membrane synthesis. *J. Mater. Chem.* 22: 25272–25276.
222. Zhang, C., Dai, Y, Johnson, J.R., Karvan, O., and Koros, W.J. 2012. High performance ZIF-8/6FDA-DAM mixed matrix membrane for propylene/propane separations. *J. Membr. Sci.* 389: 34–42.
223. Banerjee, R., Phan, A, Wang, B., Knobler, C., Furukawa, H., O'Keeffe, M., and Yaghi, O.M. 2008. High-throughput synthesis of zeolitic imidazolate frameworks and application to CO_2 capture. *Science* 319: 939–943.

224. Bergh, J.V., Gücüyener, C., Pidko, E.A., Hensen, E.J.M., Gascon, J., and Kapteijn, F. 2011. Understanding the anomalous alkane selectivity of ZIF-7 in the separation of light alkane/alkene mixtures. *Chem. Eur. J* 17: 8832–8840.
225. Dumée, L., He, L., Hill, M., Zhu, B., Duke, M., Schütz, J., She, F. et al. 2013. Seeded growth of ZIF-8 on the surface of carbon nanotubes towards self-supporting gas separation membranes. *J. Mater. Chem. A* 1: 9208–9214.
226. Hao, L., Li, P., Yang, T., and Chung, T.S. 2013. Room temperature ionic liquid/ZIF-8 mixed-matrix membranes for natural gas sweetening and post-combustion CO_2 capture. *J. Membr. Sci.* 436: 221–231.
227. Ploegmakers, J., Japip, S., and Nijmeijer, K. 2013. Mixed matrix membranes containing MOFs for ethylene/ethane separation. Part B: Effect of Cu_3BTC_2 on membrane transport properties. *J. Membr. Sci.* 428: 331–340.
228. Ploegmakers, J., Japip, S., and Nijmeijer, K. 2013. Mixed matrix membranes containing MOFs for ethylene/ethane separation. Part A: Membrane preparation and characterization. *J. Membr. Sci.* 428: 445–453.
229. Perez, E.V., Balkus Jr., K.J., Ferraris, J.P., and Musselman, I.H. 2009. Mixed-matrix membranes containing MOF-5 for gas separations. *J. Membr. Sci.* 328: 165–173.
230. Li, T., Pan, Y., Peinemann, K.-V., and Lai, Z. 2013. Carbon dioxide selective mixed matrix composite membrane containing ZIF-7 nano-fillers. *J. Membr. Sci.* 425: 235–242.
231. Bushell, A.F., Attfield, M.P., Mason, C.R., Budd, P.M., Yampolskii, Y., Starannikova, L., Rebrov, A. et al. 2013. Gas permeation parameters of mixed matrix membranes based on the polymer of intrinsic microporosity PIM-1 and the zeolitic imidazolate framework ZIF-8. *J. Membr. Sci.* 427: 48–62.
232. Song, Q., Nataraj, S.K., Roussenova, M.V., Tan, J.C., Hughes, D.J., Li, W., Bourgoin, P. et al. 2012. Zeolitic imidazolate frame-work (ZIF-8) based polymer nanocomposite membranes for gas separation. *Energy Environ. Sci.* 5: 8359–8369.
233. Wijenayake, S.N., Panapitiya, N.P., Versteeg, S.H., Nguyen, C.N., Goel, S, Balkus Jr., K.J., Musselman, I.H., and Ferraris, J.P. 2013. Surface cross-linking of ZIF-8/polyimide mixed matrix membranes (MMMs) for gas separation. *Ind. Eng. Chem. Res.* 52: 6991–7001.
234. Bae, T.-H., Lee, J.S., Qiu, W., Koros, W.J., Jones, C.W., and Nair, S. 2010. A high-performance gas-separation membrane containing sub micrometer-sized metal–organic frame work crystals. *Angew. Chem. Int. Ed.* 49: 9863–9866.
235. Yang, T.X. and Chung, T.S. 2013. Room-temperature synthesis of ZIF-90 nano crystals and the derived nano-composite membranes for hydrogen separation. *J. Mater. Chem. A* 1: 6081–6090.
236. Basu, S., Cano-Odena, A., and Vankelecom, I.F.J. 2010. Asymmetric Matrimids/$[Cu_3(BTC)_2]$ mixed-matrix membranes for gas separations. *J. Membr. Sci.* 362: 478–487.
237. Lively, R.P., Dose, M.E., Thompson, J.A., McCool, B.A., Chance, R.R., and Koros, W.J. 2011. Ethanol and water adsorption in methanol-derived ZIF-71. *Chem. Commun.* 47: 8667–8669.
238. Liu, S., Liu, G., Zhao, X., and Jin, W. 2013. Hydrophobic-ZIF-71 filled PEBA mixed matrix membranes for recovery of biobutanol via pervaporation. *J. Membr. Sci.* 446: 181–188.
239. Zhang, K., Lively, R.P., Dose, M.E., Brown, A.J., Zhang, C., Chung, J., Nair, S., Koros, W.J., and Chance, R.R. 2013. Alcohol and water adsorption in zeolitic imidazolate frameworks. *Chem. Commun.* 49: 3245–3247.

240. Hayashi, J., Mizuta, H., Yamamoto, M., Kusakabe, K., Morooka, S., and Suh, S. 1996. Separation of ethane/ethylene and propane/propylene systems with a carbonized BPDA-ODA polyimide membrane. *Ind. Eng. Chem. Res.* 35: 4176–4181.
241. Japip, S., Wang, H., Xiao, Y., and Chung, T.S. 2014. Highly permeable zeolitic imidazolate framework (ZIF)-71 nanoparticles enhanced polyimide membranes for gas separation. *J. Membr. Sci.* 467: 162–174.
242. Xu, H.Y., Yang, B.H., Wang, J.F., Guang, S.Y., and Li, C. 1973. Preparation, Tg improvement, and thermal stability enhancement mechanism of soluble poly (methyl methacrylate) nanocomposites by incorporating octavinyl polyhedral oligomeric silsesquioxanes. *J. Polym. Sci. Polym. Chem.* 41: 79.
243. Oaten, M. and Choudhury, N.R. 2005. Silsesquioxane–urethane hybrid for thin film applications. *Macromolecules* 38: 6392.
244. Capaldi, F.M., Rutledge, G.C., and Boyce, M.C. 2005. Structure and dynamics of blends of polyhedral oligomeric silsesquioxanes and polyethylene by atomistic simulation. *Macromolecules* 38: 6700.
245. Joshi, M., Butola, B.S., Simon, G., and Kukaleva, N. 2006. Rheological and viscoelastic behavior of HDPE/octamethyl–POSS nanocomposites. *Macromolecules* 39: 1839.
246. Huang, J.C., He, C.B., Xiao, Y., Mya, K.Y., Dai, J., and Siow, Y.P. 2003. Polyimide/POSS nanocomposites: Interfacial interaction, thermal properties and mechanical properties. *Polymer* 44: 4491.
247. Paul, D.R. and Robeson, L.M. 2008. Polymer nanotechnology: Nanocomposites. *Polymer* 49: 3187.
248. Li, F., Li, Y., Chung, T.-S., and Kawi, S. 2010. Facilitated transport by hybrid POSS®–Matrimid®–Zn2+ nanocomposite membranes for the separation of natural gas. *J. Membr. Sci.* 356: 14–21.
249. Rahman, M.M., Filiz, V., Shishatskiy, S., Abetz, C., Neumann, S., Bolmer, S., Khan, M.M., and Abetz, V. 2013. PEBAX with PEG functionalized POSS as nanocomposite membranes for CO_2 separation. *J. Membr. Sci.* 437: 286–297.
250. Rahman, M.M., Shishatskiy, S., Abetz, C., Georgopanos, P., Neumanna, S., Khan, M.M., Filiz, V., and Abetz, V. 2014. Influence of temperature upon properties of tailor-made PEBAX MH 1657 nanocomposite membranes for post-combustion CO_2 capture. *J. Membr. Sci.* 469: 344–354.
251. Rahman, M.M., Filiz, V., Khan, M.M., Gacal, B.N., and Abetz, V. Functionalization of POSS nanoparticles and fabrication of block copolymer nanocomposite membranes for CO_2 separation. *React. Funct. Polym.* 86: 125–133.
252. Gorgojo, P., Uriel, S., Téllez, C., and Coronas, J. 2008. Development of mixed matrix membranes based on zeolite Nu-6(2) for gas separation. *Micropor. Mesopor. Mater.* 115: 85–92.
253. Gorgojo, P., Sieffert, D., Staudt, C., Tellez, C., and Coronas, J. 2012. Exfoliated zeolite Nu-6(2) as filler for 6FDA-based copolyimide mixed matrix membranes. *J. Membr. Sci.* 411–412: 146–152.
254. Choi, S., Coronas, J., Lai, Z., Yust, D., Onorato, F., and Tsapatsis, M. 2008. Fabrication and gas separation properties of polybenzimidazole(PBI)/nanoporous silicates hybrid membrane. *J. Membr. Sci.* 316: 145–152.
255. Choi, S., Coronas, J., Jordan, E., Oh, W., Nair, S., Onorato, F., Shantz, D.F., and Tsapatsis, M. 2008. Layered silicates by swelling of AMH-3 and nanocomposite membranes. *Angew. Chem. Int. Ed.* 47: 552–555.

256. Kim, W., Lee, J.S., Bucknall, D.G., Koros, W.J., and Nair, S. 2013. Nanoporous layered silicate AMH-3/cellulose acetate nanocomposite membranes forgas separations. *J. Membr. Sci.* 441: 129–136.
257. Galve, A., Sieffert, D., Vispe, E., Tellez, C., Coronas, J., and Staudt, C. 2011. Copolyimide mixed matrix membranes with oriented microporous titanosilicate JDF-L1sheet particles. *J. Membr. Sci.* 370: 131–140.
258. Galve, A., Sieffert, D., Staudt, C., Ferrando, M., Guell, C., Tellez, C., and Coronas, J. 2013. Combination of ordered mesoporous silica MCM-41and layered titanosilicate JDF-L1 fillers for 6FDA-based copolyimide mixed matrix membranes. *J. Membr. Sci.* 431: 163–170.
259. Castarlenas, S., Gorgojo, P., Casado-Coterillo, C., Masheshwari, S., Tsapatsis, M., Tellez, C., and Coronas, J. 2013. Melt compounding of swollen titanosilicate JDF-L1with polysulfone to obtain mixed matrix membranes for H_2/CH_4 separation. *Ind. Eng. Chem. Res.* 52: 1901–1907.
260. Jeong, H.K., Krych, W., Ramanan, H., Nair, S., Marand, E., and Tsapatsis, M. 2004. Fabrication of polymer/selective-flake nanocomposite membranes and their use in gas separation. *Chem. Mater.* 16: 3838–3845.
261. Vaughan, B.R. and Marand, E. 2008. Transport properties of polymer–aluminophosphate nano-composites prepared by simple mixing. *J. Membr. Sci.* 310: 197–207.
262. Murray, H.H. 2006. Structure and composition of the clay minerals and their physical and chemical properties, *Developments in Clay Science*, Vol. 2, Elsevier, Amsterdam, the Netherlands, 7–31.
263. Hashemifard, S.A., Ismail, A.F., and Matsuura, T. 2011. Effects of montmorillonite nano-clay fillers on PEI mixed matrix membrane for CO_2 removal. *Chem. Eng. J.* 170: 316–325.
264. Liang, C.Y., Uchytil, P., Petrychkovych, R., Lai, Y.C., Friess, K., Sipek, M., Reddy, M.M., and Suen, S.Y. 2012. A comparison on gas separation between PES/MMT and PES/TiO_2 mixed matrix membranes. *Sep. Purif. Technol.* 92: 57–63.
265. Bhole, Y.S., Wanjale, S.D., Kharul, U.K., and Jog, J.P. 2007. Assessing feasibility of polyarylate–clay nanocomposites towards improvement of gas selectivity. *J. Membr. Sci.* 306: 277–286.
266. Defontaine, G., Barichard, A., Letaief, S., Feng, C., Matsuura, T., and Detellier, C. 2010. Nanoporous polymer–clay hybrid membranes for gas separation. *J. Colloid Interface Sci.* 343: 622–627.
267. Hashemifard, S.A., Ismail, A.F., and Matsuura, T. 2011. Mixed matrix membrane incorporated with large pore size halloysite nanotubes (HNT) as filler for gas separation: Experimental. *J. Colloid Interface Sci.* 359: 359–370.
268. Johnson, J.R. and Koros, W.J. 2009. Utilization of nanoplatelets inorganic–inorganic hybrid separation materials: Separation advantages and formation challenges. *J. Taiwan Inst. Chem. Eng.* 40: 268–275.
269. Zulhairun, A.K. and Ismail, A.F. 2014. The role of layered silicate loadings and their dispersion states on the gas separation performance of mixed matrix membrane. *J. Membr. Sci.* 46: 20–30.
270. Kong, Y., Du, H., Yang, J., Shi, D., Wang, Y., Zhang, Y., and Xin, W. 2002. Study on polyimide/TiO_2 nanocomposite membranes for gas separation. *Desalination* 146: 49–55.
271. Moghadam, F., Omidkhah, M.R., Vasheghani-Farahani, E., Pedram, M.Z., and Dorosti, F. 2011. The effect of TiO_2 nanoparticles on gas transport properties of Matrimid 5218-based mixed matrix membranes. *Sep. Purif. Technol.* 77: 128–136.

272. Ahmad, J. and Hågg, M.B. 2013. Polyvinylacetate/titanium dioxide nanocomposite membranes for gas separation. *J. Membr. Sci.* 445: 200–210.
273. Sun, H., Ma, C., Yuan, B., Wang, T., Xu, Y., Xue, Q., Li, P., and Kong, Y. 2014. Cardo polyimides/TiO_2 mixed matrix membranes: Synthesis, characterization, and gas separation property improvement. *Sep. Purif. Technol.* 122: 367–375.
274. Xin, Q., Wu, H., Jiang, Z., Li, Y., Wang, S., Li, Q., Li, X., Lu, X., Cao, X., and Yang, J. 2014. SPEEK/amine-functionalized TiO_2 submicrospheres mixed matrix membranes for CO_2 separation. *J. Membr. Sci.* 467: 23–35.
275. Matteucci, S., Kusuma, V.A., Sanders, D., Swinnea, S., and Freeman, B.D. 2008. Gas transport in TiO_2 nanoparticle-filled poly(1-trimethylsilyl-1-propyne). *J. Membr. Sci.* 307: 196–217.
276. Matterucci, S., Kusuma, V.A., Swinnea, S., and Freeman, B.D. 2008. Gas permeability, solubility and diffusivity in 1,2-polybutadiene containing brookite nanoparticles. *Polymer* 49: 757–773.
277. Hosseini, S.S., Li, Y., Chung, T.-S., and Liu, Y. 2007. Enhanced gas separation performance of nanocomposite membranes using MgO nanoparticles. *J. Membr. Sci.* 302: 207–217.
278. Matteucci, S., Kusuma, V.A., Kelman, S.D., and Freeman, B.D. 2008. Gas transport properties of MgO filled poly(1-trimethylsilyl-1-propyne) nanocomposites. *Polymer* 49: 1659–1675.
279. Matteucci, S., Kusuma, V.A., Kelman, S.D., and Freeman, B.D. 2008. Gas transport properties—Magnesium properties. *Macromolecules* 41: 2144–2156.
280. Joly, C., Goiset, S., Schrotter, J.C., Sanchez, J., and Escoubes, M. 1997. Sol-gel polyimide-silica composite membrane: Gas transport properties. *J. Membr. Sci.* 130: 63–74.
281. Sadeghi, M., Semsarzadeh, M.A., and Moadel, H. 2009. Enhancement of the gas separation properties of polybenzimidazole (PBI) membrane by incorporation of silica nano particles. *J. Membr. Sci.* 331: 21–30.
282. Zhang, Y., Balkus Jr., K.J., Musselman, I.H., and Ferraris, J.P. 2008. Mixed-matrix membranes composed of Matrimid® and mesoporous ZSM-5 nanoparticles. *J. Membr. Sci.* 325: 28–39.
283. Chung, T.S., Jiang, L.Y., Li, Y., and Kulprathipanja, S. 2007. Mixed matrix membranes (MMMs) comprising organic polymers with dispersed inorganic fillers for gas separation. *Prog. Polym. Sci.* 32: 483–507.
284. Goh, P.S., Ismail, A.F., Sanip, S.M., Ng, B.C., and Aziz, M. 2011. Recent advances of inorganic fillers in mixed matrix membrane for gas separation. *Sep. Purif. Technol.* 81: 243.
285. Chew, T.L., Ahmad, A.L., and Bhatia, S. 2010. Ordered mesoporous silica (OMS) as an adsorbent and membrane for separation of carbon dioxide (CO_2). *Adv. Colloid Interface Sci.* 153: 43–57.
286. Brady, R., Woonton, B., Gee, M.L., and O'Connor, A.J. 2008. Hierarchical mesoporous silica materials for separation of functional food ingredients—A review. *Innov. Food Sci. Emerg. Technol.* 9: 243–248.
287. Shen, Y. and Lua, A.C. 2012. Preparation and characterization of mixed matrix membranes based on PVDF and three inorganic fillers (fumed nonporous silica, zeolite 4A and mesoporous MCM-41) for gas separation. *Chem. Eng. J.* 192: 201–210.
288. Weng, T.H., Tseng, H.H., and Wey, M.Y. 2010. Effect of SBA-15 texture on the gas separation characteristics of SBA-15/polymer multilayer mixed matrix membrane. *J. Membr. Sci.* 369: 550–559.

289. Ferrari, M.C., Galizia, M., De Angelis, M.G., and Sarti, G.C. 2010. Gas and vapor transport in mixed matrix membranes based on amorphous Teflon AF1600 and AF2400 and fumed silica. *Ind. Eng. Chem. Res.* 49: 11920–11935.
290. Cornelius, C.J. and Marand, E. 2002. Hybrid silica-polyimide composite membranes: Gas transport properties. *J. Membr. Sci.* 202: 97–118.
291. Ahn, J., Chung, W.-J., Pinnau, I., Song, J., Du, N., Robertson, G.P., and Guiver, M.D. 2010. Gas transport behavior of mixed-matrix membranes composed of silica nanoparticles in a polymer of intrinsic microporosity (PIM-1). *J. Membr. Sci.* 346: 280–287.
292. Takahashi, S. and Paul, D.R. 2006. Gas permeation in poly(ether imide) nanocomposite membranes based on surface-treated silica. Part 2: With chemical coupling to matrix. *Polymer* 47: 7535–7547.
293. Takahashi, S. and Paul, D.R. 2006. Gas permeation in poly (ether imide) nanocomposite membranes based on surface-treated silica. Part 1: Without chemical coupling to matrix. *Polymer* 47: 7519–7534.
294. Joseph, J., Tseng, C.-Y., Pan, C.-J., Chen, H.-M., Lin, C.W., Pillai, K.S., and Hwang, B.J. 2010. Growing well-defined monodispersed silica in polyimide host membranes using a surfactant assisted sol-gel process. *Polymer* 51: 5663–5668.
295. Yu, Y., Chien, W. and Tsai, T. 2010. High transparent soluble polyimide/silica hybrid optical thin films. *Polym. Test.* 29: 33–40.
296. Sadeghi, M., Khanbabaei, G., Dehaghani, A.H.S., Sadeghi, M., Aravand, M.A., Akbarzade, M., and Khatti, S. 2008. Gas permeation properties of ethylene vinyl acetate–silica nanocomposite membranes. *J. Membr. Sci.* 322: 423–428.
297. Sadeghi, M., Semsarzadeh, M.A., and Moadel, H. 2009. Enhancement of the gas separation properties of polybenzimidazole (PBI) membrane by incorporation of silica nano particles. *J. Membr. Sci.* 331: 21–30.
298. Xing, R. and Ho, W.S.W. 2011. Crosslinked polyvinylalcohol–polysiloxane/fumed silica mixed matrix membranes containing amines for CO_2/H_2 separation. *J. Membr. Sci.* 367: 91–102.
299. Sadeghia, M., Semsarzadehb, M.A., Barikanic, M., and Chenar, M.P. 2011. Gas separation properties of polyether-based polyurethane–silica nanocomposite membranes. *J. Membr. Sci.* 376: 188–195.
300. Rafiq, S., Mana, Z., Maulud, A., Muhammada, N., and Maitra, S. 2012. Separation of CO_2 from CH_4 using polysulfone/polyimide silica nanocomposite membranes. *Sep. Purif. Technol.* 90: 162–172.
301. Khanbabaeia, G., Vasheghani-Farahania, E., and Rahmatpour, A. 2012. Pure and mixed gas CH_4 and n-C_4H_{10} permeation in PDMS-fumed silica nanocomposite membranes. *Chem. Eng. J.* 191: 369–377.
302. Shen, Y. and Lua, A.C. 2012. Structural and transport properties of BTDA-TDI/MDI co-polyimide (P84)–silica nanocomposite membranes for gas separation. *Chem. Eng. J.* 188: 199–209.
303. Sadeghi, M., Talakesh, M.M., Ghalei, B., and Shafiei, M. 2013. Preparation, characterization and gas permeation properties of a polycaprolactone based polyurethane-silica nanocomposite membrane. *J. Membr. Sci.* 427: 21–29.
304. Vu, D.Q., Koros, W.J., and Miller, S.J. 2003. Mixed matrix membranes using carbon molecular sieves. II. Modeling permeation behavior. *J. Membr. Sci.* 211: 335–348.
305. Bouma, R.H.B., Checchetti, A., Chidichimo, G., and Drioli, E. 1997. Permeation through a heterogeneous membrane: The effect of the dispersed phase. *J. Membr. Sci.* 128: 141–149.

306. Pal, R. 2008. Permeation models for mixed matrix membranes. *J. Colloid Interface Sci.* 317: 191–198.
307. Banhegyi, G. 1986. Comparison of electrical mixture rules for composites. *Colloid Polym. Sci. 264*: 1030–1050.
308. Lewis, T. and Nielsen, L. 1970. Dynamic mechanical properties of particulate-filled composites. *J. Appl. Polym. Sci.* 14: 1449–1471.
309. Nielsen, L. 1973. Thermal conductivity of particulate-filled polymers. *J. Appl. Polym. Sci.* 17: 3819–3820.
310. Pal, R. 2007. New models for thermal conductivity of particulate composites. *J. Reinf. Plast. Compos.* 26: 643–651.
311. Mahajan, R. 2000. Formation, characterization and modeling of mixed matrix membrane materials, PhD thesis, University of Texas, Austin, TX.
312. Mahajan, R. and Koros, W.J. 2002. Mixed matrix membrane materials with glassy polymers. Part 1. *Polym. Eng. Sci.* 42: 1420–1431.
313. Merkel, T.C., Freeman, B.D., Spontak, R.J., He, Z., Pinnau, I., Meakin, P., and Hill, A.J. 2003. Sorption, transport and structural evidence for enhanced free volume in poly(4-methyl-2-pentyne)/fumed silica nanocomposite membranes. *Chem. Mater.* 15: 109–123.
314. Maxwell, J.C. 1873. *Treatise on Electricity and Magnetism*, Oxford University Press, London.
315. Hu, X., Cong, H., Shen, Y., and Radosz, M. 2007. Nanocomposite membranes for CO_2 separations: Silica/brominated poly (phenylene oxide). *Ind. Eng. Chem. Res.* 46: 1547–1551.
316. Cong, H., Hu, X., Radosz, M., and Shen, Y. 2007. Brominated poly (2,6-diphenyl-1, 4-phenylene oxide) and its SiO_2 nanocomposite membranes for gas separation. *Ind. Eng. Chem. Res.* 46: 2567–2575.
317. Merkel, T.C., Freeman, B.D., Spontak, R.J., He, Z., Pinnau, I., Meakin, P., and Hill, A.J. 2002. Ultra-permeable, reverse-selective nanocomposite membranes. *Science* 296: 519–522.
318. Cohen, M.H. and Turnbull, D. 1959. Molecular transport in liquids and glasses. *J. Chem. Phys.* 31: 1164–1169.

6

Nanocomposite Membranes in Biomedical Applications

6.1 Introduction

Research on biosystems has generated dynamic interdisciplinary research and application domains. The domain includes better understanding and treatment of living systems, revolutionary biotechnology processes, synthesis of new drugs and their targeted delivery, regenerative medicine, neuromorphic engineering, and biocompatible materials for sustainable environment.

It is anticipated that nanotechnology has potential to drive innovation and play a critical role in biomedical applications, particularly in drug delivery. Advances in nanotechnology that enable drugs to preserve their efficacy while being delivered to precise therapeutic targets are creating a host of opportunities for drug developers. In addition, by combining nanotechnology-based target-specific drug therapy with methods for early diagnosis of pathologies, it is getting closer for creating the ultimate functional drug carrier. A variety of nanostructures are being investigated as functional drug carriers for treating a wide range of therapies, most notably cardiovascular defects, autoimmune diseases, and cancer. Although the concept of nanoparticles in drug delivery is not new, the number of research programs and active drug development projects in this field has escalated as funding for nanotechnology has increased. The result is the emergence of a host of novel nanotechnologies tailored to meet the physicochemical and therapeutic requirements of drug developers. With all this potential for advanced drug delivery and targeted therapy, with reduced side effects, nanotechnology-based drug delivery systems hold the promise of significantly improving the quality of life through nanomedicine.

Nanobiomedicine applies nanoscale principles and techniques for understanding and transforming inert materials and biosystems (nonliving or living) for medical purposes such as drug synthesis, brain understanding, body part replacement, visualization, and tools for medical interventions. Integration of nanotechnology with biomedicine and biology as well as with information technology and cognitive science will be accelerated with time.

Nano-biosystem technology is one of the challenging and fastest growing fields. It is essential for better understanding of living systems and for developing new tools for medicine and solutions for health care (such as synthesis of new drugs and their targeted delivery, regenerative medicine, and neuromorphic engineering). One of the important challenges is to understand the processes inside cells and neural systems. Nano-biosystems provide models for man-made nanosystems. Efforts are directed for better biocompatible materials and nano-biomaterials for industrial applications.

Though the nanostructured materials and nanocomposites provide the mankind with innumerable opportunities in the area of biomedical applications, the application of nanocomposite membranes is limited, which essentially has got a key role to play in the drug delivery system.

6.2 Potential Areas

Bioimaging technique has played a vital role in improving human health by using *imaging* technique to advance diagnosis, treatment, and prevention of diseases. The development of a wide range of imaging techniques such as magnetic resonance imaging (MRI), positron emission tomography (PET), ultrasound, and optical imaging are nowadays important tools for the early detection of disease, understanding the basic molecular aspects of living organisms, and evaluation of medical treatment. In recent years, after the development of several core/shell nanostructured materials, bioimaging techniques have been developed a lot.

Use of nanoparticles in bioimaging is a recent biomedical advancement.[1,2] It has been shown that nanoparticle-based systems are commonly used for the dual purpose of drug delivery and bioimaging.[3,4] Some of them could be collaterally used for biosensing[5] and gene delivery.[6] Biomedical applications based on material property of nanoparticles such as magnetic[7] and silica[8] and polymeric[9] nanoparticles have been reported.

Nanotechnology-based biosensors have a promising future. A *biosensor* is an analytical device for the analysis or sensing of biological samples by converting a biological response into an electrical signal. It is essentially a biocompatible diagnostic device able to respond to a signal generated because of biochemical reaction (e.g., enzyme–substrate reactions) or bimolecular interactions (antigen–antibody, receptor–ligand, nucleic acid–protein, nucleic acid–nucleic acid, metal–macromolecule), convert the signal into an electronic mode such that it can be quantified and discretized, and generate amplified, comprehendible, and feasible output. A biosensor has a highly specific and stable biocatalyst, capable of analyzing on the basis of a reaction independent of physical parameters such as agitation, pH, and temperature, giving an accurate, precise, reproducible, and linear response over the required range without dilution or concentration.

Due to different breakthroughs in the field of medical science and engineering, drugs have become more and more customized and efficient to specifically address the target problem in a more localized fashion. Most of the drugs till date have been dependent on the circulatory system of the body to be channeled into their site of action, with the penalty of heavy dilution rate, reduced efficacy, and consequent increased side effects. These drawbacks sometimes surpass the benevolence of the drug itself. Research is now focused on more sophisticated modes of drug delivery, which can be designed to channel itself and confined to its site of action with a high level of specificity, evading the immune system till its purpose is served. In targeted drug delivery, the medicinal target is concentrated on specific tissues, while reducing their relative concentration on others. Conventional targeted drug delivery can be categorized as (1) active and (2) passive targeting. In active targeting, antibodies are used which, being inherently specific, need no external targeting facility. The more important and difficult technology is the second approach of passive or enhanced permeability and retention (EPR)-dependent drug targeting which does involve immunoglobulins. Targeting mechanisms are diverse but mainly based on biomarkers and their ligand–receptor interaction with corresponding tags present on the drug carriers. Macrosystems of delivery become complex and nonspecific, due to the difficulty of combining so many types of materials (drug, adjuvant, carrier, targeting, and other molecules) and therefore have low efficacy. Also drug-carrying capacity and payload delivery remain low and slightly erratic.

Gene is essentially the functional component of the DNA and incorporation of a DNA sequence into another both *in vitro* and *in vivo* is a common technique of genetic engineering. The oldest technique of gene transfer from one living cell to another was by using viral vector, plasmids, and other inherently infectious cells or their components—a process termed as *transduction*. Over the years, nonviral gene transfer methods in prokaryotic cells were developed which are known as *transformation*. Therefore, basically gene delivery can occur via two vectors: viral and nonviral. The nonviral vectors have added advantages over the viral vectors, by being easier in production and storage, and inducing less cytotoxic responses. Such safer nonviral methods when modified and implemented on the more complicated and advanced eukaryotes with the potential of application in medical treatment of a wide variety of diseases form the basis of *gene transfection*. Typically gene transfection is done through chemical methods (using polyethylene glycol [PEG]), electroporation, liposomes, or coprecipitation (with Ca phosphate) techniques. Nanoparticle-based systems as the carriers of DNA into the cell are the recent tools developed in this field.

The interaction mechanism of nanoparticles with human cells is a subject of research interest, as it holds the key for future developments in the field of biodiagnostic and therapeutics, among other fields. The range of nanoparticles between 50 and 200 nm appears very effective for uptake in cells and has opened up new avenues of applications. Gold nanoparticles are commonly used for the detection of DNA. Spectroscopic and electrophoretic techniques have been applied to evaluate the interaction of Au with calf thymus DNA.

Tissue engineering involves the use of cells, engineering materials, and suitable biochemical factors to improve or replace biological functions. Although the tissue engineering covers a broad range of applications, in practice the term has come to represent applications that repair or replace structural tissues (bone, cartilage, blood vessels, bladder, etc.). These are tissues that function by virtue of their mechanical properties. A closely related field known as cell transplantation is concerned with the transplantation of cells that perform a specific biochemical function (e.g., an artificial pancreas or an artificial liver). Tissue engineering deals with problems by using living cells as engineering materials. These could be artificial skin that includes living fibroblasts, cartilage repaired with living chondrocytes, or other types of cells used in other ways.

6.3 Core–Shell Nanoparticles and Polymeric Nanohybrid Devices

The field of biomedical engineering has been substantially influenced by the advent of nanoparticles. The major advantages of nanoparticles over larger sized particles are their high surface-to-volume ratio and hence higher surface energy, unique optical, electronic, and excellent magnetic properties. The high surface area also allows them to be modified adequately so as to improve their pharmacokinetic properties and increase vascular circulation lifetime, along with improving bioavailability, especially for biomedical applications. The most important property that has attracted the attention of researchers worldwide is their ability to have better surface modifications, which help in targeted drug delivery as well as can solve the dual purpose of monitoring of drug release as well. In general, the size-dependent properties of nanoparticles (mainly optical, electronic, and magnetic) have been observed to be very much useful for biomedical applications. Other added features of nanoparticles include enhanced target specificity and permeability across semipermeable biomembranes. These properties make nanoparticles an attractive drug delivery vehicle along with the possibility of monitoring drug release.

Core–shell nanoparticles have a core made of a material coated with another material on top of it. In biological applications, core–shell nanoparticles have major advantages over simple nanoparticles leading to the improvement of properties such as less cytotoxicity, increase in dispersibility, bio- and cytocompatibility, better conjugation with other bioactive molecules, and increased thermal and chemical stability. The coating of a benign material on top of the core makes the nanoparticles much less toxic and biocompatible. The shell layer acts as a nontoxic layer. At times, it also improves the core material property. Hydrophilicity of nanoparticles is important to disperse them in aqueous biological systems. The increase in biodispersivity and bio- and cytocompatibility makes it a useful alternative to the conventional drug delivery vehicle. In many cases, the material of interest may be difficult to conjugate with a particular type of biomolecules; in that case, coating of a suitable biocompatible material helps to solve this problem. However, coating of an inert material generally enhances the stability of core particles. A schematic presentation[10] of a core–shell nanoparticle for multipurpose biomedical applications is shown in Figure 6.1. Optionally functionalized and devised nanoparticles can be achieved for individualized diagnosis and treatments.

Core–shell nanogels are composed of a metal core and a hydrophilic shell such as PEG and poly(*N*-isopropylacrylamide-*co*-acrylic acid). Hybrid nanogel has been made by coating the Ag–Au bimetallic nanoparticle core with a thermoresponsive nonlinear PEG-based hydrogel as shell, which was then loaded with anticancer drug temozolomide and used for drug delivery as well as fluorescence imaging of mouse melanoma cells (B16F10 cell line).[11] Core–shell-structured

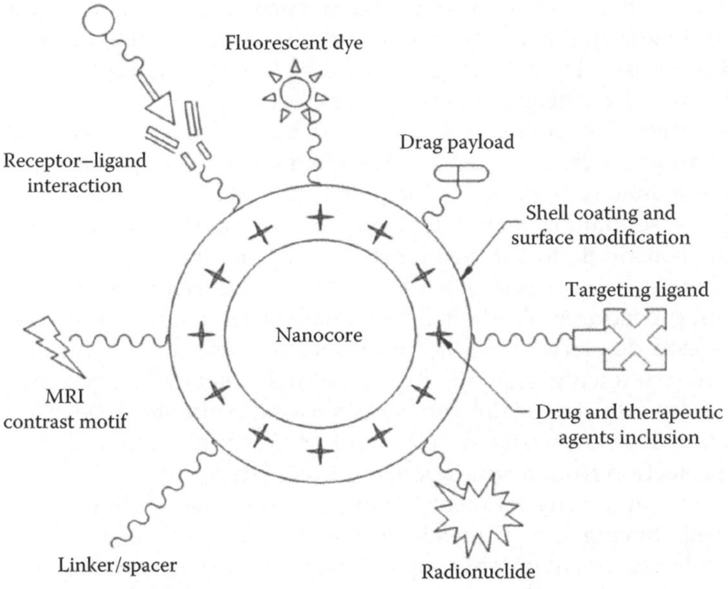

FIGURE 6.1
Scheme of a multifunctional nanoparticle for molecular imaging, drug delivery, and drug therapy. (Reprinted from *Adv. Colloid Interface Sci.*, 209, Chatterjee, K., Sarkar, S., Rao, K.J., and Paria, S., Core/shell nanoparticles in biomedical applications, 8–39, Copyright 2014, Figure 1 and Table 2, with permission from Elsevier.)

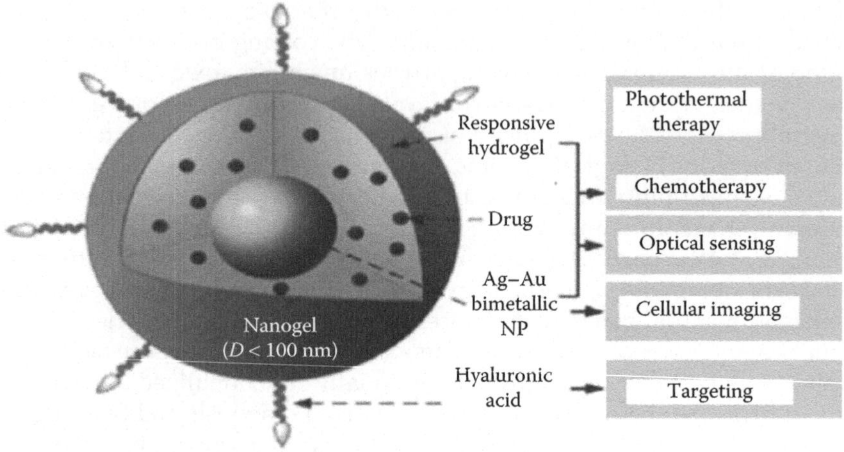

FIGURE 6.2
Schematic illustration of multifunctional core/shell hybrid nanogels.

hybrid nanogels (40–80 nm) composed of a Ag nanoparticle as a core and a smart gel of poly(*N*-isopropylacrylamide-*co*-acrylic acid) as a shell have been developed,[12] which can overcome cellular barriers to enter the intracellular region and light up the mouse melanoma cells, including the nuclear regions. The pH-responsive hybrid nanogels offer a high drug loading capacity as well as a pH-controllable drug release behavior (Figure 6.2).

The emergence of nanohybrid materials has an edge over chemotherapy and radiation therapy as cancer therapeutics. This is primarily because nanohybrid materials offer engineered particles with specific size, shape, and other essential properties. It is anticipated that these materials will significantly contribute to the next generation of medical care technology and pharmaceuticals in areas of disease diagnosis, disease prevention, and other treatment procedures. The benefits of a nanoscale drug delivery system are (1) a targeted delivery, allowing for an increased drug concentration at the desired site, reducing systemic exposure to a potentially toxic compound; (2) a constant rate of drug delivery to allow for the maintenance of a constant therapeutic dose at the site of delivery; and (3) an increase in drug stability due to protection from degradation and loss of drug.[13,14]

The tumor microenvironment comprises fast-growing, hyperproliferative cancer cells having a high metabolic rate and demand for the new vessels (neovascularization) in order to supply them with oxygen and nutrients.[15] The pathophysiologic condition of a tumor is its utilization of the glycolysis pathway to obtain additional energy, resulting in the generation of an acidic environment.[16] Tumor cells release enzymes such as matrix metalloproteinases, which lead to an imbalance of angiogenic regulators dilating the tumor vessels, resulting in large gap junctions, ranging from 200 to 2000 nm in size, between endothelial cells.[15] Moreover, the higher interstitial pressure generated by a compromised lymphatic drainage and a lower intravascular pressure limit the movement of macromolecules/particulate materials out of the tumor blood vessels into the extravascular compartment.[17] This enhanced vascular permeability and the lack of adequate lymphatic drainage at tumor sites facilitate passive targeting using polymeric nanovectors. The phenomenon is termed the enhanced permeability and retention effect.[18–20] The schematic of tumor targeting by nanohybrids via EPR is shown in Figure 6.3.[21] This effect is also attributed to the molecular weight, surface charge, and nature of the polymer. The conjugation of drugs to polymers avoids the random bioavailability of the low-molecular-weight drugs and enables target specificity. However, the molecular weight of the nano-vectors also plays an equally important role in the delivery, as nano-vectors of a molecular weight of less than 50 kDa or of a size less than 6 nm are rapidly cleared by the kidney following systemic administration.[22–25] However, the size is crucial for nonbiodegradable polymers, in order to be eliminated by the renal system, following the drug delivery.

Polymeric nanohybrid materials comprise a core material, a therapeutic *payload*, and a biological surface modification that helps in the biodistribution and selective cell targeting moieties. The schematic of multifunctional polymeric nanohybrid devices for targeted drug delivery is shown in Figure 6.4. The nanohybrid materials/nanovectors in conjunction with drugs are mostly delivered intravenously, as they bear the key characteristics of their ability to be tailored

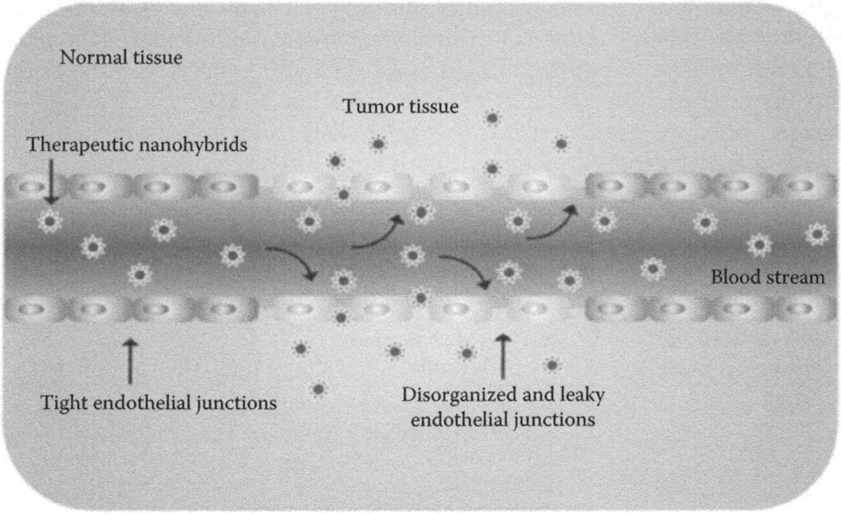

FIGURE 6.3
Schematic of tumor targeting by nanohybrids via an EPR effect. (Reprinted from *Adv. Drug Deliv. Rev.*, 63, Prakash, S., Malhotra, M., Shao, W., Tomaro-Duchesneau, C., and Abbasi, S., Polymeric nanohybrids and functionalized carbon nanotubes as drug delivery carriers for cancer therapy, 1340–1351, Copyright 2011, Figures 1 and 2, with permission from Elsevier.)

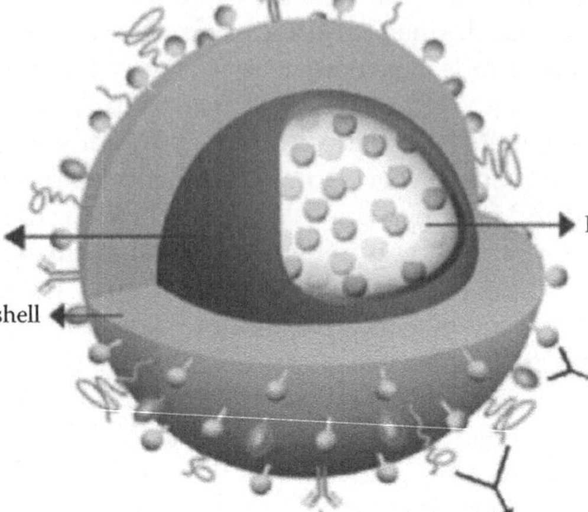

FIGURE 6.4
Schematic of multifunctional polymeric nanohybrid devices for targeted drug delivery.

to bypass the biological/physiological and immunological barriers of the body. The use of nanovector drug delivery vehicles has gained importance in biomedical applications, as they enable the encapsulation and the successful delivery of drugs with poor aqueous solubility profiles such as paclitaxel, an antitumor agent.[26–28] Another advantage of utilizing polymeric nanovectors is the potential for noninvasive targeting to the tumor. Nanohybrid materials exhibit multifunctional features that facilitate imaging, targeting, and drug delivery.[29,30] The overview of biomedical applications of core/shell nanoparticles, including the polymeric nanohybrids, is given in Table 6.1.

TABLE 6.1

Classification of Core/Shell Nanoparticles Based on Different Core Materials and Their Applications

Core/Shell (Metal or Metal Alloy as Core)	Surface Modification/ Ligands	Application	References
Fe/CNP	Polyacrylic acid (PAA), polyvinylpyrrolidone (PVP), poly(2-acetoxyethyl methacrylate), poly-*N*-hydroxyethylacrylamide	MRI	31,32
Fe/γ-Fe_2O_3	Dopamine, PEG-600, dextran	MRI	33

(Continued)

TABLE 6.1 (*Continued*)

Classification of Core/Shell Nanoparticles Based on Different Core Materials and Their Applications

Core/Shell (Metal or Metal Alloy as Core)	Surface Modification/ Ligands	Application	References
Au/polyaniline	Enzyme HRP	Amperometric sensor	34
Au/polypropyleneimine	Myoglobin	Amperometric sensor	35
Ag/Au-polypyrrole	Dopamine receptors	Amperometric sensor	36
Au/Pd	$[P(C_6)_3C_{14}][Tf_2N]$	Amperometric sensor	37
Au/citrate	Homocysteine, glutathione	Amperometric sensor	38
Au/CoFe	PNA oligomers	Optical sensors	39
Au/ polyallylamine-chlorophyllide	Cofactors	Optical sensors	40
Fe/Au	Assorted ligands and bioconjugation molecules	Drug delivery	41
Fe/*N*-isopolyacrylamide	Assorted ligands and bioconjugation molecules	Drug delivery, MRI	42
Au/PEG-amino acid	Radioactive iodine and other targeting molecules	Drug targeting	43
Au/Os	Enzyme glucose oxidase and cofactors	Amperometric sensor	44
Ag/titanium	Chromophores and enzymes	Optical sensors	45
Au/oleic acid/*N*-isoproprylacrylamide	Stabilizer molecules and target receptors		46
Au-ssDNA	Assorted ligands		47
FeCo/graphite		MRI, optical imaging	48
Co(Fe)/Au	Possible attachment of organic molecules with thiol-terminations	MRI, medical labeling	49
FePt/ZnO		Piezoelectric sensor	50
Au/Ag + silica/polymer dual conjugate	Monoclonal antibodies onto Au/Ag NPs ligated to polymer surface using trimethoxysilane	Immunosensor chip	51

(*Continued*)

TABLE 6.1 (*Continued*)

Classification of Core/Shell Nanoparticles Based on Different Core Materials and Their Applications

Core/Shell (Metal or Metal Alloy as Core)	Surface Modification/ Ligands	Application	References
Fe_3O_4/SiO_2	Fluorescein isothiocyanate (FTIC) dye, chelated Gd(III)	MRI	52
Fe_3O_4/poly(allylamine hydrochloride)/Au		MRI	53
Fe_3O_4/CS or oleic acid-entrapped curcumin		MRI, optical imaging, drug delivery vehicle	54
γ-Fe_2O_3/polymers (PEG; d-glucuronic acid, PEI, PEG-PEI)		Potential application in MRI	55
γ-Fe_2O_3/poly(2-methacryloyloxyethyl (2,3,5-triiodobenzoate)		MRI, X-ray	56
γ-Fe_2O_3/SiO_2	PEG, amino acid, FTIC	MRI, biolabeling	57
SiO_2/Au	Dye-functionalized monomer 1-pyrenebutyl acrylate and a trimethoxysilane-carrying one, (3-acryloxypropyl)-trimethoxysilane, antibody conjugation	Optical imaging, drug delivery	58
$SiO_2/NaYF_4$		Optical imaging	59
$Fe_3O_4/CaCO_3$/PMMA	Assorted ligands and bioconjugation molecules	Drug delivery	60
MnO	PEG, d-glucuronic acid, lactobionic acid	Potential application in MRI	61
Fe oxide or Fe_3O_4/Au	DNA ligase enzyme/ desthiobiotin	Piezometric sensor, piezometric and optical sensors	62
SiO_2/polyaniline	Enzyme HRP	Amperometric sensor	63
Fe_3O_4/SiO_2	Assorted biomarkers	Amperometric sensor	64
$CoFe_2O_4$/Au	Biomarkers and enzymes	Amperometric sensor	65
Fe_3O_4/CS	Hemoglobin for H_2O_2 detection	Amperometric sensor	66
Fe_3O_4/silica/Au	Enzymes/nucleotides	Optoelectric sensors	67

(*Continued*)

TABLE 6.1 (*Continued*)

Classification of Core/Shell Nanoparticles Based on Different Core Materials and Their Applications

Core/Shell (Metal or Metal Alloy as Core)	Surface Modification/ Ligands	Application	References
MnO/ssDNA	Biomarkers and assorted molecules	MRI imaging, drug delivery, tumor sensor	68
CdSe/CdS/SiO_2		Biolabeling	69
GaP/$GaPO_4$		Piezoelectric sensor	70
β-$NaGdF_4$:Yb^{3+}/Tm^{3+}	PVP octylamine-PAA	MRI, optical imaging, biolabeling	71
α-$NaGdF_4$:Yb^{3+}:Er^{3+}/ $NaGdF_4$		MRI	72
$NaYF_4$/$NaGdF_4$		MRI	73
$NaGdF_4$:Tm^{3+}/Er^{3+}/Yb^{3+}	Azelaic acid	MRI, optical imaging	74
$KGdF_4$:Yb^{3+},Er^{3+},Ho^{3+},Tm^{3+}		Optical–magnetic dual modal nanoprobes	75
Gd_2O_3/D-glucuronic acid	D-Glucuronic acid	MRI	76
GdF_3/citrate GdF_3/LaF_3	Citrate, 2-aminoethyl phosphate	MRI	77
Gd_2O_3	PEG, D-glucuronic acid, and lactobionic acid	MRI	78
Gd_2O_3/MnO	Lactobionic acid	MRI	78
$Gd_2(OH)_5NO_3nH_{2O}$	PEG	MRI	79
Gd_2O_3/SiO_2	Poly(2-methacryloyloxyethyl phosphorylcholine), poly(lactic-*co*-glycolic acid) (PLGA)		80
$NaYF_4$/Si-DTTA (3-aminopropyl (trimethoxysilyl) diethylenetriamine tetraacetic acid)-loaded Gd(III)		MRI	81
Gd/SiO_2, Gd-DTPA/SiO_2	Multilayered silica, PEI, 3-hydroxypicolinate	MRI, biolabeling	82
Fe_3O_4 embedded in poly(D,L-lactide)/polyvinyl alcohol		MRI, ultrasound	83
Cyanine dye/SiO_2		Optical imaging	84
TRITC dye/SiO_2		Optical imaging	85

(*Continued*)

TABLE 6.1 (*Continued*)

Classification of Core/Shell Nanoparticles Based on Different Core Materials and Their Applications

Core/Shell (Metal or Metal Alloy as Core)	Surface Modification/ Ligands	Application	References
Rhodamine B or (6G) dye/ SiO_2		Optical imaging	86
Cy5 dye/SiO_2	PEG	Optical imaging	87
Alexa Fluor 700 (or 750) dye/SiO_2		Optical imaging	87
DY730 (or 780)/SiO_2		Optical imaging	87
Fluorescein isothiocyanate dye/SiO_2		Cell labeling, Optical imaging	88
Coumarin 7 dye/SiO_2		Optical imaging	89
Chitosan/β-lactoglobulin	Biomarkers	Drug delivery	90
PEG/PCL	MPEG	Drug delivery	91
Cholesterol/CS	MPEG	Drug delivery	92
PLGA/PEG	Folate and assorted biomarkers and stabilizers	Drug targeting	93
Polystyrene/ polybutyl-2-cyanoacrylate	Assorted biomarkers and stabilizers	Drug targeting	94
Polycaprolactone/ dextran	Lectin, biomarkers	Drug delivery	95
Chitosan/cholesterol	Folate and PEG	Drug delivery	96
Ferrite impregnated acrylonitrile/acrylamide	Assorted biomarkers and stabilizers	Drug delivery	97
PMMA/PEI	Assorted biomarkers and stabilizers	Drug delivery	98
PLGF-PLAF/PLEOF	Assorted biomarkers and stabilizers	Drug delivery	99
Polystyrene/ polybutyl-2-cyanoacrylate	Thioflavin receptors	Drug delivery	100
PMMA/PEI	Lactate, aspartate, and biomarkers	Drug delivery	101
N-Isopropylmethacrylamide/ *N,N′*-methylenebisacrylamide	Peptide recognition tag and EGFR	Gene silencing and drug targeting	102
PEG/CS	Gene insert and associated targeting molecules	Gene transfection	102
PEG/polyglycerol	Gene insert and associated targeting molecules	Gene transfection	103
Hyperbranched polyglycerol/PE	Gene insert and associated targeting molecules	Gene transfection	103

(*Continued*)

TABLE 6.1 (*Continued*)

Classification of Core/Shell Nanoparticles Based on Different Core Materials and Their Applications

Core/Shell (Metal or Metal Alloy as Core)	Surface Modification/ Ligands	Application	References
PEG/PBLG	Gene insert and associated targeting molecules	Gene transfection	104
l-Aspartate/PEI	Gene insert and associated targeting molecules	Gene transfection	105
PLGA/folate-coated PEG-cholesterol	Gene insert and associated targeting molecules	Gene transfection	106
DNA-PMMA/CS	Gene insert and associated targeting molecules	Gene transfection	107
Histidine and lysine oligopeptide/DNA	Gene insert and associated targeting molecules	Gene transfection	108
Oligopeptide/DNA-PEG	Gene insert and associated targeting molecules	Gene transfection	109
Polylysine/PELGE	Gene insert and associated targeting molecules	Gene transfection	110
PAEA128-b-PS40/DNA	Gene insert and associated targeting molecules	Gene transfection	111
PGLA/DNA-functionalized glycol/chitosan	Folate and other biomarkers	Gene transfection	112
CS/polyacrylamide	Gene insert and associated targeting molecules	Gene transfection	113

Sources: Reprinted from *Adv. Colloid Interface Sci.*, 209, Chatterjee, K., Sarkar, S., Rao, K.J., and Paria, S., Core/shell nanoparticles in biomedical applications, 8–39, Copyright 2014, Figure 1 and Table 2; *Carbon*, 49, Vashist, S.K., Zheng, D., Pastorin, G., AlRubeaan, K., Luong, J.H.T., and Sheu, F.-S., Delivery of drugs and biomolecules using carbon nanotubes, 4077–4097, Copyright 2011, Table 1; *Prog. Polym. Sci.*, 38, Fernandes, E.M., Piresa, R.A., Manoa, J.F., and Reis, R.L., Bionanocomposites from lignocellulosic resources: Properties, applications and future trends for their use in the biomedical field, 1415–1441, Copyright 2013, Table 5, with permission from Elsevier.

Notes: CNP, carbon nano-particles; HRP, horseradish peroxidase; PNA, peptide nucleic acid; ssDNA, single stranded DNA; PMMA, polymethyl methacrylate; DTPA, diethylenetriamine pentaacetic acid; TRITC, tetramethylrhodamine isothiocyanate; Rhodamine 6G, a highly fluorescent rhodamine family dye; PCL, poly(ε-caprolactone); MPEG, modular polyethylene glycol; PLGF-PLAF/PLEOF, poly(lactide-co-glycolide fumarate)-poly(lactide fumarate)/poly(lactide-ethylene oxide fumarate); EGFR, epidermal growth factor receptor; PE, phosphatidylethanolamine; PBLG, poly(γ-benzyl *L*-glutamate); PELGE, monomethoxy (poly ethylene glycol)-poly(lactide-*co*-glycolide)-monomethoxy (poly ethylene glycol).

6.4 Chitin and Chitosan

Chitin (CT) is a well-known natural biopolymer, whereas chitosan (CS) is a bio-based or artificial polymer. Both of the polymers are nontoxic, biodegradable, and biocompatible in nature. These biomaterials can be easily processed into different forms such as membranes, sponges, gels, scaffolds, microparticles, nanoparticles, and nanofibers for a variety of biomedical applications such as drug delivery, gene therapy, tissue engineering, and wound healing.

CT and CS are linear polysaccharides, comprising two monomeric units, namely, *N*-acetyl-2-amino-2-deoxy-D-glucose (N-acetylated groups) and 2-amino-2-deoxy-D-glucose residues (N-deacetylated groups, amino groups). CT samples contain low amount of 2-amino-2-deoxy-D-glucose, and hence, it is less soluble in acidic solvents. CS samples contain lesser number of *N*-acetyl-2-amino-2-deoxy-D-glucose, and hence, it is soluble in acidic solvents. The structure of CT and CS is represented in Figure 6.5.

CT and CS have versatile applications in tissue engineering[114–119] and wound healing.[120–126] They are used as excipients for drug delivery[127–131] and gene delivery.[132,133] They offer the advantage of being easily processed into gels,[134] membranes,[135,136] nanofibers,[137] nanofibrils,[138] beads,[139] microparticles,[140] nanoparticles,[141,142] scaffolds,[143–145] and spongelike forms,[146] as shown in Figure 6.6.[147]

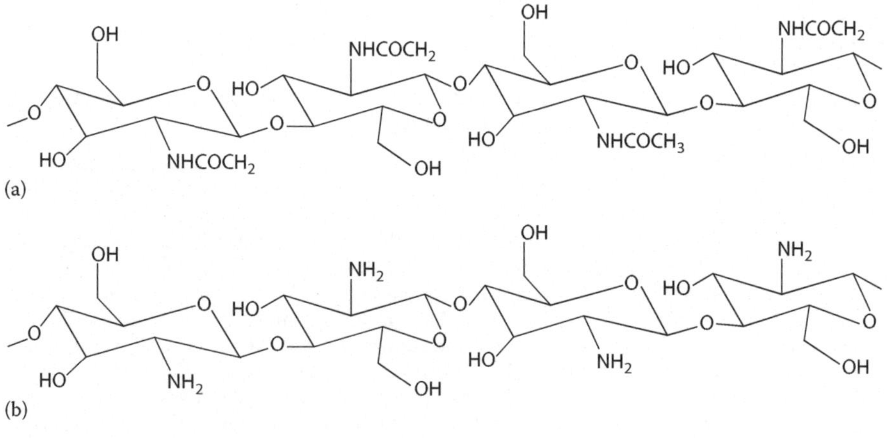

FIGURE 6.5
Structure of (a) CT and (b) CS.

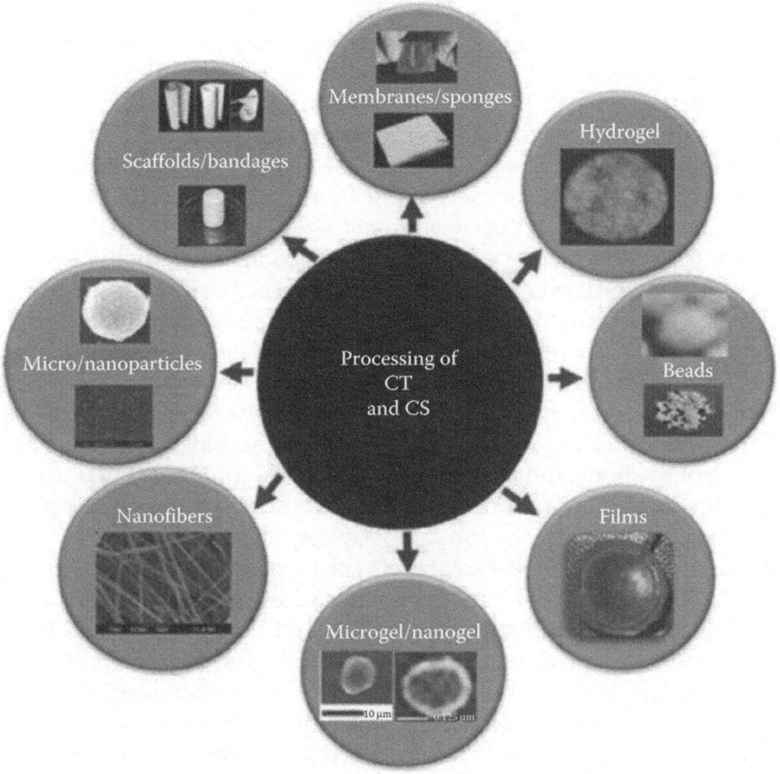

FIGURE 6.6
Schematic representation on the possibilities of processing CT and CS into different forms. (Reprinted from *Prog. Polym. Sci.*, 39, Anitha, A., Sowmyaa, S., Sudheesh Kumara, P.T., Deepthia, S., Chennazhia, K.P., Ehrlichb, H., Tsurkanc, M., and Jayakumar, R., Chitin and chitosan in selected biomedical applications, 1644–1667, Copyright 2014, Figure 3, with permission from Elsevier.)

6.5 Bio-Nanocomposites from Lignocellulosic Resources

Bio-nanocomposites are usually defined as a combination of two or more materials or phases in which one of the phases has at least one dimension in the nanometer range (1–100 nm). Matrices may be biodegradable polymers (e.g., chemically modified cellulose systems), ideally derived from renewable resources (e.g., plants). In terms of reinforcements, they might include plant fibers and by-products from lignocellulosic renewable resources or synthetic inorganic materials, as well as natural or modified clays. Plant-based nanocellulose and bacterial cellulose (BC)[148] are included in this definition. The interest in lignocellulosic polymers, in particular the cellulose-based ones, is due to their reinforcement capacity and biodegradability. Moreover, cellulose-based materials have proven to present an excellent biocompatibility.[149,150]

Biomaterials from lignocellulosic sources have potential to play an important role in human health. Methodologies are available to control the physical and chemical structure targeting their applications such as scaffolds and wound healing systems, as well as tissue engineering[151,152]; and lignocellulosic-based hydrogels, including biodegradation of some cellulose derivatives.[153–155] The selection of biomaterials is important in most of the tissue engineering strategies. Their primary function is to act as a support for the colonization by cells in the intervened area, promoting the formation of new tissue. Hence, biomaterial should have adequate mechanical, thermal, morphological, chemical, and biological characteristics.[156–158] Depending on the target tissue, the biomaterial and support structure should possess certain biological requirements. In the case of the bone, the important biological requirements are bioactivity; osteoproduction, osteoconduction, and osteoinduction; angiogenic potential; and capacity to minimize the foreign body response.[159,160]

The sources of lignocellulosic materials can be plant based or bacterial based. Wood is a natural composite material with a hierarchical architecture where biopolymers such as cellulose, hemicellulose, and lignin form a highly porous anisotropic cellular microstructure, which exhibits a unique combination of high strength, stiffness, toughness, and low density.[161,162] Bone and wood present similar hierarchical structure from the nanoscale to the macroscale. Collagen is a fibrous structural protein which is found in all vertebrates. It is present in the bone, cartilage, skin, tendons, and other tissues.[163,164] Collagen in combination with hydroxylapatite offers mechanical support in the bone; similarly, cellulose has the same function in wood. Lignocellulosic materials and bone present porous networks in both biological materials; although different in size and connectivity, they provide channels for nutrient transport and exchange.[165] Moreover, lignocellulosic materials from plant fibers such as bamboo, sisal, and hemp, among others, also present an anisotropic structure and they are the relevant source of cellulose and hemicellulose derivatives in the biomedical field. BC sometimes referred to as bacterial nanocellulose or microbial cellulose is a natural polymer whose properties are similar to the hydrogels produced from synthetic polymers; for example, it displays high water content (98%–99%) and good sorption of liquids. It is nonallergenic and can be safely sterilized. BC is synthesized by the acetic bacterium *Acetobacter xylinum* (or *Gluconacetobacter xylinus*), a gram-negative strain of acetic acid-producing bacteria using a fermentation process.[166,167] It is a nanomaterial, having several characteristics that make it valuable for biomedical applications such as a highly pure and crystalline structure, high mechanical properties, ultrafine network, high hydrophilicity, and biocompatibility, and it has the advantage of *in situ* mouldability.[168–171]

BC and cellulosic nanofibers are widely used as biobased nano-reinforcements in several polymeric matrices due to their superior mechanical properties.[172,173] The high surface area-to-volume ratio of the nanofibers combined with their microporous structure favors cell adhesion, proliferation, migration, and differentiation, all of which are highly desirable properties for tissue engineering applications.[174] Cellulose acetate (CA) has been used as an isolated component or as a part of composites or blends in the development of TE supports. Blending CA with starch produces starch CA, a biocompatible material that exhibits biodegradability and biocompatibility.[175] A series of different CA combinations have been attempted to produce TE scaffolds,[176–184] bone cements,[185–188] and drug delivery vehicles.[189–191] Inspired by the Nature and following a biomimetic approach, natural wood cellular structures have been selected to serve as a 3D porous scaffold for bone TE by the researchers.[192–194]

Composite hydrogels from lignocellulosic materials are a class of biomaterials that have demonstrated great potential for biological and medical applications.[195–197] These structures can absorb up to 1000 times their dry weight.[198,199] They appear very promising for biomedical applications because they are usually biocompatible. The aqueous environment of the hydrogel can protect cells and fragile drugs (such as peptides). They provide good transport of nutrients to cells and metabolic products from the cells. They can be easily modified with cell adhesion ligands. They can be injected *in vivo* as a liquid that gelifies at body temperature.

6.6 Polymer–Bioactive Glass Nanocomposites

Bioactive glasses of silicate composition, which were first developed by Hench and coworkers in 1969,[200] represent a group of surface reactive materials which are able to bond to the bone in a physiological environment. Bioactive glasses widely used in biomedical applications consist of a silicate network incorporating sodium, calcium, and phosphorus in different relative proportions.

Early applications of bioactive glasses were in the form of solid pieces for small bone replacement, that is, in middle ear surgery. Later, other clinical applications of bioactive glasses were proposed, for example, in periodontology[201] and endodontology[202,203] or as a coating on metallic orthopedic implants.[204,205] More recently, great potential has been attributed to the application of bioactive glasses in tissue engineering and regenerative medicine.[206,207] Bone tissue engineering is one of the possible most exciting future clinical applications of bioactive glasses, for example, to fabricate optimal scaffolds with osteogenic and angiogenic potential.[208,209] Bioactive glass/biodegradable polymer composite materials have emerged recently as a new family of bioactive materials with applications ranging from structural implants to tissue engineering scaffolds. These composites exploit the flexibility of polymers with the stiffness, strength, and bioactive character of the bioactive glass fillers. So far, the work on this class of composites has been carried out using conventional (micron-sized) bioactive glass particles as fillers (or coatings).[210] However, all the specific effects and advantages of bioactive glasses mentioned above, including surface bioreactivity, can be enhanced or modified and controlled to a greater extent, if nanoparticles (or nanofibers) are available, compared to conventional micron-sized powders.

Fabrication techniques for bioactive glasses involve melting methods such as microemulsion techniques,[211–217] gas-phase synthesis,[218–220] sol–gel techniques,[221–223] and laser spinning techniques.[224–227] The typical feature common to all bioactive glasses, being melt or sol–gel derived, is the ability to interact with living tissues forming strong bonds to the bone (and in some cases soft) tissue, a property commonly termed bioreactivity or bioactivity.

The combination of biodegradable polymers and bioactive ceramics (and glasses) results in a new group of composite materials for applications such as temporary orthopedic implants, bone filler materials, or 3D biocompatible scaffolds in the field of tissue engineering. The goal of these composite materials is to impart strength and bioactivity by using inorganic bioactive filler while keeping the positive properties of the polymer intact such as flexibility and capacity to deform under loads.[228] Nanocomposite systems comprising nanoscale bioactive glass (nanoparticles or nanofibres) and biodegradable polymers can be the following:

- Poly(3-hydroxybutyrate)/nanoparticle-based bioactive glass composites[229,230]
- Poly(L-lactic acid)/bioactive glass nanocomposites[231–233]
- Natural polymer/bioactive glass nanocomposites[234–235]
- Bioactive nanocomposites containing bioactive glass nanofibers[236,237]

Nanoscale particulate and nanofiber bioactive glasses have shown advantages over conventional (micron-sized) bioactive glasses due to their large surface area and enhanced solubility as well as reactivity coupled with the possibility to induce nanotopographic surface features in composite materials. These nanomaterials have also inspired researchers to investigate new applications of bioactive glasses in biomedical engineering. Their clinical effectiveness, however, still needs to be tested and validated.

6.7 Gold Nanoparticles

Inertness, nontoxic nature toward cell, and biocompatibility of gold (Au) make gold nanoparticle a promising material for biological and biomedical applications as well as a useful material in therapy and imaging.[238–249] In 1978, cisplatin was approved by the Food and Drug Administration as an antitumor agent, which opens up a new gate to investigate and explore using gold nanoparticles as an antitumor agent. It is found that the combination of Au(I) and Au(III) complexes enhances the antitumor activity of the known antitumor compound.[250,251] Gold nanoparticles have the tendency to absorb and scatter visible and near-infrared (NIR) light resonantly upon excitation of surface plasmon oscillation. The sense of light scattering signal in gold nanoparticles is more profound (highly intense) and comparatively much brighter than the chemical fluorophores. This property can be useful in imaging techniques, that is, single-molecule imaging can be achieved.[252,253]

The gold nanoparticles can be synthesized by the following:

- Physical methods such as microwave irradiation, sonochemical method, ultraviolet radiation, laser ablation, thermolytic process, and photochemical process[254–258]
- Chemical methods employing citrate or sodium borohydride as a reducing agent[259–261]
- Biological methods such as the use of fungus or bacteria as a source[262,263]

Gold nanoparticles in different shapes such as nanospheres,[264,265] nanorods,[266,267] nanoshells,[268] and nanocages[269,270] are used for biomedical applications. Surface modification or functionalized gold nanoparticles for biological and biomedical applications include bioimaging, single-molecule tracking, biosensing, drug delivery, transfection, and diagnostic. Through proper functionalization, the particles can be engineered to accumulate preferentially in tumor cells using targeting ligands, providing a tool for cancer diagnosis and gene therapy.[271,272] In order to obtain high specificity and accuracy to target tumor cells, nanoparticles can be conjugated with tumor-targeting ligands such as peptides, small organic molecules, and antibodies that help them specifically target the tumor tissues or cells, thus giving an early detection.[273–275] Versatile nature of gold nanoparticles has enabled to make use of them for optical imaging of cells[276–278] and phantoms.[279,280] Nanoshells have been designed for diagnostic imaging and at higher light intensity using optical coherence tomography.[281] An enhancement of optical contrast in a mouse colon tumor model has been reported when gold nanoshells are injected. Xiao et al.[282] developed and characterized multifunctional gold nanorod using an anticancer drug doxorubicin to target tumor cell and imaging by PET. Kim et al.[283] prepared PEGylated gold nanoparticles (87 nm in diameter) in aqueous dissolvable microneedles for controlled drug delivery into hamster oral tissue *in vivo*. Popovtzer et al.[284] demonstrated the use of gold nanoprobes that will selectively target tumor-selective antigens to detect head and neck cancer, while inducing distinct contrast in computed tomography imaging.

Currently, radioactive isotopes have been widely used for medical treatment such as cancer therapy with continually growing interest. Diagaradjane et al.[285] investigated the gold nanoshells for noninvasive modulation of *in vivo* tumor radiation response. Use of gold (Au 198 colloid) has been reported for the treatment of pleural and peritoneal effusions due to neoplastic infiltration.[286–288] Gold nanoparticles can be a promising candidate for antioxidant action, as they inhibit the formation of reactive oxygen species, scavenging free radicals, thus increasing the antioxidant defense enzymes of the body and creating a sustained control over hyperglycemic conditions which may prove to be beneficial in the treatment of diabetes.[289–292]

Nanotechnology plays a key role in the development of biosensors. Because of their small size, nanosensors appear to be a promising tool for simple and rapid analysis. Inorganic nanoparticles bound with biological moieties (proteins and peptides) have been developed in order to detect and amplify signals. Gold-based electrochemical immunosensors enhance the electrochemical signal transduction of the binding event between the antigen and the antibody, which in turn provide better surface for immune reagent stability upon immobilization.[293] A variety of small carcinogenic substances such as Aflatoxin B_1,[294] Ochratoxin A,[295] naphthalene,[296] herbicides picloram,[297] and hormones such as human chorionic gonadotrophin have been detected by gold-based electrochemical immunosensors.

6.8 Graphene

Apart from other carbon allotropes (fullerenes, carbon nanotubes [CNTs], and graphite), graphene exhibits unique chemical and physical properties. Because of its honeycomb lattice with two carbon atoms per unit cell, the valence and conduction bands touch the Brillouin zone corners, giving rise to a linear dispersion of the energy spectrum.[298] Single-layer graphene is highly transparent toward visible light (2.3% absorption).[299] It has high mechanical strength with a Young's modulus of 1.1 TPa.[300] High thermal conductivity of about 5000 Wm K^{-1} and large surface area (2630 m^2/g) have also been reported.[301,302] Both sides of a graphene plane can be used as a substrate for addition or adsorption of molecules and functional groups in a controlled manner. Covalent and noncovalent surface modifications can been performed to improve biocompatibility and colloidal stability. Generally, covalent modifications include oxidation by Hummers' method to make graphene oxide (GO) or reduced GO (rGO), conjugation of hydrophilic polymers, 1,3-dipolar cycloaddition, or amine coupling to carboxylic groups. Most of noncovalent modifications are achieved using hydrophobic forces or π–π interactions on the pristine graphene surface or unmodified graphenic compounds lying between functional groups on GO surfaces.[303]

Graphenes have been proposed as a good material for attachment and delivery of drugs, such as anticancer agents.[304,305] Covalent attachment of chitosan,[306] folic acid,[307] or PEG[308] to GO thus produces a platform for the delivery of anti-inflammatory and water-insoluble anticancer drugs such as doxorubicin (Dox).[309,310] The rGO can be successfully used to develop highly efficient electrochemical and biological sensors due to their different functionalities, which can be designed to be very sensitive to small changes in the chemical or biological environment.[311–313]

Functionalization chemistry techniques have been developed to increase solubility and biocompatibility of GO by preparing different sizes of PEGylated GO sheets that are soluble in buffers and stable in serum without agglomeration. The GO sheets are photoluminescent in the visible and infrared spectral regions. This photoluminescence of GO has been used for live cell imaging in the NIR region. The use of nanographene has been explored for *in vivo* tumor targeting.[314] Pharmacokinetics and tumor targeting efficacy have been quantitatively evaluated by noninvasive PET imaging.

Graphene and GO sheets can be used as biocompatible, transferable, and implantable platforms for stem cell culture.[315] Graphene is an excellent biocompatible scaffold that does not hamper the proliferation of human mesenchymal stem cells and accelerates their specific differentiation into osteoblasts.[316] The differentiation rate is comparable to the one achieved with common growth factors, demonstrating the potential of graphene for stem cell research. Electrical properties of graphene allow various therapeutic applications such as neural stimulating electrodes for brain disease treatments.[317–319]

Photothermal therapy as a physical treatment approach to treat cancer has emerged as an alternative of currently used cancer therapies. Hybrid materials have also been successfully developed based on GO or graphene modified by magnetic nanoparticles. This composition is employed for simultaneous cancer therapy by loading an anticancer drug and MRI.[320–322] *In vivo* study of GO tumor uptake shows graphene as a potential candidate for photothermal therapy.[323,324]

Sufficient efforts should be directed for the synthesis and functionalization of graphene with desirable physical and chemical properties. Novel approaches with ease for mass production have to be developed for the separation of graphene layers and to prevent them from agglomeration during the biomedical applications. It is important to develop technologically and economically feasible approaches to functionalize graphene with desirable electrical, chemical, physical, and biological properties, tuning the toxicity (containing cyto- and genotoxicity) of graphene to bacteria or cells, improving the specific selectivity of graphene for biosensing and biomedical imaging and other biomedical applications.

6.9 Carbon Nanotubes

CNTs possess unique and excellent structural, optical, and electrical properties for the development of advanced drug delivery systems. Their very large surface area allows multi-conjugation of various molecules on the sidewalls.

Molecules containing aromatic groups can be easily bound to CNTs noncovalently by strong p–p interactions. One-dimensional functionalized CNTs could improve the binding to a single cell by interacting through multiple binding sites due to their flexibility.[325] Intrinsic optical and electrical properties of CNTs are specifically utilized in imaging and therapeutic applications. Single-walled CNTs (SWCNTs) absorb light strongly in the NIR range (800–1600 nm), which contains the tissue transparent region of electromagnetic wavelengths (800–1400 nm). Therefore, they are extensively employed in photothermal therapy[326–328] and photoacoustic imaging.[329] Optical properties of SWCNTs can also be used for Raman detection and imaging.[330,331]

Pristine CNTs are intrinsically hydrophobic and cannot disperse uniformly in most solvents and biological media, that is, they cannot be employed directly for drug or biomolecular delivery. Thus, functionalization must be developed for improving their biocompatibility and solubility, which allow further modification of CNTs with drugs and biomolecules. These methods include (1) noncovalent functionalization outside CNTs (e.g., on external walls), (2) defect functionalization at the opened tips and sidewalls of CNTs, (3) covalent functionalization (also outside CNTs on their sidewalls), and (4) encapsulation of bioactive molecules or drugs inside CNTs. The most common method for noncovalent modification is to absorb functional moieties containing aromatic groups onto the external wall of CNTs through p–p interactions.[332–335]

CNT serves as a potential candidate in designing drug delivery system for anticancer drugs such as Dox,[336–342] platinum-based anticancer drugs,[343–345] and antimicrobial and anti-inflammatory drugs.[346–349] In addition, it is a suitable pathway for delivery of biomolecules.[350–355] DNA can be attached to the amino groups of functionalized multiwalled CNT (f-MWCNT).[356] The linkage of DNA to f-MWCNT is utilized for improving dispersibility of nanotubes in aqueous media as well as for efficient gene transfection without the use of viral genes. Polyethylenimine (PEI) can be grafted onto MWCNT to form (PEI–*g*-MWCNT) complex, which is used for the immobilization and release of DNA.[357]

CNT membranes can be employed as the active element of a switchable transdermal drug delivery device, which can facilitate more effective treatments of drug abuse and addiction. These membranes are integrated with a nicotine formulation to obtain switchable transdermal nicotine delivery rates on human skin (*in vitro*). Nicotine is electrophoretically pumped at low power (12 days battery life) across the CNT membrane at rates necessary for nicotine replacement therapy. CNT membranes are fabricated using a microtome-cutting method[357] with high CNT loadings (5%–10%) of either MWCNTs or double-walled CNTs for an increased pore area. To obtain efficient electroosmosis pumping, MWCNT membranes were further functionalized with negatively charged dye molecules that have four sulfonate ($-SO^{-3}$) groups using a two-step process as shown in Figure 6.7c.[358]

High and low flux values of nicotine through CNT membrane are found as 1.1 and 0.19 μmol/h cm^2, respectively, for a 35 mg/ml nicotine donor solution

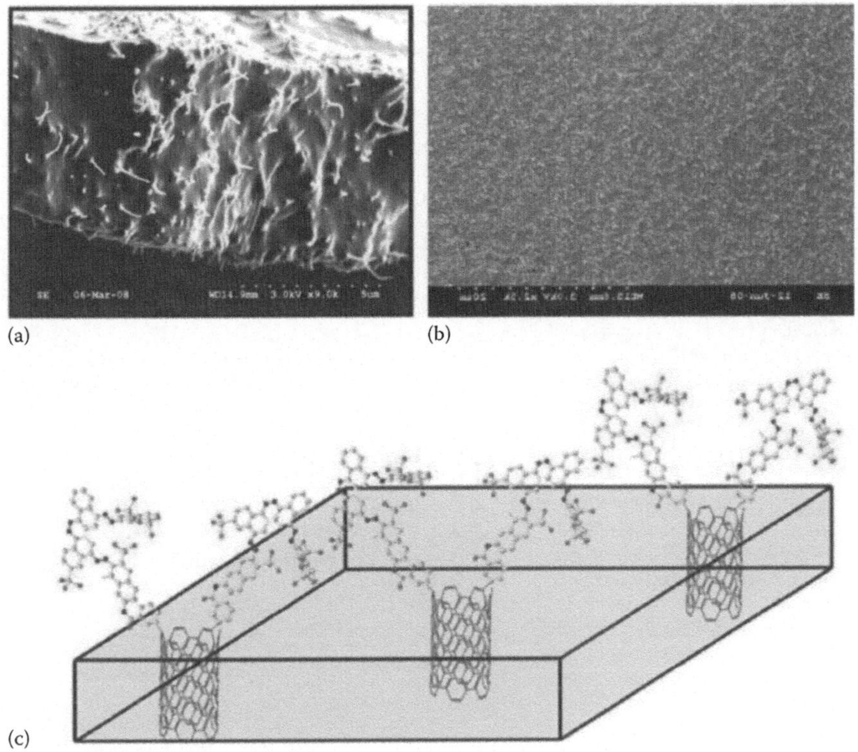

FIGURE 6.7
Scanning electron micrographic images of microtome-cut CNT membrane: (a) cross-sectional view, and (b) top view. (c) Schematic shows the molecular structure of the anionic dye covalently functionalized on the surface of CNTs.

of 8 pH value. The ratio of nicotine flux is 5.5 for −300 and 0 mV. Donor solution is a 35 mg/ml concentration nicotine aqueous solution with a pH value of 8. Switchable transdermal nicotine delivery has been carried out using a modified Franz cell installed with three electrodes for applying bias as shown in Figure 6.8. A diffusion in series model is created to simulate switchable transdermal nicotine delivery using CNT membrane as the rate-limiting component. In this model, nicotine has to pass three layers of diffusion barrier before entering the saline receiver compartment (Figure 6.8).[359] These three layers consist of 5 μm-thick CNT membrane, 140 μm-thick 2% hydroxyethylcellulose gel, and 200 μm-thick human skin. The active area of CNT membrane is 0.07 cm^2. The gel is mainly composed of water (98%). Hence, the diffusion coefficient is near bulk water, thus, there is a minimal concentration gradient of nicotine within the gel.

The unique physical properties of CNT membranes allow highly efficient electrophoretic pumping. It may be utilized to control the delivery rates of an

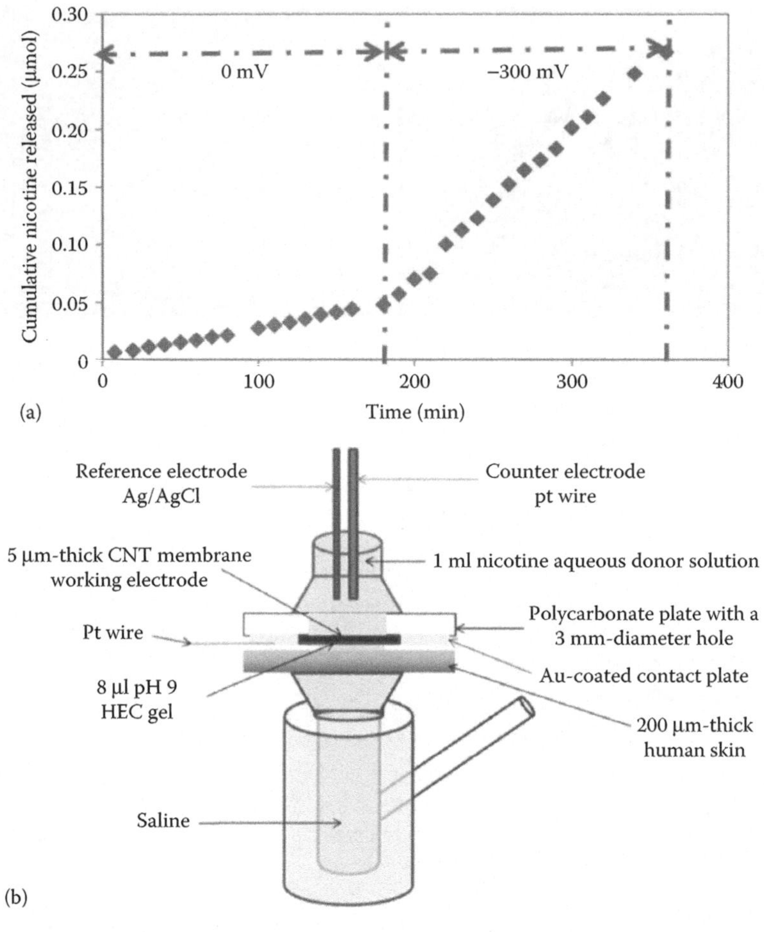

FIGURE 6.8
(a) Flux of nicotine through CNT membrane with/without applying a −300 mV bias. (b) Schematic for switchable transdermal drug delivery (skin/gel/CNT membrane). HEC, hydroxyethylcellulose.

addictive drug for a programmable treatment device. The CNT membrane acts as the rate-limiting component and observed flux values are consistent with a simple diffusion in series model based on Fick's law. Energy requirement and power usage in the CNT membranes are reduced considerably.

Due to the good electric properties, CNTs can provide the nanocomposites with electrical transduction mechanisms, such as piezoresistivity and capacitive detection.[360] A new approach for the fabrication of flexible pressure sensors based on aligned CNTs (A-CNTs)[361] has been developed with inspiration from micro-electro-mechanical system (MEMS)-scale pressure sensors,[362] which are devices of great interest for biomedical applications

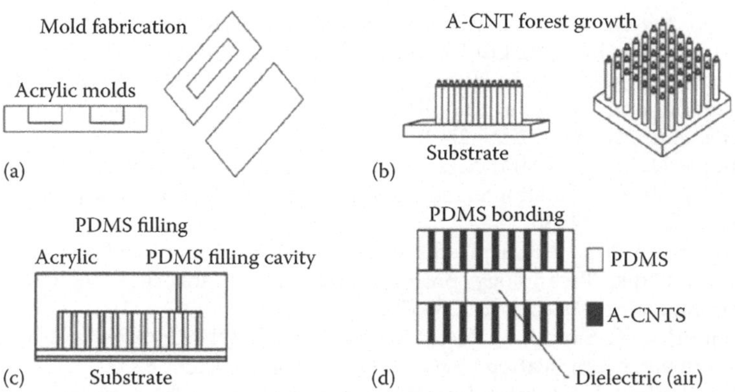

FIGURE 6.9
Fabrication steps for a flexible pressure sensor with A-CNTs/PDMS nanocomposites: (a) acrylic molds fabricated by CNC milling, (b) A-CNT forest growth, (c) cross-sectional view of PDMS filling, and (d) cross-sectional view of the three bonded thin membranes. (Reprinted from *Procedia Eng.*, 25, Sepúlvedaa, A.T., Fachinb, F., Guzmán de Villoriab, R., Wardleb, B.L., Vianaa, J.C., Pontesa, A.J., and Rochaa, L.A., Nanocomposite flexible pressure sensor for biomedical applications, 140–143, Copyright 2011, Figure 1, with permission from Elsevier.)

due to the small size (usually planar areas below 4 mm^2). A-CNTs embedded in a flexible substrate of polydimethylsiloxane (PDMS) are used to fabricate elements of the capacitive sensors. PDMS is a transparent, nontoxic, and biocompatible silicone elastomer. The technology process flow for fabricating the flexible pressure sensor is given in Figure 6.9.

The technology is suitable for blood pressure sensors that can be attached to a stent graft and deployed during an endovascular aneurysm repair procedure. The device is foldable (extremely flexible) and characterized by a (very small) profile that integrates with the stent-graft cross section.

References

1. Gunasekera, U.A., Pankhurst, Q.A., and Douek, M. 2009. Imaging applications of nanotechnology in cancer. *Target. Oncol.* 4: 169–181.
2. Sharma, P., Brown, S, Walter, G., Santra, S., and Moudgil, B. 2006. Nanoparticles for bioimaging. *Adv. Colloid Interface Sci.* 123: 471–485.
3. Mader, H.S., Kele, P., Saleh, S.M., and Wolfbeis, O.S. 2010. Upconverting luminescent nanoparticles for use in bioconjugation and bioimaging. *Curr. Opin. Chem. Biol.* 14: 582–596.
4. Janib, S.M., Moses, A.S., and MacKay, J.A. 2010. Imaging and drug delivery using theranostic nanoparticles. *Adv. Drug. Deliv. Rev.* 62: 1052–1063.
5. Selvan, S.T., Tan, T.T.Y., Yi, D.K., and Jana, N.R. 2009. Functional and multifunctional nanoparticles for bioimaging and biosensing. *Langmuir* 26: 11631–11641.

6. Liu, G., Swierczewska, M., Lee, S., and Chen, X. 2010. Functional nanoparticles for molecular imaging guided gene delivery. *Nano Today* 5: 524–539.
7. Hao, R., Xing, R., Xu, Z., Hou, Y., Gao, S., and Sun, S. 2010. Synthesis, functionalization, and biomedical applications of multifunctional magnetic nanoparticles. *Adv. Mater.* 22: 2729–2742.
8. Knopp, D., Tang, D., and Niessner, R. 2009. Review: Bioanalytical applications of biomolecule functionalized nanometer-sized doped silica particles. *Anal. Chim. Acta* 647: 14–30.
9. Haidar, Z.S. 2010. Bio-inspired/functional colloidal core-shell polymeric-based nanosystems: Technology promise in tissue engineering, bioimaging and nanomedicine. *Polymers* 2: 323–352.
10. Chatterjee, K., Sarkar, S., Rao, K.J., and Paria, S. 2014. Core/shell nanoparticles in biomedical applications. *Adv. Colloid Interface Sci.* 209: 8–39.
11. Wu, W., Shen, J., Banerjee, P., and Zhou, S. 2010. Core–shell hybrid nanogels for integration of optical temperature-sensing, targeted tumor cell imaging and combined chemophotothermal treatment. *Biomaterials* 31: 7555–7566.
12. Wu, W., Zhou, T., Berliner, A., Banerjee, P., and Zhou, S. 2010. Smart core-shell hybrid nanogels with Ag nanoparticle core for cancer cell imaging and gel shell for pH-regulated drug delivery. *Chem. Mater.* 22: 1966–1976.
13. Malam, Y., Loizidou, M., and Seifalian, A.M. 2009. Liposomes and nanoparticles: Nanosized vehicles for drug delivery in cancer. *Trends Pharmacol. Sci.* 30: 592–599.
14. Hoffman, A.S. 2008. The origins and evolution of "controlled" drug delivery systems. *J. Control. Release* 132: 153–163.
15. Carmeliet, P. and Jain, R.K. 2000. Angiogenesis in cancer and other diseases. *Nature* 407: 249–257.
16. Pelicano, H., Martin, D.S., Xu, R.H., and Huang, P. 2006. Glycolysis inhibition for anticancer treatment. *Oncogene* 25: 4633–4646.
17. Jain, R.K. 2001. Delivery of molecular medicine to solid tumors: Lessons from in vivo imaging of gene expression and function. *J. Control. Release* 74: 7–25.
18. Maeda, H. 2001. SMANCS and polymer-conjugated macromolecular drugs: Advantages in cancer chemotherapy. *Adv. Drug Deliv. Rev.* 46: 169–185.
19. Maeda, H. and Matsumura, Y. 1989. Tumoritropic and lymphotropic principles of macromolecular drugs. *Crit. Rev. Ther. Drug Carrier Syst.* 6: 193–210.
20. Matsumura, Y. and Maeda, H. 1986. A new concept for macromolecular therapeutics in cancer chemotherapy: Mechanism of tumoritropic accumulation of proteins and the anti-tumor agent. *Cancer Res.* 46: 6387–6392.
21. Prakash, S., Malhotra, M., Shao, W., Tomaro-Duchesneau, C., and Abbasi, S. 2011. Polymeric nanohybrids and functionalized carbon nanotubes as drug delivery carriers for cancer therapy. *Adv. Drug Deliv. Rev.* 63: 1340–1351.
22. Brenner, B.M., Hostetter, T.H., and Humes, H.D. 1978. Glomerular permselectivity: Barrier function based on discrimination of molecular size and charge. *Am. J. Physiol.* 234: F455–F460.
23. Maack, T., Johnson, V., Kau, S.T., Figueiredo, J., and Sigulem, D. 1979. Renal filtration, transport, and metabolism of low-molecular-weight proteins: A review. *Kidney Int.* 16: 251–270.
24. Mihara, K., Hojo, T., Fujikawa, M., Takakura, Y., Sezaki, H., and Hashida, M. 1993. Disposition characteristics of protein drugs in the perfused rat kidney. *Pharm. Res.* 10: 823–827.

25. Mihara, K., Mori, M., Hojo, T., Takakura, Y., Sezaki, H., and Hashida, M. 1993. Disposition characteristics of model macromolecules in the perfused rat kidney. *Biol. Pharm. Bull.* 16: 158–162.
26. Torchilin, V.P. 2005. Lipid-core micelles for targeted drug delivery. *Curr. Drug Deliv.* 2: 319–327.
27. Zeng, F., Liu, J., and Allen, C. 2004. Synthesis and characterization of biodegradable poly (ethylene glycol)-block-poly(5-benzyloxy-trimethylene carbonate) copolymers for drug delivery. *Biomacromolecules* 5: 1810–1817.
28. Roby, A., Erdogan, S., and Torchilin, V.P. 2006. Solubilization of poorly soluble PDT agent, mesotetraphenylporphin, in plain or immunotargeted PEG-PE micelles results in dramatically improved cancer cell killing in vitro. *Eur. J. Pharm. Biopharm.* 62: 235–240.
29. Duncan, R. 2006. Polymer conjugates as anticancer nanomedicines. *Nat. Rev. Cancer* 6: 688–701.
30. Maria, V.J. and Duncan, R. 2006. Polymer conjugates: Nanosized medicines for treating cancer. *Trends Biotechnol.* 24: 39–47.
31. Mu, Q., Yang, L., Davis, J.C., Vankayala, R., Hwang, K.C., and Zhao, J. 2010. Biocompatibility of polymer grafted core/shell iron/carbon nanoparticles. *Biomaterials* 31: 5083–5090.
32. Miguel, O.B., Gossuin, Y., Morales, M., Gillis, P., Muller, R., and Veintemillas-Verdaguer S. 2007. Comparative analysis of the 1H NMR relaxation enhancement produced by iron oxide and core-shell iron–iron oxide nanoparticles. *Magn. Reson. Imaging* 25: 1437–1441.
33. Wu, W., Zhou, T., Berliner, A., Banerjee, P., and Zhou, S. 2010. Smart core-shell hybrid nanogels with Ag nanoparticle core for cancer cell imaging and gel shell for pH-regulated drug delivery. *Chem. Mater.* 22: 1966–1976.
34. Li, Z., Zhang, Y., Shuter, B., and Muhammad Idris, N. 2009. Hybrid lanthanide nanoparticles with paramagnetic shell coated on upconversion fluorescent nanocrystals. *Langmuir* 25: 12015–12018.
35. Zhang, H. and Hu, N. 2007. Assembly of myoglobin layer-by-layer films with poly(propyleneimine) dendrimer-stabilized gold nanoparticles and its application in electrochemical biosensing. *Biosens. Bioelectron.* 23: 393–399.
36. Feng, X., Huang, H., Ye, Q., Zhu, J.J., and Hou, W. 2007. Ag/polypyrrole core-shell nanostructures: Interface polymerization, characterization, and modification by gold nanoparticles. *J. Phys. Chem. C* 111: 8463–8468.
37. Chen, X., Pan, H., Liu, H., and Du, M. 2010. Nonenzymatic glucose sensor based on flower-shaped Au@Pd core–shell nanoparticles–ionic liquids composite film modified glassy carbon electrodes. *Electrochim. Acta* 56: 636–643.
38. Stobiecka, M. and Hepel, M. 2010. Rapid functionalization of metal nanoparticles by moderator-tunable ligand-exchange process for biosensor designs. *Sens. Actuators B Chem.* 149: 373–380.
39. Pita, M., Abad, J.M., Vaz-Dominguez, C., Briones, C., Mateo-Martí, E., and Martín-Gago, J.A. 2008. Synthesis of cobalt ferrite core/metallic shell nanoparticles for the development of a specific PNA/DNA biosensor. *J. Colloid Interface Sci.* 321: 484–492.
40. Hamer, M., Carballo, R., and Rezzano, I. 2010. Polyallylamine-chlorophyllide derivatized gold and silver nanoparticles as optical probes for sensor applications. *Sens. Actuators B Chem.* 145: 250–253.

41. Jafari, T., Simchi, A., and Khakpash, N. 2010. Synthesis and cytotoxicity assessment of superparamagnetic iron–gold core–shell nanoparticles coated with polyglycerol. *J. Colloid Interface Sci.* 345: 64–71.
42. Nattama, S., Rahimi, M., Wadajkar, A.S., Koppolu, B., Hua, J., and Nwariaku, F. 2007. Characterization of polymer coated magnetic nanoparticles for targeted treatment of cancer. Engineering in Medicine and Biology Workshop, IEEE, Dallas, TX, 35–38.
43. Kim, T., Momin, E., Choi, J., Yuan, K., Zaidi, H., and Kim, J. 2011. Mesoporous silica-coated hollow manganese oxide nanoparticles as positive T 1 contrast agents for labeling and MRI tracking of adipose-derived mesenchymal stem cells. *J. Am. Chem. Soc.* 133: 2955–2961.
44. Scodeller, P., Flexer, V., Szamocki, R., Calvo, E., Tognalli, N., and Troiani, H. 2008. Wired enzyme core–shell Au nanoparticle biosensor. *J. Am. Chem. Soc.* 130: 12690–12697.
45. Sakai, H., Kanda, T., Shibata, H., Ohkubo, T., and Abe, M. 2006. Preparation of highly dispersed core/shell-type titania nanocapsules containing a single Ag nanoparticle. *J. Am. Chem. Soc.* 128: 4944–4945.
46. Kim, J.H. and Lee, T. 2006. Discrete thermally responsive hydrogel-coated gold nanoparticles for use as drug-delivery vehicles. *Drug Dev. Res.* 67: 61–69.
47. Sun, L., Zhang, Z., Wang, S., Zhang, J., Li, H., and Ren, L. 2009. Effect of pH on the interaction of gold nanoparticles with DNA and application in the detection of human gene mutation. *Nanoscale Res. Lett.* 4: 216–220.
48. Kosuge, H., Sherlock, S.P., Kitagawa, T., Terashima, M., Barral, J.K., and Nishimura, D.G. 2011. FeCo/graphite nanocrystals for multi-modality imaging of experimental vascular inflammation. *PLoS One* 6: 14523.
49. Xu, Y.H., Bai, J., and Wang, J.P. 2007. High-magnetic-moment multifunctional nanoparticles for nanomedicine applications. *J. Magn. Magn. Mater.* 311: 131–134.
50. Zhou, T., Lu, M., Zhang, Z., Gong, H., Chin, W.S., and Liu, B. 2010. Synthesis and characterization of multifunctional FePt/ZnO core/shell nanoparticles. *Adv. Mater.* 22: 403–406.
51. Ji, X., Xu, S., Wang, L., Liu, M., Pan, K., and Yuan, H. 2005. Immunoassay using the probelabeled Au/Ag core-shell nanoparticles based on surface-enhanced Raman scattering. *Colloids Surf. A. Physicochem. Eng. Asp.* 257: 171–175.
52. Tanaka, K., Narita, A., Kitamura, N., Uchiyama, W., Morita, M., and Inubushi, T. 2010. Preparation for highly sensitive MRI contrast agents using core/shell type nanoparticles consisting of multiple SPIO cores with thin silica coating. *Langmuir* 26: 11759–11762.
53. Wang, L., Bai, J., Li, Y., and Huang, Y. 2008. Multifunctional nanoparticles displaying magnetization and near-IR absorption. *Angew. Chem. Int. Ed* 47: 2439–2442.
54. Tran, L.D., Hoang, N.M.T., Mai, T.T., Tran, H.V., Nguyen, N.T., and Tran, T.D. 2010. Nanosized magnetofluorescent Fe_3O_4–curcumin conjugate for multimodal monitoring and drug targeting. *Colloids Surf. A. Physicochem. Eng. Asp.* 371: 104–112.
55. Schweiger, C., Pietzonka, C., Heverhagen, J., and Kissel, T. 2011. Novel magnetic iron oxide nanoparticles coated with poly(ethylene imine)-g-poly(ethylene glycol) for potential biomedical application: Synthesis, stability, cytotoxicity and MR imaging. *Int. J. Pharm.* 408: 130–137.
56. Galperin, A. and Margel, S. 2007. Synthesis and characterization of radiopaque magnetic core-shell nanoparticles for X-ray imaging applications. *J. Biomed. Mater. Res. B. Appl. Biomater.* 83: 490–498.

57. Pinho, S.L., Pereira, G.A., Voisin, P., Kassem, J., Bouchaud, V., and Etienne, L. 2010. Fine tuning of the relaxometry of γ-Fe_2O_3@SiO_2 nanoparticles by tweaking the silica coating thickness. *ACS Nano* 4: 5339–5349.
58. Loo, C., Lowery, A., Halas, N., West, J., and Drezek, R. 2005. Immunotargeted nanoshells for integrated cancer imaging and therapy. *Nano Lett.* 5: 709–711.
59. Li, Z., Zhang, Y., and Jiang, S. 2008. Multicolor core/shell-structured upconversion fluorescent nanoparticles. *Adv. Mater.* 20: 4765–4769.
60. Wang, C., Yan, J., Cui, X., Cong, D., and Wang, H. 2010. Preparation and characterization of magnetic hollow PMMA nanospheres via in situ emulsion polymerization. *Colloids Surf. A. Physicochem. Eng. Asp.* 363: 71–77.
61. Park, J.Y., Choi, E.S., Baek, M.J., Lee, G.H., Woo, S., and Chang, Y. 2009. Water-soluble ultra-small paramagnetic or superparamagnetic metal oxide nanoparticles for molecular MR imaging. *Eur. J. Inorg. Chem.* 2009: 2477–2481.
62. Wang, X.J., Wang, L.L., Huang, W.Q., Tang, L.M., Zou, B., and Chen, K.Q. 2006. A surface optical phonon assisted transition in a semi-infinite superlattice with a cap layer. *Semicond. Sci. Technol.* 21: 751.
63. Luo, X., Vidal, G.D., Killard, A.J., Morrin, A., and Smyth, M.R. 2007. Nanocauliflowers: A nanostructured polyaniline-modified screen-printed electrode with a self-assembled polystyrene template and its application in an amperometric enzyme biosensor. *Electroanalysis* 19: 876–883.
64. Qiu, J., Peng, H., and Liang, R. 2007. Ferrocene-modified Fe_3O_4@SiO_2 magnetic nanoparticles as building blocks for construction of reagentless enzyme-based biosensors. *Electrochem. Commun.* 9: 2734–2738.
65. Jimenez, J., Sheparovych, R., Pita, M., Narvaez García, A., Dominguez, E., and Minko, S. 2008. Magneto-induced self-assembling of conductive nanowires for biosensor applications. *J. Phys. Chem. C* 112: 7337–7344.
66. Tan, X.C., Zhang, J.L., Tan, S.W., Zhao, D.D., Huang, Z.W., and Mi, Y. 2009. Amperometric hydrogen peroxide biosensor based on immobilization of hemoglobin on a glassy carbon electrode modified with Fe_3O_4/chitosan core–shell microspheres. *Sensors* 9: 6185–6199.
67. Stoeva, S.I., Huo, F., Lee J.S., and Mirkin, C.A. 2005. Three-layer composite magnetic nanoparticle probes for DNA. *J. Am. Chem. Soc.* 127: 15362–15363.
68. Shukoor, M.I., Natalio, F., Tahir, M.N., Wiens, M., Tarantola, M., and Therese H.A. 2009. Pathogen-mimicking MnO nanoparticles for selective activation of the TLR9 pathway and imaging of cancer cells. *Adv. Funct. Mater.* 19: 3717–3725.
69. Zhu, M.Q., Han, J.J., and Li, A.D. 2007. CdSe/CdS/SiO_2 core/shell/shell nanoparticles. *J. Nanosci. Nanotechnol.* 7: 2343–2348.
70. Zhang, C., Sun, L., Zhang, Y., and Yan, C. 2010. Rare earth upconversion nanophosphors: Synthesis, functionalization and application as biolabels and energy transfer donors. *J. Rare Earths* 28: 807–819.
71. Johnson, N.J., Oakden, W., Stanisz, G.J., Scott Prosser, R., and van Veggel, F.C. 2011. Size-tunable, ultrasmall NaGdF4 nanoparticles: insights into their T1 MRI contrast enhancement. *Chem. Mater.* 23: 3714–3722.
72. Park, Y.I., Kim, J.H., Lee, K.T., Jeon, K.S., Na, H.B., Yu, J.H., Kim, H.M et al. 2009. Nonblinking and nonbleaching upconverting nanoparticles as an optical imaging nanoprobe and T1 magnetic resonance imaging contrast agent. *Adv. Mater.* 21: 4467–4471.
73. Abel, K.A., Boyer, J.C., and Veggel, F.C.V. 2009. Hard proof of the $NaYF_4$/$NaGdF_4$ nanocrystal core/shell structure. *J. Am. Chem. Soc.* 131: 14644–14645.

74. Zhou, J., Sun, Y., Du, X., Xiong, L., Hu, H., and Li, F. 2010. Dual-modality in vivo imaging using rare-earth nanocrystals with near-infrared to near-infrared (NIR-to-NIR) upconversion luminescence and magnetic resonance properties. *Biomaterials* 31: 3287–3295.
75. Yang, L., Zhang, Y., Li, J., Li, Y., Zhong, J., and Chu, P.K. 2010. Magnetic and upconverted luminescent properties of multifunctional lanthanide doped cubic $KGdF_4$ nanocrystals. *Nanoscale* 2: 2805–2810.
76. Park, J.Y., Baek, M.J., Choi, E.S., Woo, S., Kim, J.H., Kim, T.J., Jung, J.C., Chae, K.S., Chang, Y., and Lee, G.H. 2009. Paramagnetic ultrasmall gadolinium oxide nanoparticles as advanced T 1 MRI contrast agent: account for large longitudinal relaxivity, optimal particle diameter, and in vivo T 1 MR images. *ACS Nano* 3: 3663–3669.
77. Evanics, F., Diamente, P., Van Veggel, F., Stanisz, G., and Prosser, R. 2006. Water-soluble GdF_3 and GdF_3/LaF_3 nanoparticles physical characterization and NMR relaxation properties. *Chem. Mater.* 18: 2499–2505.
78. Choi, E.S., Park, J.Y., Baek, M.J., Xu, W., Kattel, K., Kim, J.H., Lee, J.J. et al. 2010. Water-soluble ultra-small manganese oxide surface doped gadolinium oxide (Gd_2O_3@ MnO) nanoparticles for MRI contrast agent. *Eur. J. Inorg. Chem.* 28: 4555–4560.
79. Yoon, Y.-s., Lee, B., Lee, K.S., Im, G.H., Byeon, S.-H., Lee, J.H., and Lee, I.S. 2009. Surface modification of exfoliated layered gadoliniumhydroxide for the development of multimodal contrast agents for MRI and fluorescence imaging. *Adv. Funct. Mater.* 19: 3375–3380.
80. Shao, Y.Z., Liu, L.Z., Song, S.Q., Cao, R.H., Liu, H., Cui., C.Y., Li, X., Bie, M.J., and Li, L. 2011. A novel one-step synthesis of Gd3+-incorporated mesoporous SiO2 nanoparticles for use as an efficient MRI contrast agent. *Contrast. Media Mol. Imaging* 6: 110–118.
81. Li, Z., Zhang, Y., Shuter, B., and Muhammad Idris N. 2009. Hybrid lanthanide nanoparticles with paramagnetic shell coated on upconversion fluorescent nanocrystals. *Langmuir* 25: 12015–12018.
82. Kobayashi, Y., Imai, J., Nagao, D., Takeda, M., Ohuchi, N., Kasuya, A., and Konno, M. 2007. Preparation of multilayered silica–Gd–silica core-shell particles and their magnetic resonance images. *Colloids Surf. A Physicochem. Eng. Asp.* 308: 14–19.
83. Yang, F., Li, Y., Chen, Z., Zhang, Y., Wu, J., and Gu, N. 2009. Superparamagnetic iron oxide nanoparticle-embedded encapsulated microbubbles as dual contrast agents of magnetic resonance and ultrasound imaging. *Biomaterials* 30: 3882–3890.
84. Chen, G., Song, F., Wang, X., Sun, S., Fan, J., and Peng, X. 2012. Bright and stable Cy3-encapsulated fluorescent silica nanoparticles with a large Stokes shift. *Dyes Pigments* 93: 1532–1537.
85. Choi, J., Burns, A.A., Williams, R.M., Zhou, Z., Flesken-Nikitin, A., Zipfel, W.R., Wiesner, U., and Nikitin, A.Y. 2007. Coreshell silica nanoparticles as fluorescent labels for nanomedicine. *J. Biomed. Opt.* 12: 064007–064011.
86. Gao, X., He, J., Deng, L., and Cao, H. 2009. Synthesis and characterization of functionalized rhodamine B-doped silica nanoparticles. *Opt. Mater.* 31: 1715–1719.
87. Herz, E., Ow, H., Bonner D, Burns A, and Wiesner, U. 2009. Dye structure–optical property correlations in near-infrared fluorescent core-shell silica nanoparticles. *J. Mater. Chem.* 19: 6341–6347.

88. He, X., Duan, J., Wang, K., Tan, W., Lin, X., and He, C. 2004. A novel fluorescent label based on organic dye-doped silica nanoparticles for HepG liver cancer cell recognition. *J. Nanosci. Nanotechnol.* 4: 585–589.
89. Ethiraj, A.S., Kharrazi, S., Hebalkar, N., Urban, J., Sainkar, S., and Kulkarni, S. 2007. Highly photostable dye entrapped core–shell particles. *Mater. Lett.* 61: 4738–4742.
90. Chen, L. and Subirade, M. 2005. Chitosan/β-lactoglobulin core–shell nanoparticles as nutraceutical carriers. *Biomaterials* 26: 6041–6053.
91. Li, Z., Zhang, Y., and Jiang, S. 2008. Multicolor core/shell-structured upconversion fluorescent nanoparticles. *Adv. Mater.* 20: 4765–4769.
92. Jang, M.K., Jeong, Y.I., and Nah, J.W. 2010. Characterization and preparation of core–shell type nanoparticle for encapsulation of anticancer drug. *Colloids Surf. B Biointerfaces* 81: 530–536.
93. Wang, H., Zhao, P., Su, W., Wang, S., Liao, Z., Niu, R., and Chanjet, J. 2010. PLGA/polymeric liposome for targeted drug and gene co-delivery. *Biomaterials* 31: 8741–8748.
94. Siegemund, T., Paulke, B.R., Schmiedel, H., Bordag, N., Hoffmann, A., Harkany, T., Tanila, H., Kacza, J., and Härtig, W. 2006. Thioflavins released from nanoparticles target fibrillar amyloid β in the hippocampus of APP/PS1 transgenic mice. *Int. J. Dev. Neurosci.* 24: 195–201.
95. Rodrigues, J., Santos-Magalhaes, N., Coelho, L., Couvreur, P., Ponchel, G., and Gref, R. 2003. Novel core (polyester)-shell (polysaccharide) nanoparticles: Protein loading and surface modification with lectins. *J. Control. Release* 92: 103–112.
96. Wang, H., Zhao, P., Liang, X., Gong, X., Song, T., Niu, R., and Chang, J. 2010. Folate-PEG coated cationic modified chitosan–cholesterol liposomes for tumor-targeted drug delivery. *Biomaterials* 31: 4129–4138.
97. Sahiner, N. and Ilgin, P. 2010. Synthesis and characterization of soft polymeric nanoparticles and composites with tunable properties. *J. Polym. Sci. A. Polym. Chem.* 48: 5239–5246.
98. Feng, M. and Li, P. 2007. Amine-containing core-shell nanoparticles as potential drug carriers for intracellular delivery. *J. Biomed. Mater. Res. A* 80: 184–193.
99. He, X., Ma, J., Mercado, A.E., Xu, W., and Jabbari E. 2008. Cytotoxicity of paclitaxel in biodegradable self-assembled core–shell poly(lactide-co-glycolide ethylene oxide fumarate) nanoparticles. *Pharm. Res.* 25: 1552–1562.
100. Dickerson, E., Blackburn, W., Smith, M., Kapa, L., Lyon, L.A., and McDonald, J. 2010. Chemosensitization of cancer cells by siRNA using targeted nanogel delivery. *BMC Cancer* 10: 10.
101. Pimpha, N., Rattanonchai, U., Surassmo, S., Opanasopit, P., Rattanarungchai, C., and Sunintaboon, P. 2008. Preparation of PMMA/acid-modified chitosan core–shell nanoparticles and their potential as gene carriers. *Colloid Polym. Sci.* 286: 907–916.
102. Zhang, K., Fang, H., Wang, Z., Li, Z., Taylor, J.-S.A., and Wooley, K.L. 2010. Structure-activity relationships of cationic shell-crosslinked knedel-like nanoparticles: Shell composition and transfection efficiency/cytotoxicity. *Biomaterials* 31: 1805–1813.
103. Zhang, K., Fang, H., Wang, Z., Taylor, J.-S.A., and Wooley, K.L. 2009. Cationic shell-crosslinked knedel-like nanoparticles for highly efficient gene and oligonucleotide transfection of mammalian cells. *Biomaterials* 30: 968–977.
104. Yu, D., Wang, A., Huang, H., and Chen, Y. 2008. PEG-PBLG nanoparticle-mediated HSV-TK/GCV gene therapy for oral squamous cell carcinoma. *Nanomedicine* 3: 813–821.

105. Yu, J., Quan, J., Kwon, J., Xu, C., Sun, B., Jiang, H.-L., and Nah, J.W. 2009. Fabrication of a novel core–shell gene delivery system based on a brush-like polycation of α, β-poly(laspartate-graft-PEI). *Pharm. Res.* 26: 2152–2163.
106. Wiradharma, N., Khan, M., Tong, Y.W., Wang, S., and Yang, Y.Y. 2008. Self-assembled cationic peptide nanoparticles capable of inducing efficient gene expression in vitro. *Adv. Funct. Mater.* 18: 943–951.
107. Harada-Shiba, M., Yamauchi, K., Harada, A., Takamisawa, I., Shimokado, K., and Kataoka, K. 2002. Polyion complex micelles as vectors in gene therapy: Pharmacokinetics and in vivo gene transfer. *Gene. Ther.* 9: 407–414.
108. Nie, Y., Yuan, W.M., Gong, T., Lu, J., Fu, Y., and Zhang, Z.R. 2007. Investigation on characterization and transfection of a novel multi-polyplex gene delivery system. *J. Appl. Polym. Sci.* 106: 1028–1033.
109. Lee, P.W., Hsu, S.H., Tsai, J.S., Chen, F.R., Huang, P.J., Ke, C.J., Liao, Z.X., Hsiao, C.W., Lin, H.J., and Sung, H.W. 2010. Multifunctional core-shell polymeric nanoparticles for transdermal DNA delivery and epidermal Langerhans cells tracking. *Biomaterials* 31: 2425–2434.
110. Chen, Q., Hu, Y., Chen, Y., Jiang, X., and Yang, Y. 2005. Microstructure formation and property of chitosan-poly(acrylic acid) nanoparticles prepared by macromolecular complex. *Macromol. Biosci.* 5: 993–1000.
111. Vacanti, C.A. 2006. The history of tissue engineering. *J. Cell Mol. Med.* 10: 569–576.
112. Krajewska, B. 2005. Membrane-based processes performed with use of chitin/chitosan materials. *Sep. Purif. Technol.* 41: 305–312.
113. Venkatesan, J. and Kim, S.K. 2010. Chitosan composites for bone tissue engineering—An overview. *Mar. Drugs* 8: 2252–2266.
114. Jayakumar, R., New, N., Tokura, S., and Tamura, H. 2007. Sulfated chitin and chitosan as novel biomaterials. *Int. J. Biol. Macromol.* 40: 175–181.
115. Tigli, R.S. and Gumusderelioglu, M. 2009. Evaluation of alginate-chitosan semi-IPNs as cartilage scaffolds. *J. Mater. Sci. Mater. Med.* 20: 699–709.
116. Ragetly, G.R., Slavik, G.J., Cunningham, B.T., Schaeffer D.J., and Griffon, D.J. 2010. Cartilage tissue engineering on fibrous chitosan scaffolds produced by a replica molding technique. *J. Biomed. Mater. Res. A* 93: 46–55.
117. Ragetly, G.R., Griffon, D.J., Lee, H.B., Fredericks, L.P., Gordon-Evans, W., and Chung, Y.S. 2010. Effect of chitosan scaffold microstructure on mesenchymal stem cells chondrogenesis. *Acta Biomater.* 6: 1430–1436.
118. Mourya, V.K. and Inamdar, N.N. 2008. Chitosan-modifications and applications: Opportunities galore. *React. Funct. Polym.* 68: 1013–1051.
119. Suh, J.K.F. and Matthew, H.W.T. 2000. Application of chitosan-based polysaccharide biomaterials in cartilage tissue engineering: A review. *Biomaterials* 21: 2589–2598.
120. Muzzarelli, R.A.A. 2009. Chitins and chitosans for the repair of wounded skin, nerve, cartilage and bone. *Carbohydr. Polym.* 76: 167–182.
121. Dai, M., Zheng, X.L., Xu, X., Kong, X.Y., Li, X.Y., Guo, G., Luo, F., Zhao, X., Wei, Y.Q., and Qian, Z. 2009. Chitosan–alginate sponge: Preparation and application in curcumin delivery for dermal wound healing in rat. *J. Biomed. Biotechnol.* 2009: 595126.
122. Sudheesh Kumar, P.T., Abhilash, S., Manzoor, K., Nair, S.V., Tamura, H., and Jayakumar, R. 2010. Preparation and characterization of novel chitin/nano silver composite scaffolds for wound dressing applications. *Carbohydr. Polym.* 80: 761–777.
123. Kofuji, K., Huang, Y., Tsubaki, K., Kokido, F., Nishikaw, K., Isobe, T., and Murata, M. 2010. Preparation and evaluation of a novel wound dressing sheet comprised of glucan-chitosan complex. *React. Funct. Polym.* 70: 784–789.

124. Cai, Z.X., Mo, X.M., Zhang, K.H., Fan, L.P., Yin, A.L., He, C.L., and Wang, H.S. Fabrication of chitosan/silk fibroin composite nanofibers for wound-dressing applications. *Int. J. Mol. Sci.* 11: 3529–3539.
125. Kang, Y.O., Yoon, I.S., Lee, S.Y., Kim, D.D., Lee, S.J., Park, W.H., and Hudson, S.M. 2010. Chitosan coated poly(vinyl alcohol) nanofibers for wound dressings. *J. Biomed. Mater. Res. B* 92: 568–576.
126. Dong, Y., Liu, H.Z., Xu, L., Li, G., Ma, Z.N., Han, F., Yao, H.M., Sun, Y.H., and Li, S.M. 2010. A novel CHS/ALG bi-layer composite membrane with sustained antimicrobial efficacy used as wound dressing. *Chin. Chem. Lett.* 21: 1011–1014.
127. Radhakumary, C., Antonty, M., and Sreenivasan, K. 2011. Drug loaded thermoresponsive and cytocompatible chitosan based hydrogel as a potential wound dressing. *Carbohydr. Polym.* 83: 705–713.
128. Dev, A., Binulal, N.S., Anitha, A., Nair, S.V., Furuike, T., Tamura, H., and Jayakumar, R. 2010. Preparation of novel poly(lactic acid)/chitosan nanoparticles for anti-HIV drug delivery applications. *Carbohydr. Polym.* 80: 833–838.
129. Dev, A., Mohan, J.C., Sreeja, V., Tamura, H., Patzke, G.R., Hussain, F., Weyeneth, S., Nair, S.V., and Jayakumar, R. 2010. Novel carboxymethyl chitinnanoparticles for cancer drug delivery applications. *Carbohydr. Polym.* 79: 1073–1079.
130. Ravi Kumar, M.N.V. 2000. A review of chitin and chitosan applications. *React. Funct. Polym.* 46: 1–27.
131. Sashiwa, H. and Aiba, S. 2004. Chemically modified chitin and chitosan as biomaterials. *Prog. Polym. Sci.* 29: 887–888.
132. Ravi Kumar, M.N.V., Muzzarelli, R.A.A., Muzzarelli, C., Sashiwa, H., and Domb, A.J. 2004. Chitosan chemistry and pharmaceutical perspectives. *Chem. Rev.* 104: 6017.
133. Yi, H., Wu, L.Q., Bentley, W.E., Ghodssi, R., Rubloff, G.W., Culver, J.N., and Payne, G.F. 2005. Biofabrication with chitosan. *Biomacromolecules* 6: 2881–2894.
134. Roughley, P., Hoemann, C., Desrosiers, E., Mwale, F., Antoniou, J., and Alini, M. 2006. The potential of chitosan-based gels containing invertebral disc cells for nucleus pulposus supplementation. *Biomaterials* 27: 388–396.
135. Ehrlich, H., Krajewska, B., Hanke, T., Born, R., Heinemann, S., Knieb, C., and Worch, H. 2006. Chitosan membrane as a template for hydroxyapatite crystal growth in a model dual membrane diffusion system. *J. Membr. Sci.* 273: 124–128.
136. Jayakumar, R., Divya, R.V.V., Shalumon, K.T., Sudheesh Kumar, P.T., Nair, S.V., Furuike, T., and Tamura, H. 2009. Bioactive and osteoblast cell attachment studies of novel and chitin membranes for tissue-engineering applications. *Int. J. Biol. Macromol.* 45: 260–264.
137. Jayakumar, R., Prabaharan, M., Nair, S.V., and Tamura, H. 2010. Novel chitin and chitosan nanofibers in biomedical applications. *Biotechnol. Adv.* 28: 142–150.
138. Morganti, P. and Morganti, G. 2008. Chitin nanofibrils for advanced cosmeceuticals. *Clin. Dermatol.* 26: 334–340.
139. Jayakumar, R., Reis, R.L., and Mano, J.F. 2007. Synthesis and characterization of pH-sensitive thiol-containing chitosan beads for controlled drug delivery applications. *Drug Deliv.* 14: 9–17.
140. Prabaharan, M. and Mano, J.F. 2005. Chitosan-based particles as controlled drug delivery systems. *Drug. Deliv.* 12: 41–57.

141. Anitha, A., Divya, R.V.V., Krishna, R, Sreeja, V., Selvamurugan, N., Nair, S.V., Tamura, H., and Jayakumar, R. 2009. Synthesis, characterization, cytotoxicity and antibacterial studies of chitosan, O-carboxymethyl, N,O-carboxymethyl chitosan nanoparticles. *Carbohydr. Polym.* 78: 672–677.
142. Anitha, A., Deepagan, V.G., Divya, R.V.V., Deepthy, M., Nair, S.V., and Jayakumar, R. 2011. Preparation, characterization, in vitro drug release and biological studies of curcumin loaded dextran sulphate-chitosan nanoparticles. *Carbohydr. Polym.* 84: 1158–1164.
143. Drury, J.L. and Mooney, D.J. Hydrogels for tissue engineering: Scaffold design variables and applications. *Biomaterials* 24: 4337–4351.
144. Jayakumar, R., Prabaharan, M., Sudheesh Kumar, P.T., Nair, S.V., and Tamura, H. Biomaterials based on chitin and chitosan in wound dressing applications. *Biotechnol. Adv.* 29: 322–337.
145. Prabaharan, M., Jayakumar, R., and Nair, SV. 2012. Electrospun nanofibrous scaffolds: Current status and prospects in drug delivery. *Adv. Polym. Sci.* 246: 241–262.
146. Muramatsu, K., Masuda, S., Yoshihara, S., and Fujisawa, A. 2003. In vitro degradation behavior of freeze-dried carboxymethyl-chitin sponges processed by vacuum-heating and gamma irradiation. *Polym. Degrad. Stabil.* 81: 327–332.
147. Anitha, A., Sowmyaa, S., Sudheesh Kumara, P.T., Deepthia, S., Chennazhia, K.P., Ehrlichb, H., Tsurkanc, M., and Jayakumar, R. 2014. Chitin and chitosan in selected biomedical applications. *Prog. Polym. Sci.* 39: 1644–1667.
148. Klemm, D., Kramer, F., Moritz, S., Lindstrom, T., Ankerfors, M., Gray, D., and Dorris, A. 2011. Nanocelluloses: A new family of nature-based materials. *Angew. Chem. Int. Ed.* 50: 5438–5466.
149. Entcheva, E., Bien, H., Yin, L.H., Chung, C.Y., Farrell, M., and Kostov, Y. 2004. Functional cardiac cell constructs on cellulose-based scaffolding. *Biomaterials* 25: 5753–5762.
150. Martson, M., Viljanto, J., Hurme, T., Laippala, P., and Saukko, P. 1999. Is cellulose sponge degradable or stable as implantation material? An in vivo subcutaneous study in the rat. *Biomaterials* 20: 1989–1995.
151. Fernandes, E.M., Piresa, R.A., Manoa, J.F., and Reis, R.L. 2013. Bionanocomposites from lignocellulosic resources: Properties, applications and future trends for their use in the biomedical field. *Prog. Polym. Sci.* 38: 1415–1441.
152. Petersen, N. and Gatenholm, P. 2011. Bacterial cellulose-based materials and medical devices: Current state and perspectives. *Appl. Microbiol. Biotechnol.* 91: 1277–1286.
153. Czaja, W.K., Young, D.J., Kawecki, M., and Brown Jr., R.M. 2007. The future prospects of microbial cellulose in biomedical applications. *Biomacromolecules* 8: 1–12.
154. Sannino, A., Demitri, C., and Madaghiele, M. 2009. Biodegradable cellulose-based hydrogels: Design and applications. *Materials* 2: 353–373.
155. Chang, C. and Zhang, L. 2011. Cellulose-based hydrogels: Present status and application prospects. *Carbohyd. Polym.* 84: 40–53.
156. Li, X. and Pan, X. 2010. Hydrogels based on hemicellulose and lignin from lignocelluloses biorefinery: A mini-review. *J. Biobased Mater. Bioenergy* 4: 289–297.
157. Langer, R. and Vacanti, J.P. 1993. Tissue engineering. *Science* 260: 920–926.
158. Hutmacher, D.W. 2001. Scaffold design and fabrication technologies for engineering tissues—State of the art and future perspectives. *J. Biomater. Sci. Polym.* 12: 107–124.

159. Dvir, T., Timko, B.P., Kohane, D.S., and Langer, R. 2011. Nanotechnological strategies for engineering complex tissues. *Nat. Nanotechnol.* 6: 13–22.
160. Hench, L.L. 1998. Bioactive materials: The potential for tissue regeneration. *J. Biomed. Mater. Res.* 41: 511–518.
161. Hench, L.L., Jones, J.R., and Sepulveda, P. 2002. Bioactive materials for tissue engineering scaffolds, *Future Strategies for Tissue and Organ Replacement*, J.M. Polak, L.L. Hench, and P. Kemp (eds.), Imperial College Press, London, 3–24.
162. Gibson, L.J. 2003. Cellular solids. *MRS Bull.* 28: 270–271.
163. Silva, S.P., Sabino, M.A., Fernandes, E.M., Correlo, V.M., Boesel, L.F., and Reis, R.L. 2005. Cork: Properties, capabilities and applications. *Int. Mater. Rev.* 50: 345–365.
164. Ratner, B.D., Hoffman, A.S., Schoen, F.J., and Lemons, J.E. (eds.). 2004. *Biomaterials Science: An Introduction to Materials in Medicine*, Elsevier Press, San Diego, CA, 851.
165. Cheng, Z.Y. and Teoh, S.H. 2004. Surface modification of ultra thin poly (epsilon-caprolactone) films using acrylic acid and collagen. *Biomaterials* 25: 1991–2001.
166. Gross, K.A. and Ezerietis, E. 2003. Juniper wood as a possible implant material. *J. Biomed. Mater. Res. Part A* 64: 672–683.
167. Klemm, D., Schumann, D., Kramer, F., Heßler, N., Hornung, M., Schmauder, H.P., and Marsch, S. 2006. Nanocelluloses as innovative polymers in research and application polysaccharides. II. *Adv. Polym. Sci.* 205: 49–96.
168. Millon, L.E. and Wan, W.K. 2006. The polyvinyl alcohol–bacterial cellulose system as a new nanocomposite for biomedical applications. *J. Biomed. Mater. Res. Part B* 79: 245–253.
169. Klemm, D., Schumann, D., Kramer, F., Hessler, N., Koth, D., and Sultanova, B. 2009. Nanocellulose materials—Different cellulose, different functionality. *Macromol. Res.* 280: 60–71.
170. Tahara, N., Tabuchi, M., Watanabe, K., Yano, H., Morinaga, Y., and Yoshinaga, F. 1997. Degree of polymerization of cellulose from *Acetobacter xylinum* BPR2001 decreased by cellulase produced by the strain. *Biosci. Biotechnol. Biochem.* 61: 1862–1865.
171. Miyamoto, T., Takahashi, S.I., Ito, H., Inagaki, H., and Noishiki, Y. 1989. Tissue biocompatibility of cellulose and its derivatives. *J. Biomed. Mater. Res.* 23: 125–133.
172. Svensson, A., Nicklasson, E., Harrah, T., Panilaitis, B., Kaplan, D.L., Brittberg, M., and Gatenholm, P. 2005. Bacterial cellulose as a potential scaffold for tissue engineering of cartilage. *Biomaterials* 26: 419–431.
173. Yano, H., Sugiyama, J., Nakagaito, A.N., Nogi, M., Matsuura, T., Hikita, M., and Handa, K. 2005. Optically transparent composites reinforced with networks of bacterial nanofibers. *Adv. Mater.* 17: 153–175.
174. Oksman, K., Mathew, A.P., Bondeson, D., and Kvien, I. 2006. Manufacturing process of cellulose whiskers/polylactic acid nanocomposites. *Compos. Sci. Technol.* 66: 2776–2784.
175. Vasita, R. and Katti, D.S. 2006. Nanofibers and their applications in tissue engineering. *Int. J. Nanomed.* 1: 15–30.
176. Marques, A.P., Reis, R.L., and Hunt, J.A. 2002. The biocompatibility of novel starch-based polymers and composites: In vitro studies. *Biomaterials* 23: 1471–1478.
177. Alves, N.M., Saiz-Arroyo, C., Rodriguez-Perez, M.A., Reis, R.L., and Mano, J.F. 2007. Microhardness of starch based biomaterials in simulated physiological conditions. *Acta Biomater.* 3: 69–76.
178. Marques, A.P. and Reis, RL. 2005. Hydroxyapatite reinforcement of different starch-based polymers affects osteoblast-like cells adhesion/spreading and proliferation. *Mater. Sci. Eng. C* 25: 215–229.

179. Marques, A.P., Cruz, H.R., Coutinho, O.P., and Reis, R.L. 2005. Effect of starch based biomaterials on the in-vitro proliferation and viability of osteoblast-like cells. *J. Mater. Sci. Mater. Med.* 16: 833–842.
180. Leonor, I.B. and Reis, R.L. 2003. An innovative auto-catalytic deposition route to produce calcium–phosphate coatings on polymeric biomaterials. *J. Mater. Sci. Mater. Med.* 14: 435–441.
181. Gouma, P., Xue, R., Goldbeck, C.P., Perrotta, P., and Balazsi, C. 2012. Nanohydroxyapatite–cellulose acetate composites for growing of bone cells. *Mater. Sci. Eng. C* 32: 607–612.
182. Zhang, K., Ma, Y., and Francis, L.F. 2002. Porous polymer/bioactive glass composites for soft-to-hard tissue interfaces. *J. Biomed. Mater. Res.* 61: 551–563.
183. Rodriguez, K., Renneckar, S., and Gatenholm, P. 2011. Biomimetic calcium phosphate crystal mineralization on electrospun cellulose-based scaffolds. *ACS Appl. Mater. Interfaces* 3: 681–689.
184. Kim, E.J., Boehm, C.A., Fleischman, A.J., Muschler, G.F., Kostov, Y.V., and Roy, S. 2009. Modulating human connective tissue progenitor cell behavior on cellulose acetate scaffolds by surface microtextures. *J. Biomed. Mater. Res. Part A* 90: 1198–1205.
185. Ye, H., Xia, Z.D., Ferguson, D.J.P., Triffitt, J.T., and Cui, Z.F. 2007. Studies on the use of hollow fibre membrane bioreactors for tissue generation by using rat bone marrow fibroblastic cells and a composite scaffold. *J. Mater. Sci. Mater. Med.* 18: 641–648.
186. Boesel, L.F., Mano, J.F., and Reis, R.L. 2004. Optimization of the formulation and mechanical properties of starch based partially degradable bone cements. *J. Mater. Sci. Mater. Med.* 15: 73–83.
187. Boesel, L.F., Fernandes, M.H.V., and Reis, R.L. 2004. The behavior of novel hydrophilic composite bone cements in simulated body fluids. *J. Biomed. Mater. Res. Part B* 70: 368–377.
188. Boesel, L.F., Cachinho, S.C.P., Fernandes, M.H.V., and Reis, R.L. 2007. The in-vitro bioactivity of two novel hydrophilic, partially degradable bone cements. *Acta Biomater.* 3: 175–182.
189. Boesel, L.F. and Reis, R.L. 2008. A review on the polymer properties of hydrophilic, partially degradable and bioactive acrylic cements (HDBC). *Prog. Polym. Sci.* 33: 180–190.
190. Jeon, J.H. and Puleo, D.A. 2008. Formulations for intermittent release of parathyroid hormone (1–34) and local enhancement of osteoblast activities. *Pharm. Dev. Technol.* 13: 505–512.
191. Jeon, J.H., Piepgrass, W.T., Lin, Y.L., Thornas, M.V., and Puleo, D.A. 2008. Localized intermittent delivery of simvastatin hydroxyacid stimulates bone formation in rats. *J. Periodontol.* 79: 1457–1464.
192. Jeon, J.H., Thomas, M.V., and Puleo, D.A. 2007. Bioerodible devices for intermittent release of simvastatin acid. *Int. J. Pharm.* 340: 6–12.
193. De Arellano-López, A.R., Martínez-Fernández, J., González, P., Domínguez, C., Fernández-Quero, V., and Singh, M. 2004. Biomorphic SiC: A new engineering ceramic material. *Int. J. Appl. Ceramic Technol.* 1: 56–67.
194. Tampieri, A., Sprio, S., Ruffini, A., Celotti, G., Lesci, I.G., and Roveri, N. 2009. From wood to bone: Multi-step process to convert wood hierarchical structures into biomimetic hydroxyapatite scaffolds for bone tissue engineering. *J. Mater. Chem.* 19: 4973–4980.

195. Gonzalez, P., Serra, J., Liste, S., Chiussi, S., Leon, B., Perez-Amor, M., Martinez-Fernandez, J., de Arellano-Lopez, A.R., and Varela-Feria, F.M. 2003. New biomorphic SiC ceramics coated with bioactive glass for biomedical applications. *Biomaterials* 24: 4827–4832.
196. Peppas, N.A., Hilt, J.Z., Khademhosseini, A., and Langer, R. 2006. Hydrogels in biology and medicine: From molecular principles to bionanotechnology. *Adv. Mater.* 18: 1345–1360.
197. Allan, S.H. 2002. Hydrogels for biomedical applications. *Adv. Drug Deliv. Rev.* 54: 3–12.
198. Eyholzer, C., de Courac, A.B., Duc, F., Bourban, P.E., Tingaut, P., Zimmermann, T., Månson, J.A.E., and Oksman, K. 2011. Biocomposite hydrogels with carboxymethylated, nanofibrillated cellulose powder for replacement of the nucleus pulposus. *Biomacromolecules* 12: 1419–1427.
199. Elvira, C., Abraham, G.A., Gallardo, A., and Román, J.S. 2005. Smart biodegradable hydrogels with applications in drug delivery and tissue engineering, *Biodegradable Systems in Tissue Engineering and Regenerative Medicine,* R.L. Reis and J.S. Román (eds.), CRC Press, Boca Raton, FL, 493–508.
200. Hench, L.L., Splinter, R.J., Allen, W.C., and Greenlee, T.K. 1998. Bonding mechanisms at the interface of ceramic prosthetic materials. *J. Biomed. Mater. Res.* 5(6): 117–141.
201. Hench, L.L. 1998. Bioceramics. *J. Am. Ceram. Soc.* 81: 1705–1728.
202. Zamet, J.S., Darbar, U.R., Griffiths, G.S., Burgin, W., and Newman, H.N. 1997. Particulate bioglass (perioglas(R)) in the treatment of periodontal intrabony defects. *J. Dent. Res.* 76: 2219.
203. Gatti, A.M., Simonetti, L.A., Monari, E., Guidi, S., and Greenspan, D. 2006. Bone augmentation with bioactive glass in three cases of dental implant placement. *J. Biomater. Appl.* 20: 325–339.
204. Zehnder, M., Soderling, E., Salonen, J., and Waltimo, T. 2004. Preliminary evaluation of bioactive glass S53P4 as an endodontic medication in vitro. *J. Endodont.* 30: 220–224.
205. Waltimo, T., Mohn, D., Paque, F., Brunner, T.J., Stark, W.J., Imfeld, T., Schätzle, M., and Zehnder, M. 2009. Fine-tuning of bioactive glass for root canal disinfection. *J. Dent. Res.* 88(3): 235–238.
206. Gomez-Vega, J.M., Saiz, E., Tomsia, A.P., Marshall, G.W., and Marshall, S.J. 2000. Bioactive glass coatings with hydroxyapatite and bioglass (R) particles on Ti-based implants. 1. Processing. *Biomaterials* 21: 105–111.
207. Kitsugi, T., Nakamura, T., Oka, M., Senaha, Y., Goto, T., and Shibuya, T. 1996. Bone-bonding behavior of plasma-sprayed coatings of bioglass (R), AW-glass ceramic, and tricalcium phosphate on titanium alloy. *J. Biomed. Mater. Res.* 30(2): 261–269.
208. Rezwan, K., Chen, Q.Z., Blaker, J.J, and Boccaccini, A.R. 2006. Biodegradable and bioactive porous polymer/inorganic composite scaffolds for bone tissue engineering. *Biomaterials* 27(18): 3413–3431.
209. Xynos, I.D., Hukkanen, M.V.J., Batten, J.J, Buttery, L.D, Hench, L.L, and Polak, J.M. 2000. Bioglass (R) 45S5 stimulates osteoblast turnover and enhances bone formation in vitro: Implications and applications for bone tissue engineering. *Calcified Tissue Int.* 67: 321–329.
210. Yao, J., Radin, S., Leboy, S., and Ducheyne, P. 2005. The effect of bioactive glass content on synthesis and bioactivity of composite poly(lactic-co-glycolic acid)/bioactive glass substrate for tissue engineering. *Biomaterials* 26(14): 1935–1945.

211. Hench, L.L. and Polak, J.M. 2002. Third-generation biomedical materials. *Science* 295: 1014–1017.
212. Chen, Q.Z.Z., Thompson, I.D., and Boccaccini, A.R. 2006. 45S5 Bioglass (R)-derived glass–ceramic scaffolds for bone tissue engineering. *Biomaterials* 27: 2414–2425.
213. Pileni, M.P. 2003. The role of soft colloidal templates in controlling the size and shape of inorganic nanocrystals. *Nat. Mater.* 2: 145–150.
214. Sun, Y., Guo, G., Tao, D., and Wang, Z. 2007. Reverse microemulsion-directed synthesis of hydroxyapatite nanoparticles under hydrothermal conditions. *J. Phys. Chem. Solids* 68: 373–377.
215. Singh, S., Bhardwaj, P., Singh, V., Aggarwal, S., and Mandal, U.K. 2008. Synthesis of nanocrystalline calcium phosphate in microemulsion—Effect of nature of surfactants. *J. Colloid Interface Sci.* 319: 322–329.
216. Lim, G.K., Wang, J., Ng, S.C., and Gan, L.M. 1996. Processing of fine hydroxyapatite powders via an inverse microemulsion route. *Mater. Lett.* 28: 431–436.
217. Karagiozov, C. and Momchilova, D. 2005. Synthesis of nano-sized particles from metal carbonates by the method of reversed mycelles. *Chem. Eng. Process.* 44: 115–119.
218. Stark, W.J., Madler, L., Maciejewski, M., Pratsinis, S.E., and Baiker, A. 2003. Flame synthesis of nanocrystalline ceria–zirconia: Effect of carrier liquid. *Chem. Commun.* 5: 588–589.
219. Stark, W.J., Mädler, L., and Pratsinis, S.E. 2004. Metal oxides prepared by flame spray pyrolysis. WO 2004/005184.
220. Stark, W.J. and Pratsinis, S.E. 2004. Metal delivery system for nanoparticle manufacture. WO 2004/103900A1.
221. Sepulveda, P., Jones, J.R., and Hench, L.L. 2001. Characterization of melt-derived 45S5 and sol–gel-derived 58S bioactive glasses. *J. Biomed. Mater. Res.* 58: 734–740.
222. Saravanapavan, P., Jones, J.R., Pryce, R.S., and Hench, L.L. 2003. Bioactivity of gel–glass powders in the $CaO–SiO_2$ system: A comparison with ternary ($CaO–P_2O_5–SiO_2$) and quaternary glasses ($SiO_2–CaO–P_2O_5–Na_2O$). *J. Biomed. Mater. Res. A* 66A: 110–119.
223. Li, R, Clark, A.E., and Hench, L.L. 1991. An investigation of bioactive glass powders by sol–gel processing. *J. Appl. Biomater.* 2: 231–239.
224. Quintero, F., Mann, A.B., Pou, J., Lusquinos, F., and Riveiro, A. 2007. Rapid production of ultra-long amorphous ceramic nanofibers by laser spinning. *Appl. Phys. Lett.* 90: 153109.
225. Quintero, F., Pou, J., Lusquinos, F., and Riveiro, A. 2007. Experimental analysis of the production of micro- and nanofibres by laser spinning. *Appl. Surf. Sci.* 254: 1042–1047.
226. Quintero, F., Dieste, O., Pou, J., Lusquinos, F., and Riveiro, A. 2009. On the conditions to produce micro- and nanofibres by laser spinning. *J. Phys. D Appl. Phys.* 42: 065501.10.
227. Quintero, F., Pou, J., Comesana, R., Lusquinos, F., Riveiro, A., Mann, A.B., Hill, R.G., Wu, Z.Y., and Jones, J.R. 2009. Laser spinning of bioactive glass nanofibers. *Adv. Funct. Mater.* 19: 1–7.
228. Guarino, V., Causa, F., and Ambrosio, L. 2007. Bioactive scaffolds for bone and ligament tissue. *Expert Rev. Med. Dev.* 4(3): 405–418.
229. Misra, S.K., Mohn, D., Brunner, T.J., Stark, W.J., Philip, S.E., Roy I, Salih, V., Knowles, J.C., and Boccaccini, A.R. 2008. Comparison of nanoscale and microscale bioactive glass on the properties of P(3HB)/Bioglass(R) composites. *Biomaterials* 29(12): 1750–1761.

230. Misra, S.K., Ansari, T., Mohn, D., Valappil, S.P., Brunner, T.J, Stark, W.J., Roy, I. et al. 2010. Effect of nanoparticulate bioactive glass particles on bioactivity and cytocompatibility of poly(3-hydroxybutyrate) composites. *J. Roy. Soc. Interface* 7: 453–465.
231. Hong, Z., Reis, R.L., and Mano, J.F. 2008. Preparation and in vitro characterization of scaffolds of poly(l-lactic acid) containing bioactive glass ceramic nanoparticles. *Acta Biomater.* 4: 1297–1306.
232. El-Kady, A.M., Ali, A.F., and Farag, M.M. 2010. Development, characterization, and in vitro bioactivity studies of sol–gel bioactive glass/poly(L-lactide) nanocomposite scaffolds. *Mater. Sci. Eng. C* 30(1): 120–131.
233. Liu, A., Hong, Z., Zhuang, X., Chen, X., Cui, Y., Liu, Y., and Jing, X. 2008. Surface modification of bioactive glass nanoparticles and the mechanical and biological properties of poly(L-lactide) composites. *Acta Biomater.* 4: 1005–1015.
234. Peter, M., Kumar, P.T.S., Binulal, N.S., Nair, S.V., Tamura, H., and Jayakumar, R. 2009. Development of novel a-chitin/nanobioactive glass ceramic composite scaffolds for tissue engineering applications. *Carbohyd. Polym.* 78(4): 926–931.
235. Peter, M., Binulal, N.S., Soumya, S., Nair, S.V., Furuike, T., Tamura, H., and Jayakumar, R. 2010. Nanocomposite scaffolds of bioactive glass ceramic nanoparticles disseminated chitosan matrix for tissue engineering applications. *Carbohyd. Polym.* 79(2): 284–289.
236. Kim, H.W., Kim, H.E., and Knowles, J.C. 2006. Production and potential of bioactive glass nanofibers as a next-generation biomaterial. *Adv. Funct. Mater.* 16: 1529–1535.
237. Kim, H.W., Lee, H.H., and Chun, G.S. 2008. Bioactivity and osteoblast responses of novel biomedical nanocomposites of bioactive glass nanofiber filled poly(lactic acid). *J. Biomed. Mater. Res. A* 85A: 651–663.
238. Faraji, A.H. and Wipf, P. 2009. Nanoparticles in cellular drug delivery. *Bioorg. Med. Chem.* 17: 2950–2962.
239. Anker, N., Hall, W.P., Lyandres, O., Shah, N.C., Zhao, J., and Van Duyne, R.P. 2008. Biosensing with plasmonic nanosensors. *Nat. Mater.* 7: 442–453.
240. Jain, P.K., Huang, X.H., El-Sayed, I.H., and El-Sayed, M.A. 2008. Noble metals on the nanoscale: Optical and photothermal properties and some applications in imaging, sensing, biology, and medicine. *Acc. Chem. Res.* 41: 578–586.
241. Stewart, M.E., Anderton, C.R., Thompson, L.B., Maria, J., Gray, S.K., Rogers, J.A., and Nuzzo, R.G. 2008. Nanostructured plasmonic sensors. *Chem. Rev.* 108: 494–521.
242. Ghosh, P., Han, G., De, M., Kim, C.K., and Rotello, V.M. 2008. Gold nanoparticles in delivery applications. *Adv. Drug. Deliv. Rev.* 60: 1307–1315.
243. Boisselier, E. and Astruc, D. 2009. Gold nanoparticles in nanomedicine: Preparations, imaging, diagnostics, therapies and toxicity. *Chem. Soc. Rev.* 38: 1759–1782.
244. Chourpa, I., Lei, F.H., Dubois, P., Manfait, M., and Sockalingum, G.D. 2008. Intracellular applications of analytical SERS spectroscopy and multispectral imaging. *Chem. Soc. Rev.* 37: 993–1000.
245. Eustis, S. and El-Sayed, MA. 2006. Why gold nanoparticles are more precious than pretty gold: Noble metal surface plasmon resonance and its enhancement of the radiative and nonradiative properties of nanocrystals of different shapes. *Chem. Soc. Rev.* 35: 209–217.
246. Gao, J.H. and Xu, B. 2008. Applications of nanomaterials inside cells. *Nano Today* 4: 37–51.

247. Mayhew, T.M., Muhlfeld, C., Vanhecke, D., and Ochs, M. 2009. A review of recent methods for efficiently quantifying immunogold and other nanoparticles using TEM sections through cells, tissues and organs. *Ann. Anat.* 91: 153–170.
248. Nel, A.E., Madler, L., Velegol D, Xia, T., Hoek, E.M., Somasundaran, P., Klaessig, F., Castranova, V., and Thompson, M. 2009. Understanding biophysicochemical interactions at the nano-bio interface. *Nat. Mater.* 8: 543–557.
249. Sperling, R.A., Rivera, P., Zhang, F., Zanella, M., and Parak, W.J. 2008. Biological applications of gold nanoparticles. *Chem. Soc. Rev.* 37: 1896–1908.
250. Berners-Price, S.J., Norman, R.E., and Sadler, P.J. 1987. The autoxidation and proton dissociation constants of tertiary diphosphines: Relevance to biological activity. *J. Inorg. Biochem.* 31: 197–209.
251. Haiduc, I. and Silvestru, C. 1989. Rhodium, iridium, copper and gold antitumor organometallic compounds. *In Vivo* 3: 285–294.
252. Sonnichsen, C., Franzl, T., Wilk, T, von Plessen, G., Feldmann, J., Wilson, O., and Mulvaney, P. 2002. Drastic reduction of plasmon damping in gold nanorods. *Phys. Rev. Lett.* 88: 402.
253. Yguerabide, J. and Yguerabide, E.E. 1998. Light-scattering submicroscopic particles as highly fluorescent analogs and their use as tracer labels in clinical and biological applications. *Anal. Biochem.* 262: 137–156.
254. Gedanken, T. 2007. A microwave-assisted polyol method for the deposition of silver nanoparticles on silica spheres. *Nanotechnology* 18: 255601–255608.
255. Jin, Y., Wang, P., Yin, D., Liu, J., Qin, L., Yu, N., Xie, G., and Li, B. 2007. Gold nanoparticles prepared by sonochemical method in thiol-functionalized ionic liquid. *Colloid Surf. A Physicochem. Eng. Asp.* 302: 366–370.
256. Abyaneh, M.K., Paramanik, D., Varma, S., Gosavi, S.W., and Kulkarni, S.K. 2007. Formation of gold nanoparticles in polymethylmethacrylate by UV irradiation. *J. Phys. D Appl. Phys.* 40: 3771–3779.
257. Giorgetti, E., Giusti, A., Laza, S.C., Marsili, P., and Giammanco, F. 2007. Production of colloidal gold nanoparticles by picosecond laser ablation in liquids. *Phys. Status Solid A* 204: 1693–1698.
258. Nakamoto, M., Yamamoto, M., and Fukusumi, M. 2002. Thermolysis of gold(I) thiolate complexes producing novel gold nanoparticles passivated by alkyl groups. *Chem. Commun. (Camb)* 2002: 1622–1623.
259. Mandal, T.K., Fleming, M.S., and Walt, D.R. 2002. Preparation of polymer coated gold nanoparticles by surface-confined living radical polymerization at ambient temperature. *Nano Lett.* 2: 3–7.
260. Brust, M., Walker, M., Bethell, D., Schiffrin, D.J., and Whyman, R. 1994. Synthesis of thiol-derivatised gold nanoparticles in a two-phase liquid-liquid system. *J. Chem. Soc. Chem. Commun.* 7: 801–802.
261. Brust, M., Fink, J., Bethell, D., Schiffrin, E.J., and Kiely, C.J. 1995. Synthesis and reactions of functionalised gold nanoparticles. *J. Chem. Soc. Chem. Commun.* 21: 1655–1656.
262. Alivisatos, P. 2004. The use of nanocrystals in biological detection. *Nat. Biotechnol.* 22: 47–52.
263. Katzand, I. and Willner, E. 2004. Integrated nanoparticle-biomolecule hybrid systems: Synthesis, properties, and applications. *Angew. Chem. Int. Ed.* 43: 6042–6108.
264. Turkevich, J., Stevenson, P.C., and Hillier, J. 1951. A study of the nucleation and growth processes in the synthesis of colloidal gold. *Discuss Faraday Soc.* 11: 55–75.

265. Frens, G. 1973. Controlled nucleation for the regulation of the particle size in monodisperse gold suspensions. *Nature* 241: 20–22.
266. Martin, C.R. 1994. Nanomaterials—A membrane-based synthetic approach. *Science* 266: 1961–1966.
267. Van der Zande, B.M.I., Boehmer, M.R., and Fokkink, L.G.J. 1997. Aqueous gold Sols of rod-shaped particles. *J. Phys. Chem. B* 101: 852–854.
268. Loo, C., Lin, A., Hirsch, L., Lee, M.H., Barton, J., Halas, N., West, J., and Drezek, R. 2004. Nanoshell-enabled photonics-based imaging and therapy of cancer. *TCRT* 3: 33–40.
269. Sara, E.S., Jingyi, C., Yugang, S., Xianmao, L., Leslie, A., Larie, C., and Younan, X. 2008. Gold nanocages: Synthesis, properties and applications. *Acc. Chem. Res.* 41: 1587–1595.
270. Xia, Y., Li, W., Cobley, C.M., Chen, J., Xia, X., Zhang, Q., Yang, M., Cho, E,C., and Brown, P.K. 2011. Gold nanocages: From synthesis to theranostic applications. *Acc. Chem. Res.* 44: 914–924.
271. Nel, A., Xia, T., Madler, L., and Li, N. 2006. Toxic potential of materials at the nanolevel. *Science* 311: 622–627.
272. Evaldas, S., Nicklas, R.J., Gorm, D., Meredin, S., Ulla, V., Agnete, L., Wolfgang, K., and Håkan, W. 2009. Biodistribution of gold nanoparticles in mouse lung following intratracheal instillation. *Chem. Cent. J.* 3: 1–7.
273. Greish, K. J. 2007. Enhanced permeability and retention of macromolecular drugs in solid tumors: A royal gate for targeted anticancer nanomedicines. *Drug Target.* 15: 457–464.
274. Hawley, A.E., Illum, L., and Davis, S.S. 1997. Preparation of biodegradable, surface engineered PLGA nanospheres with enhanced lymphatic drainage and lymph node uptake. *Pharm. Res.* 14: 657–661.
275. Davis, M.E., Chen, Z., and Shin, D.M. 2008. Nanoparticle therapeutics: An emerging treatment modality for cancer. *Nat. Rev. Drug Discov.* 7: 771–782.
276. Yang, P.H., Sun, X., Chiu, J.F., Sun, H., and He, Q. 2005. Transferrin-mediated gold nanoparticle cellular uptake. *Bioconjug. Chem.* 16: 494–496.
277. Cai, W., Gao, T., Hong, H., and Sun, J. 2008. Applications of gold nanoparticles in cancer nanotechnology. *Nanotechnol. Sci. Appl.* 1: 17–32.
278. Durr, N.J., Larson, T., Smith, D.K., Korgel, B.A., Sokolov, K., and BenYakar, A. 2007. Two-photon luminescence imaging of cancer cells using molecularly targeted gold nanorods. *Nano Lett.* 7: 941–945.
279. Li, P.C., Wei, C.W., Liao, C.K., Chen, C.D., Pao, K.C., Wang, C.R., Wu, Y.N., and Shieh, D.B. 2007. Photoacoustic imaging of multiple targets using gold nanorods. *Trans. Ultrason. Ferroelectr. Freq. Control* 54: 1642–1647.
280. Oyelere, A.K., Chen, P.C., Huang, X., El-Sayed, I.H., El-Sayed, M.A. 2007. Peptide-conjugated gold nanorods for nuclear targeting. *Bioconjug. Chem.* 18: 1490–1497.
281. Gobin, A.M., Lee, M.H., Halas, N.J., James, W.D., Drezek, R.A., and West, J.L. 2007. Near-infrared resonant nanoshells for combined optical imaging and photothermal cancer therapy. *Nano Lett.* 7: 1929–1934.
282. Xiao, Y., Hong, H., and Matson, V.Z. 2012. Gold nanorods conjugated with doxorubicin and cRGD for combined anticancer drug delivery and PET imaging. *Theranostics* 2(8): 757–768.
283. Kim, C.S., Wilder, P.S., Ahn, Y.C., Liaw, L.H.L., Chen, Z., and Kwon, Y.J. 2010. Enhanced detection of early-stage oral cancer in vivo by optical coherence tomography using multimodal delivery of gold nanoparticles. *J. Biomed. Opt.* 1(1): 106–113.

284. Popovtzer, R., Agrawal, A., Kotov, N.A., Popovtzer, A., Balter, J., Carey, T.E., and Kopelman, R. 2008. Targeted gold nanoparticles enable molecular CT imaging of cancer. *Nano Lett.* 8: 4593–4596.
285. Diagaradjane, P., Shetty, A, Wang, J.C., Elliott, A.M., Schwartz, J., and Shentu, S. 2008. Modulation of in vivo tumor radiation response via gold nanoshell-mediated vascular-focused hyperthermia: Characterizing an integrated anti-hypoxic and localized vascular disrupting targeting strategy. *Nano Lett.* 8: 1492–1500.
286. Muller, J.H. 1950. Further progress in treatment of peritoneal carcinosis in ovarian cancer with artificial radioactivity; Au198. *Gynaecoligia* 129: 289–294.
287. Walton, R.J. and Sinclair, W.K. 1952. Intracavitary irradiation with radioactive colloidal gold in the palliative treatment of malignant pleural and peritoneal effusions. *Br. Med. Bull.* 8: 165–172.
288. Mackay, N.R. 1957. Radioactive colloidal gold in the treatment of pleural and peritoneal effusions of malignant origin: Review of 235 cases. *Lancet* 2: 761.
289. Yakimovich, N.O., Ezhevskii, A.A., Guseinov, D.V., Smirnova, L.A., Gracheva, T.A., and Klychkov, K.S. 2008. Antioxidant properties of gold nanoparticles studied by ESR spectroscopy. *Russ. Chem. Bull.* 57(3): 520–523.
290. Nie, Z., Liu, K.J., Zhong, C.J., Wang, L.F., Yang, Y., Tian, Q., and Liu, Y. 2007. Enhanced radical scavenging activity by antioxidant-functionalized gold nanoparticles: A novel inspiration for development of new artificial antioxidants. *Free Radic. Biol. Med.* 43: 1243–1254.
291. Yin, J.J., Lao, F., Meng, J., Fu, P.P., Zhao, Y., Xing, G., Gao, X., Sun, B., Wang, P.C., Chen, C., and Liang, X.J. 2008. Inhibition of tumor growth by endohedral metallofullerenol nanoparticles optimized as reactive oxygen species scavenger. *Mol. Pharmacol.* 74: 1132–1140.
292. Barath Mani Kanth, S., Kalishwaralal, K., Sriram, M., Pandian, S.R., Youn, H.S., and Eom, S. 2010. Anti-oxidant effect of gold nanoparticles restrains hyperglycemic conditions in diabetic mice. *J. Nanobiotechnol.* 8: 6.
293. De la Escosura-Muniz, A., Parolo, C., and Merkoci, A. 2010. Immunosensing using nanoparticles. *Mater. Today* 13: 17.
294. Liu, Y., Qin, Z.H., Wu, X.F., and Jiang, H. 2006. Immune-biosensor for aflatoxin B1 based bio-electrocatalytic reaction on micro-comb electrode. *Biochem. Eng. J.* 32: 211–217.
295. Bone, L., Vidal, J.C., Duato, P., and Castillo, J.R. 2010. Ochratoxin a nanostructured electrochemical immunosensors based on polyclonal antibodies and gold nanoparticles coupled to the antigen. *Anal. Methods* 2: 335–341.
296. Zhang, Y. and Zhuang, H.S. 2010. Amperometric Immunosensor based on layer-by-layer assembly of thiourea and nano-gold particles on gold electrode for determination of naphthalene. *Chinese J. Anal. Chem.* 38: 153.
297. Tang, L., Zeng, G.M., Shen, G.L., Li, Y.P., Zhang, Y., and Huang, D.L. 2008. Rapid detection of picloram in agricultural field samples using a disposable immunomembrane-based electrochemical sensor. *Environ. Sci. Technol.* 42: 1207.
298. Novoselov, K.S., Geim, A.K., Morozov, S.V., Jiang, D., Katsnelson, M.I., Grigorieva, I.V., Dubonos, S.V., and Firsov, A.A. 2005. Two-dimensional gas of massless Dirac fermions in grapheme. *Nature* 438: 197–200.
299. Nair, R.R., Blake, P., Grigorenko, A.N., Novoselov, K.S., Booth, T.J., Stauber, T., Peres, N.M.R., and Geim, A.K. 2008. Fine structure constant defines visual transparency of graphene. *Science* 320: 1308.

300. Lee, C., Wei, X., Kysar, J.W., and Hone, J. 2008. Measurement of the elastic properties and intrinsic strength of monolayer graphene. *Science* 321: 385–388.
301. Balandin, A.A., Ghosh, S., Bao, W., Calizo, I., Teweldebrhan, D., Miao, F., and Lau, C. 2008. Superior thermal conductivity of single-layer graphene. *Nano. Lett.* 8: 902–907.
302. Stoller, M.D., Park, S., Zhu, Y., An, J., and Ruoff, R.S. 2008. Graphene-based ultracapacitors. *Nano Lett.* 8: 3498–3502.
303. Sanchez, V.C., Jachak, A., Hurt, R.H., and Kane, A.B. 2011. Biological interactions of graphene-family nanomaterials: An interdisciplinary review. *Chem. Res. Toxicol.* 25: 15–34.
304. Hong, B.J., Compton, O.C., An, Z., Eryazici, I., and Nguyen, S.T. 2012. Successful stabilization of graphene oxide in electrolyte solutions: Enhancement of biofunctionalization and cellular uptake. *ACS Nano* 6: 63–73.
305. Artiles, M.S., Rout, C.S., and Fisher, T.S. 2011. Graphene-based hybrid materials and devices for biosensing. *Adv. Drug Deliv. Rev.* 63: 1352.
306. Rana, V.K., Choi, M.-C., Kong, J.-Y., Kim, G.Y., Kim, M.J., Kim, S.-H., Mishra, S., Singh, R.P., and Ha, C.-S. 2011. Synthesis and drug-delivery behavior of chitosan-functionalized graphene oxide hybrid nanosheets. *Macromol. Mater. Eng.* 296: 131.
307. Depan, D., Shah, J., and Misra, R.D.K. 2011. Controlled release of drug from folate-decorated and graphene mediated drug delivery system: Synthesis, loading efficiency, and drug release response. *Mater. Sci. Eng. C* 31: 1305–1312.
308. Liu, Z., Robinson, J.T., Sun, X., and Dai, H.J. 2008. PEGylated nanographene oxide for delivery of water-insoluble cancer drugs. *Am. Chem. Soc.* 130: 10876–10877.
309. Sun, X., Liu, Z., Welsher, K., Robinson, J., Goodwin, A., Zaric, S., and Dai, H. 2008. Nano-graphene oxide for cellular imaging and drug delivery. *Nano Res.* 1: 203–212.
310. Yang, X., Zhang, X., Liu, Z., Ma, Y., Huang, Y., and Chen, Y.J. 2008. High-efficiency loading and controlled release of doxorubicin hydrochloride on graphene oxide. *Phys. Chem. C* 112: 17554–17558.
311. Jung, I., Dikin, D., Park, S., Cai, W., Mielke, S.L., and Ruoff, R.S.J. 2008. Effect of water vapor on electrical properties of individual reduced graphene oxide sheets. *Phys. Chem. C.* 112: 20264–20268.
312. Fowler, J.D., Allen, M.J., Tung, V.C., Yang, Y., Kaner, R.B., and Weiller, B.H. 2009. Practical chemical sensors from chemically derived graphene. *ACS Nano* 3: 301–306.
313. Lu, G., Ocola, L.E., and Chen, J. 2009. Gas detection using low-temperature reduced graphene oxide sheets. *Appl. Phys. Lett.* 94: 083111–083113.
314. Hong, H., Yang, K., Zhang, Y., Engle, J.W., Feng, L., Yang, Y., Nayak, T.R. et al. 2012. In vivo targeting and imaging of tumor vasculature with radiolabeled, antibody-conjugated nanographene. *ACS Nano* 6: 2361–2370.
315. Lee, W.C., Lim, C.H.Y.X., Shi, H., Tang, L.A.L., Wang, Y., Lim, C.T., and Loh, K.P. 2011. Origin of enhanced stem cell growth and differentiation on graphene and graphene oxide. *ACS Nano* 5: 7334–7341.
316. Nayak, T.R., Andersen, H., Makam, V.S., Khaw, C., Bae, S., Xu, X., Ee, P.-L.R. et al. 2011. Graphene for controlled and accelerated osteogenic differentiation of human mesenchymal stem cells. *ACS Nano* 5: 4670–4678.
317. Grill, W.M., Norman, S.E., and Bellamkonda, R.V. 2009. Implanted neural interfaces: Biochallenges and engineered solutions. *Annu. Rev. Biomed. Eng.* 11: 1–24.
318. Cogan, S.F. 2008. Neural stimulation and recording electrodes. *Annu. Rev. Biomed. Eng.* 10: 275–309.
319. Wallace, G.G., Moulton, S.E., and Clark, G.M. 2009. Electrode-cellular interface. *Science* 324: 185–186.

320. Yang, X., Zhang, X., Ma, Y., Huang, Y., Wang, Y., and Chen, Y.J. 2009. Superparamagnetic graphene oxide–Fe_3O_4 nanoparticles hybrid for controlled targeted drug carriers. *Mater. Chem.* 19: 2710–2714.
321. Cong, H.P., He, J.J., Lu, Y., and Yu, S.H. 2010. Water-soluble magnetic-functionalized reduced graphene oxide sheets: In situ synthesis and magnetic resonance imaging applications. *Small* 6: 169–173.
322. Liu, K., Zhang, J.J., Cheng, F.F., Zheng, T.T., Wang, C., and Zhu, J.J.J. 2011. Green and facile synthesis of highly biocompatible graphene nanosheets and its application for cellular imaging and drug delivery. *Mater. Chem.* 21: 12034–12040.
323. Yang, K., Zhang, S., Zhang, G., Sun, X., Lee, S.-T., and Liu, Z. 2010. Graphene in mice: Ultrahigh in vivo tumor uptake and efficient photothermal therapy. *Nano Lett.* 10: 3318.
324. Markovic, Z.M., Harhaji-Trajkovic, L.M., Todorovic-Markovic, B.M., Kepić, D.P., Arsikin, K.M., Jovanović, S.P., Pantovic, A.C., Dramićanin, M.D., and Trajkovic, V.S. 2011. In vitro comparison of the photothermal anticancer activity of graphene nanoparticles and carbon nanotubes. *Biomaterials* 32: 1121–1129.
325. Liu, Z., Tabakman, S., Welsher, K., and Dai, H. 2009. Carbon nanotubes in biology and medicine: In vitro and in vivo detection, imaging and drug delivery. *Nano Res* 2: 85–120.
326. Chakravarty, P., Marches, R., Zimmerman, N.S., Swafford, A.D.E., Bajaj, P., Musselman, I.H., Pantano, P., Draper, R.K., and Vitetta, E.S. 2008. Thermal ablation of tumor cells with antibody-functionalized single-walled carbon nanotubes. *Proc. Natl. Acad. Sci. USA* 105: 8697–8702.
327. Xiao, Y., Gao, X., Taratula, O., Treado, S., Urbas, A., Holbrook, R.D., Cavicchi, R.E., Avedisian, C.T., Mitra, S., Savla, R., Wagner, P.W., Srivastava, S., and He, H. 2009. Anti-HER2 IgY antibody-functionalized single-walled carbon nanotubes for detection and selective destruction of breast cancer cells. *BMC Cancer* 9: 351.
328. Kam, N.W.S., O'Connell, M., Wisdom, J.A., and Dai, H. 2005. Carbon nanotubes as multifunctional biological transporters and near-infrared agents for selective cancer cell destruction. *Proc. Natl. Acad. Sci. USA* 102: 11600–11605.
329. De la Zerda, A., Zavaleta, C., Keren, S., Vaithilingam, S., Bodapati, S., Liu, Z., Khuri-Yakub, B.T., Chen, X., Dai, H., and Gambhir, S.S. 2008. Carbon nanotubes as photoacoustic molecular imaging agents in living mice. *Nat. Nanotechnol.* 3(9): 557–562.
330. Heller, D.A., Baik, S., Eurell, T.E., and Strano, M.S. 2005. Single-walled carbon nanotube spectroscopy in live cells: Towards long-term labels and optical sensors. *Adv. Mater.* 17: 2793–2799.
331. Rao, A.M., Richter, E., Bandow, S., Chase, B., Eklund, P.C., Williams, K.A., Fang, S., Subbaswamy, K.R., Menon, M., Thess, A., Smalley, R.E., Dresselhaus, G., and Dresselhaus, M.S. 1997. Diameter-selective Raman scattering from vibrational modes in carbon nanotubes. *Science* 275: 187–190.
332. Liu, Z., Sun, X., Nakayama-Ratchford, N., and Dai, H. 2007. Supramolecular chemistry on water-soluble carbon nanotubes for drug loading and delivery. *ACS Nano* 1: 50–56.
333. Liu, Y., Wu, D.C., Zhang, W.D., Jiang, X., He, C.B., Chung, T.S., Goh, S.H., and Leong, K.W. 2005. Polyethylenimine-grafted multiwalled carbon nanotubes for secure noncovalent immobilization and efficient delivery of DNA. *Angew. Chem. Int. Ed.* 44: 4782–4785.

334. Kam, N.W.S. and Dai, H. 2005. Carbon nanotubes as intracellular protein transporters: Generality and biological functionality. *J. Am. Chem. Soc.* 127: 6021–6026.
335. Yinghuai, Z., Peng, A.T., Carpenter, K., Maguire, J.A., Hosmane, N.S., and Takagaki, M. 2005. Substituted carborane-appended water-soluble single-wall carbon nanotubes: New approach to boron neutron capture therapy drug delivery. *J. Am. Chem. Soc.* 127: 9875–9880.
336. Heister, E., Neves, V., Tilmaciu, C., Lipert, K., Beltran, V.S., Coley, H.M., Silva, S.R.P., and McFadden, J. 2009. Triple functionalisation of single-walled carbon nanotubes with doxorubicin, a monoclonal antibody, and a fluorescent marker for targeted cancer therapy. *Carbon* 47: 2152–2160.
337. Zhang, X., Meng, L., Lu, Q., Fei, Z., and Dyson, P.J. 2009. Targeted delivery and controlled release of doxorubicin to cancer cells using modified single wall carbon nanotubes. *Biomaterial* 30: 6041–6047.
338. Liu, Z., Fan, A.C., Rakhra, K., Sherlock, S., Goodwin, A., Chen, X., Yang, Q., Felsher, D.W., and Dai, H. 2009. Supramolecular stacking of doxorubicin on carbon nanotubes for in vivo cancer therapy. *Angew. Chem. Int. Ed.* 48: 7668–7672.
339. Ali-Boucetta, H., Al-Jamal, K.T., McCarthy, D., Prato, M., Bianco, A., and Kostarelos, K. 2008. Multiwalled carbon nanotube-doxorubicin supramolecular complexes for cancer therapeutics. *Chem. Commun.* 8: 459–461.
340. Park, S., Yang, H.S., Kim, D., Jo, K., and Jon, S. 2008. Rational design of amphiphilic polymers to make carbon nanotubes waterdispersible, anti-biofouling, and functionalizable. *Chem. Commun.* 25: 2876–2878.
341. Li, R., Wu, R., Zhao, L., Hu, Z., Guo, S., Pan, X., and Zou, H. 2011. Folate and iron difunctionalized multiwall carbon nanotubes as dual-targeted drug nanocarrier to cancer cells. *Carbon* 49: 1797–1805.
342. Chen, Z., Pierre, D., He, H., Tan, S., Pham-Huy, C., Hong, H., and Huang, J. 2011. Adsorption behavior of epirubicin hydrochloride on carboxylated carbon nanotubes. *Int. J. Pharm.* 405: 153–161.
343. Tripisciano, C., Kraemer, K., Taylor, A., and Borowiak-Palen, E. 2009. Single-wall carbon nanotubes based anticancer drug delivery system. *Chem. Phys. Lett.* 478: 200–205.
344. Bhirde, A.A., Patel, V., Gavard, J., Zhang, G., Sousa, A.A., Masedunskas, A., Leapman, R.D., Weigert, R., Gutkind, J.S., and Rusling, J.F. 2009. Targeted killing of cancer cells in vivo and in vitro with EGF-directed carbon nanotube-based drug delivery. *ACS Nano* 3(2): 307–316.
345. Feazell, R.P., Nakayama-Ratchford, N., Dai, H., and Lippard, S.J. 2007. Soluble single-walled carbon nanotubes as longboat delivery systems for platinum(IV) anticancer drug design. *J. Am. Chem. Soc.* 129: 8438–8439.
346. Vukovic, G.D., Tomic, S.Z., Marinkovic, A.D., Radmilovic, V., Uskokovic, P.S., and Colic, M. 2011. The response of peritoneal macrophages to dapsone covalently attached on the surface of carbon nanotubes. *Carbon* 48: 3066–3078.
347. Naficy, S., Razal, J.M., Spinks, G.M., and Wallace, G.G. 2009. Modulated release of dexamethasone from chitosan-carbon nanotube films. *Sens. Actuators A* 155: 120–124.
348. Im, J.S., Bai, B.C., and Lee, Y.S. 2010. The effect of carbon nanotubes on drug delivery in an electro-sensitive transdermal drug delivery system. *Biomaterial* 31: 1414–1419.

349. Wu, W., Wieckowski, S., Pastorin, G., Benincasa, M., Klumpp, C., Briand, J.P., Gennaro, R., Prato, M., and Bianco, A. 2005. Targeted delivery of amphotericin B to cells by using functionalized carbon nanotubes. *Angew. Chem. Int. Ed.* 44: 6358–6362.
350. Jia, N., Lian, Q., Shen, H., Wang, C., Li, X., and Yang, Z. 2007. Intracellular delivery of quantum dots tagged antisense oligodeoxynucleotides by functionalized multiwalled carbon nanotubes. *Nano Lett.* 7: 2976–2980.
351. Cai, D., Mataraza, J.M., Qin, Z.H., Huang, Z., Huang, J., Chiles, T.C., Carnahan, D., Kempa, K., and Ren, Z. 2005. Highly efficient molecular delivery into mammalian cells using carbon nanotube spearing. *Nat. Methods* 2: 449–454.
352. Gao, L., Nie, L., Wang, T., Qin, Y., Guo, Z., Yang, D., and Yan, X. 2006. Carbon nanotube delivery of the GFP gene into mammalian cells. *ChemBioChem.* 7: 239–242.
353. Kam, N.W.S., Liu, Z., and Dai, H. 2006. Carbon nanotubes as intracellular transporters for proteins and DNA: An investigation of the uptake mechanism and pathway. *Angew. Chem. Int. Ed.* 45(4): 577–581.
354. Ahmed, M., Jiang, X., Deng, Z., and Narain, R. 2009. Cationic glycofunctionalized single-walled carbon nanotubes as efficient gene delivery vehicles. *Bioconjugate Chem.* 20: 2017–2022.
355. Hazani, M., Naaman, R., Hennrich, F., and Kappes, M.M. 2003. Confocal fluorescence imaging of DNA-functionalized carbon nanotubes. *Nano Lett.* 3: 153–155.
356. Awasthi, K., Singh, D.P., Singh, S., Dash, D., and Srivastava, O.N. 2009. Attachment of biomolecules (protein and DNA) to aminofunctionalized carbon nanotubes. *New Carbon Mater.* 24: 301–306.
357. Liu, Y., Wu, D.C., Zhang, W.D., Jiang, X., He, C.B., Chung, T.S., Goh, S.H., and Leong, K.W. 2005. Polyethylenimine-grafted multiwalled carbon nanotubes for secure noncovalent immobilization and efficient delivery of DNA. *Angew. Chem. Int. Ed.* 44: 4782–4785.
358. Wua, J., Paudelb, K.S., Strasingerb, C., Hammellb, D., Stinchcombb, A.L., and Bruce, J.H. 2010. Programmable transdermal drug delivery of nicotine using carbon nanotube membranes. *PNAS* 107: 11698–11702.
359. Sun, L. and Crooks, R.M. 2000. Single carbon nanotube membranes: A well-defined model for studying mass transport through nanoporous materials. *J. Am. Chem. Soc.* 122: 12340–12345.
360. Mahar, B., Laslau, C., Yip, R., and Sun, Y. 2007. Development of carbon nanotube-based sensors—A review. *IEEE Sens. J.* 7: 266–284.
361. Sepúlvedaa, A.T., Fachinb, F., Guzmán de Villoriab, R., Wardleb, B.L., Vianaa, J.C., Pontesa, A.J., and Rochaa, L.A. 2011. Nanocomposite flexible pressure sensor for biomedical applications. *Procedia Eng.* 25: 140–143.
362. Senturia, S.D. 2000. *Microsystem Design*. Kluwer Academic Publishers, New York.
363. Vashist, S.K., Zheng, D., Pastorin, G., AlRubeaan, K., Luong, J.H.T., and Sheu, F.-S. 2011. Delivery of drugs and biomolecules using carbon nanotubes. *Carbon* 49: 4077–4097.

7

Challenges in Processing of Nanocomposite Membranes

7.1 Introduction

Nanocomposite materials have the potential to redefine the field of traditional composite materials in terms of both performance and potential applications. Polymer nanocomposites have tremendous market potential both as replacement for current composites and in the creation of new markets through their outstanding properties. Availability of nanomaterials, integration of nanomaterials into membrane systems, and societal implications because of health and environment risks posed by nanomaterials are the key challenges to the development of integrated nano-based membrane systems.[1–16] Development of the process-manufacturing technologies in terms of quantity and quality for commercialization is one of the biggest challenges. Despite the bright future of nanocomposites, there are a few issues that warrant concern about the mass commercialization of the nano-based systems. There are four main issues dealing with the production and use of nanocomposites:

- Exfoliation
- Orientation
- Compatibility
- Reaggregation

When using clay fillers, it is necessary to separate the particles into the right shape and layer structure which is called *exfoliation*. They need to be about 1 nm thin and 500 nm wide for achieving optimal gas permeability without affecting the optical quality. Particle orientation also has an effect on the success of a nanocomposite. Nanoparticles need to be dispersed throughout polymer so that they are parallel to the material's surface. This position ensures a maximum *torturous path* for the gases when passing through the polymer. Compatibility between the nanofillers and the polymer substrate may cause issues as well, depending on how they interact with each other. Certain nanofillers need to be prepared so that they can perform well with the substrate. Another concern is during the processing stage. There is a possibility of reaggregation where the particles clump together. At the same time, if it is subjected to force, there is a possibility of agglomerated nanoparticles getting split. Therefore, premature failure takes place in the final product. The alignment of nanoparticles in the composite matrix is critical to maximize unidirectional properties such as strength and toughness. As in the case for traditional composites, it is even more challenging to determine the strength, composition, and functionality of the interfacial region. In addition to the composite integrity, the nature of the nanoparticle is also critical for property improvement. It is observed that single-walled nanotubes (SWNTs) are relatively defect-free whereas, multiwalled nanotubes (MWNTs) typically have more defects, such as topological defects (pentagon, heptagon) and structural defects (discontinuous, cone-shaped walls, or bamboo structure). To improve dispersion and compatibility in polymer matrices, nanotube is functionalized. There are still some concerns such as whether functionalization of a nanotube will affect the properties to improve the final product. Scale-up is needed to produce large quantities of nanomaterials for manufacturing purposes. There is still a lack of real-time characterization methods, instrumentation, tools, as well as a lack of affordable infrastructure (facilities, equipment, design tools, skilled personnel). The mechanical and dispersion properties, and the alignment of nanotubes are mainly involved in enhancing the properties of nanotube/polymer composites. However, it is also hard to achieve this without a good interfacial bonding between nanotubes and matrix. It is yet to be established: (1) whether the chemical bonding between nanotubes and matrix exists or not and (2) do the nanotubes still maintain their extraordinary mechanical, electrical, and thermal properties if chemical bonding exists between the nanotube and the matrix?

7.2 Material Selection

Proper material selection for both the matrix and the dispersed phase is fundamentally important in the development of nanocomposite membrane. Polymer properties as well as inorganic phase properties can affect the mixed matrix membrane morphology and separation performance.[17] Usually highly selective polymers can result in mixed matrix membranes with better separation performance.[18,19] Therefore, glassy polymers with superior gas selectivity are preferred to highly permeable but poorly selective rubbery polymers.[20,21] Although glassy polymers are better than rubbery polymers because of their rigid structure, adhesion between the polymer phase and the external surface of the particles is a major problem when glassy polymers are used in the preparation of mixed matrix membranes. In these cases, weak organic–inorganic interaction causes voids to form in the polymer–filler interface.[22–24] Therefore, in the selection of the matrix phase, gas/liquid separation properties and adhesion between the two phases are quite challenging. The dispersed inorganic phase as well as the continuous phase can affect mixed matrix membrane separation properties and morphology. As mentioned earlier, porous and nonporous fillers are the two major inorganic phase materials that have been used for nanocomposite membrane preparation.

When a porous material is used as a filler in the polymer matrix, its pore size distribution, surface chemistry, and functional groups must be consistent with the gas molecule pairs. For example, activated carbon is suitable for carbon dioxide/methane separation because it has a higher adsorption selectivity for CO_2 (polar gas) than for CH_4 (nonpolar compound), but this filler is not suitable for oxygen/nitrogen separation.

By contrast, because the effect of the nonporous material on nanocomposite membrane separation potential is different from porous inorganic materials with sieving function, interaction between polymer chain segments and nanofillers as well as functional groups on the surface of the inorganic phase must be evaluated when these materials are added to a polymer matrix. For instance, adding silica to a polyimide matrix can disrupt polymer chain packing and thus increase the oxygen and nitrogen permeation rates.[25] However, adding TiO_2 to the polyimide matrix can increase the CO_2/CH_4 and H_2/CH_4 selectivity because interactions of CO_2 and H_2 with TiO_2 are stronger than TiO_2–CH_4 interactions.[26]

7.3 Interface Defects

The performance of composite materials is largely governed by how intimately the filler material interacts with the surrounding polymer matrix. Nanocomposites are no different in this regard. The high surface area of nanoparticles leads to significant agglomeration, which is difficult to overcome with traditional composite processing methods. The addition of surfactants to composite materials significantly increases particle dispersion but also diminishes material properties. Usually, because of the difference between the properties of polymer and inorganic phases and the strong aggregation tendency of the fillers, making an ideal nanocomposite membrane, that is, a mixed matrix with no defects in the polymer–particle interface, is very difficult. Weak polymer–particle adhesion results in the formation of the defects at the polymer–inorganic interface. These organic–inorganic interface defects can affect the overall membrane properties. Interface defects can be classified into the following major categories[27]:

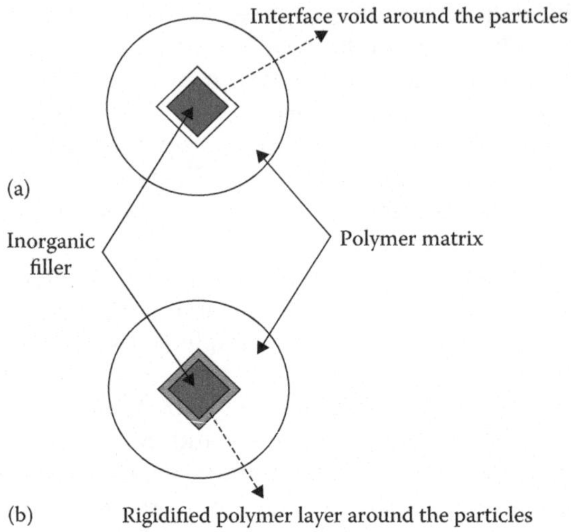

FIGURE 7.1
Interface defects at polymer–particle interface: (a) interface void around the particle, and (b) rigidified polymer layer around the particle.

1. Interface voids or sieves-in-a-cage
2. Rigid polymer layer around the particles
3. Pore blockage

Stress arising during membrane making due to solvent evaporation leads to rigidified polymer layer formation and interface void formation. Rigidified polymer layer formation near the particle is due to uniform stress around the particles. In this case, polymer chain mobility in the vicinity of particle surface is less than polymer bulk. If the stress directions are not uniform around the inorganic particles, interface voids will be formed in the particle–polymer interface. Figure 7.1 shows the interface voids and the rigidified polymer layer in the polymer–particle interface.[28]

The interface defects occur due to nucleation of nonsolvent and/or polymer lean phase around the inorganic phase during the phase separation process. For successful mixed matrix asymmetric membranes, it appears necessary that nucleation of solvents and nonsolvents at the nanoparticle (zeolite) surface be restricted. One such approach is via increasing the hydrophobicity of the zeolite surface by capping surface hydroxyls with hydrophobic organic molecules. Poor polymer–particle adhesion, polymer-packing disruption in the vicinity of the inorganic phase, repulsive force between continuous and dispersed phases, different thermal expansion coefficients for polymer and particles, and the effects of an elongation stress during fiber spinning are the other causes for interface voids formation.[29,30] The interface defects may take place in the porous and impermeable nonporous particle interface when fillers are added to the polymer matrix in mixed matrix membranes. Pore blockage is unique to porous fillers. In this condition, particle pores are clogged with a sorbent, a solvent, a contaminant, or a minor component in a feed gas or polymer chains, before, during, or after membrane preparation.[31–33] Depending on the pore size of inorganic fillers, one can classify them into total and partial pore blockages. It is obvious that when inorganic filler pores are plugged completely, the gas/water molecules cannot pass through the particle pores and the porous inorganic particles can act as an impermeable filler.[34] Whereas when pore blockage occurs partially, depending on the gas molecule dimension and blockage degree, penetrants of interest enter or pass through the pores more slowly than usual.

As mentioned above, when polymer–particle interfacial adhesion is good, a reduction in free volume occurs near the filler surface, known as polymer rigidification because of uniform stresses that arise during membrane formation.[35] In this case, the rigidified polymer layer around the particles has lower polymer chain mobility than the bulk polymer, and thus, diffusive selectivity in this region is higher than the bulk polymer, whereas permeability is lower. Therefore, lower permeability and improved selectivity in the presence of the rigidified polymer layer can be expected in the resultant mixed matrix membrane.[36,37] It is reported that permeability reduction in the rigidified region is one-third or one-fourth of the bulk polymer.[27]

Usually, for confirmation of these phenomena in the mixed matrix, one can use the glass transition temperature (T_g) analysis. If the glass transition temperature of the mixed matrix membrane is higher than that of the pure polymer or if a second higher T_g is observed, polymer chain rigidification has occurred. However, nonselective interfacial voids may be formed because of poor adhesion at the organic–inorganic interface.[38,39] These voids can reduce the apparent selectivity of the mixed matrix membrane and increase the permeability.[40–42] In this case, gas molecules pass through the less resistant interfacial voids instead of passing through particle pores, and thus, the inorganic phase becomes unusable.[43,44] When the void dimension is in the molecular range (about 5 Å), permeability increment and selectivity decrement below that of the pure polymer are observed. In this case, morphology is often termed a leaky interface for convenience.

The effect of pore blockage on the mixed matrix membrane comprising porous fillers is dependent on the degree of pore blockage as well as the molecular diameter of gases. Pore blockage of porous fillers is always accompanied by a gas permeability decrease, although its effect on selectivity depends on the inorganic filler which is used as the dispersed phase. In this case, pore dimension after blockage must be compared with molecular diameter of tested gases. If the pore size of the porous filler before blockage is in the range of molecular diameter of gases after blockage, it cannot allow both gases to pass through the pore, and thus, pore blockage greatly decreases the selectivity. When the original pore size of the fillers is larger than the molecular diameter of tested gases, pore blockage may increase the selectivity.

7.4 Effect of Particle Size and Size Distribution

In addition to the interfacial defects, particle distribution in the matrix continuous phase is another important factor which affects the mixed matrix dense membrane performance. For industrial applications, asymmetric hollow-fiber membranes are normally preferred compared to flat sheet dense mixed matrix membranes due to the following properties:

- High productivity
- High mass transfer area per unit volume
- Convenient module fabrication

An integrally skinned asymmetric membrane has a very thin and dense skin layer (0.1–1 μm) with a thick and highly porous sublayer (100–200 μm with an average void size ranging from 0.01 to 1 μm), where both layers are composed of the same material and formed in a single operation.[45–47] In this kind of membrane, the skin layer acts as the selective barrier and the sublayer serves only as a mechanical support for the skin with negligible effects on separation. Therefore, inorganic fillers must lie inside the selective skin layer during the asymmetric mixed matrix membrane preparation. In other words, dispersion of particles in the dense skin region of the membranes, organic–inorganic adhesion, and skin layer pinhole and polymer-phase defects are the important parameters that must be considered in the asymmetric mixed matrix membrane development. The size of the particles in use must be in the submicron range in order to fit inside the skin layer. That is, using submicron size particles is essential in the formation of thin and ultra-thin asymmetric mixed matrix membranes. In addition, smaller particles provide more polymer/particle interfacial area and enhance polymer–filler interface contact.[48,49] The particle distribution in the asymmetric hollow-fiber mixed matrix membrane is related to shear stress in the dope induced by the spinneret during hollow-fiber fabrication. Near the spinneret die exit, particles tend to move to the central part of flow with lower energy requirement and lower shear rates. Therefore, the particles will be concentrated in the middle part of the resultant nascent fiber wall. Because the particles do not lie in the skin layer of the nascent hollow fiber, they will become unusable, as was discussed earlier. The elongation induced by the gravity of the nascent hollow fibers while they are travelling through the air gap can change the particle distribution profile and force the particles to migrate outward to the outer skin and accumulate there. When the draw ratio is increased, the particle distribution changes considerably from a convex shape to a flat shape and then to a concave shape at high elongation draw ratios. That is, a high elongation draw ratio helps particles to move radially to the bore and shell sides, and forces them to concentrate across the fiber walls. Therefore, the particle number per unit outer surface area increases when the elongation draw ratio is increased. In addition, the elongation draw ratio increment can eliminate the macrovoid formation during dry-wet spinning because of the following reasons: It may make polymer chains parallel and retard the penetration of external coagulants. Radial outflow of solvents to both the inner and external coagulant sides, induced by fiber elongation, hinders the capillary intrusion of coagulants and eliminates the formation of macrovoids. Therefore, air gap and takeup speed are critical parameters in attempts aimed at fabricating defect-free mixed matrix membranes. In sum, for fabricating defect-free hollow-fiber mixed matrix membranes, particle distribution in the skin layer and macrovoid elimination can be controlled with an optimum elongation draw ratio. Polymer–particle interfacial defects can be eliminated by employing the above-mentioned methods.

7.5 Challenges in CNT–Polymer Nanocomposite Membrane

Carbon nanotubes (CNTs) can serve as a mere reinforcement material as well as a material to serve as membrane channels in a nanocomposite membrane system. There are four main aspects of effective reinforcement in CNT composites that are mainly responsible to lead to high-strength composites. Nonetheless, the issues pertaining to the formation of a membrane structure are quite challenging. Good deagglomeration and dispersion of nanotubes within matrix are essential for high-quality nanocomposites.

CNTs are more efficient as fillers compared to conventional microparticles. Lower amount of nanotubes (frequently below 1%) is needed to strengthen, stiffen, and impart electrical conductivity to the nanocomposite membrane. Such very low filling is impractical in the case of conventional fibers/fillers giving very weak effects or even deforcement instead of reinforcement. However, it may not be possible (and not feasible) to use high CNT loadings because it will lead to the onset of agglomeration; moreover, stiffness and strength will not be enhanced.

Due to experimental difficulties, not many measurements of the interfacial shear strength have been made. The interfacial shear strength (IFSS) for noncovalently bonded phases lies in the region of 50–100 MPa, which indicates rather good stress transfer. Nucleation of transcrystallinity by nanofibers increases the IFSS.

Alignment of nanotubes in membrane is quite challenging. An additional effort is required to align the nanotubes, for example, nanotubes can become predominantly aligned in a polymer subjected to elongational flow (like in fiber spinning). External fields, such as magnetic field, can contribute in aligning the nanotubes. Alignment of nanotubes can further help in enhancing the mechanical properties in the direction of alignment. From conventional considerations for microfibers, a fivefold increase in stiffness in the alignment direction compared to stiffness of 3D randomly oriented short fiber composite is found.

Some of the critical issues[50] that are encountered in the processing of CNT–polymer composites are uniform dispersion[51–53] and alignment of the carbon nanotubes[53,54] in the polymer matrix and functionalization of CNT for the nanotube/matrix interfacial bonding. Different techniques have been used to disperse nanotubes in polymer matrices, including solution chemistry to functionalize the nanotube surface,[55–58] the use of polymers to coat the nanotube surface,[59] *in situ* polymerization of the nanocomposite,[60,61] ultrasonic dispersion in solution,[62] melt processing,[63–66] the use of surfactants,[67,68] electrospinning,[69] electrode chemistry,[70] and gelation/crystallization.[71] Nanotube surface modifications include plasma treatment or chemical oxidation to attach the functional groups. The nanotubes should bond better to the matrix and overcome the van der Waals interactions between nanotubes.[72] Good dispersion alone shows moderate property improvements, but nanotube alignment, or orientation, leads to further improvements. Using melt compounding followed by melt drawing gives a significant increase in mechanical properties.[73] In addition to mechanical properties, the inherent conductive nature of carbon nanotubes has been utilized to induce alignment. For example, the application of a magnetic field to a nanocomposite sample during or after processing has shown nanotube alignment.[74–79]

7.6 Interfacial Defects

Polymer–inorganic incompatibility, poor polymer–particle adhesion, polymer packing disruption in the vicinity of the inorganic phase, stresses that arise due to solvent evaporation during membrane formation, repulsive force between continuous and dispersed phase, and different thermal expansion coefficients for polymer and particle cause different interfacial defects and nonideal morphologies in mixed matrix materials. Preventing agglomeration resulting from intrinsic filler attraction is generally one of the major challenges in nanocomposite preparation/processing. Functionalization and organo-modification are efficient ways to achieve deagglomeration/dispersibility. Dispersibility is a function of

- Selection of solvents
- Surfactant
- Sonication
- Mixing

Whereas solvents, surfactants, and sonication refer to solution processing of nanocomposites, the mixing refers to melt processing.

For avoiding these interfacial defects and fabricating defect-free mixed matrix membranes, the following methods have been extensively tried:

- Casting at a temperature above the polymer glass transition temperature, or using a polymer with low glass transition temperature and making the membrane close to the glass transition temperature of the polymer used as the matrix.
- Annealing already formed mixed matrix membranes above the glass transition temperature. Annealing the defective membranes does not necessarily lead to any significant improvement in the membrane morphology.
- Incorporation of a plasticizer into the polymer solution, such as Fyroflex RDP, dibutyl phthalate, and 4-hydroxy benzophenone to decrease the polymer glass transition temperature and thus to maintain polymer chain mobility and flexibility during membrane making.
- Surface modification of particles using coupling agents.[80–82] Thin-film coatings on nanoparticles themselves can be used to improve dispersion into polymeric matrices. If functionally reactive, these coatings can further be used to covalently bond the nanoparticles into the polymer matrix. For example, such a technology can be used to coat carbon nanotubes with a functional thin film, which will allow for rapid dispersion and bonding into plastics. Carbon nanotubes are yet to achieve their full potential due to dispersion problems.
- Preparation of membranes using melt processing. This method is used to maintain polymer flexibility during membrane preparation. Melt processing is yet to establish itself as a feasible process for industrially relevant membranes.[83]
- Using some copolymers such as polyimide siloxane. In this case, rubbery sections (e.g., siloxane segments) enhance the interfacial polymer–particle contacts, and thus void-free membranes can be achieved.
- Modification of particles using a sizing technique. In this method, physical deposition of the sizing agent such as polyvinylpyrrolidone onto the surface of particles is employed by treating the particles in a sizing bath solution.[84] Sizing technique, which is a surface coating approach, deserves thorough consideration and can be incorporated into mixed matrix membrane development.
- Coating the surface of the inorganic particles with a dilute polymer dope prior to dispersion in the bulk polymer is known as priming method. This method can reduce stress at the polymer–particle interface. This technique minimizes agglomeration of the particles and

promotes interaction between the bulk polymer and the polymer primed particles, thereby minimizing defective interfaces.

- Adding a low-molecular-weight additive to the membrane formulation as a third component[85] for linking the polymer chain to the inorganic particle crystal.

7.7 Challenges for Integrated Systems Nanomanufacturing

Nanomanufacturing[86] encompasses all processes aimed toward the building of nanoscale (in 1D, 2D, or 3D) structures, features, devices, and systems suitable for integration across higher dimensional scales (micro-, meso-, and macroscale) to provide functional products and useful services. It includes both bottom-up and top-down processes. Table 7.1 gives examples of a few nanomanufacturing processes. For application of nanomanufacturing to critical national needs in energy, civil infrastructure, and reduction of greenhouse gases, nanomanufacturing methods must provide a paradigm shift in the way these products are designed and manufactured. They should ultimately enhance the performance such that the properties demonstrated at the nanoscale are truly scaled to large volume manufacturing lines and processes. Furthermore, nanomanufacturing holds the exceptional promise of creating completely new markets for the nascent manufacturing industries of nanoparticles, nanostructures, and nanodevices.[87,88]

7.7.1 Nanomaterials and Nanomanufacturing

Despite the rapid growth in the field of synthesis of nanomaterials, a thorough scientific understanding of how to optimize these materials is still lacking. Key short-term objectives include understanding the deformation

TABLE 7.1

Examples of Nanomanufacturing Processes

Bottom-up Processes	Top-down Processes
Contact printing, imprinting	E-beam, ion beam lithography
Template growth	Scanning probe lithography
Spinodal wetting/dewetting	Optical near-field lithography
Laser trapping/tweezer	Femto- and attosecond laser
Assembly and joining (self- and directed assembly)	Material removal processes (mechanical, chemical, and hybrids)
Electrostatic (coatings and fibers)	Electro-erosive processes (electro, chemical, and mechanical)
Colloidal aggregation	Ultrasonic material removal

mechanisms governing the interface between matrix and nanophase in nanocomposites.[89] There is need for cost-effective, high volume synthesis, and processing of nanoparticles, nanofibers, ultrathin/monolayer films, and other building block structures from a wide range of materials.

To achieve high production and low defect rates, many of these systems may require investigation into self-assembly by competing interactions (e.g., block copolymers—kinetics and thermodynamics).[90] Self-assembly has limitations for multiscale structures. Thus, processes may evolve to incorporate molecular machines or nanorobots formed from bi- and multistable molecular systems. Nanotechnology may also play a major role in next-generation catalysts. Improved catalysis is a key to better energy efficiency, reduced environmental impact, and better process economics.

7.7.2 Prototyping, Scale-Up, and Integration

Complexities involved in large-scale manufacturing of nanomaterials, integration with macroscopic manufacturing steps, and assembly is shown in Figure 7.2. At the first level, approaches are needed for the high-rate, high-volume fabrication/synthesis of building blocks (e.g., dots, wires, tubes, particles, fibers, films) from a range of materials and with better control of size, shape, and their polydispersity. Integrating these building blocks requires an understanding of issues such as substrate and building block surface modification for directed self-assembly. In addition to self-assembly, it is recognized that product realization may come about as a combination of new and traditional manufacturing processes[91] as well as bridging top-down

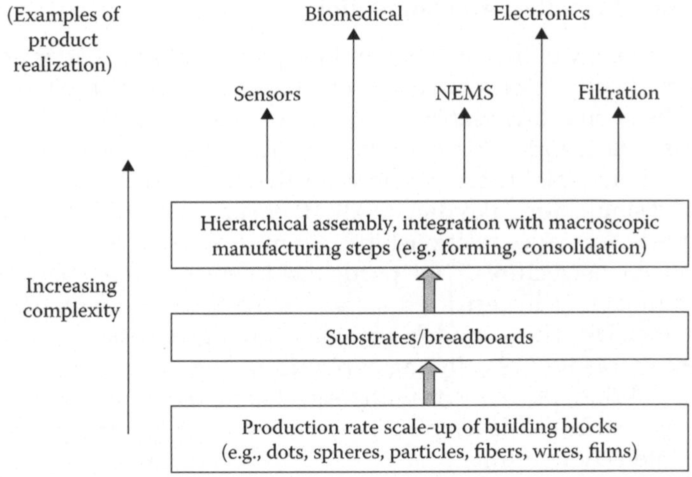

FIGURE 7.2
Schematic of levels of complexity for nanomanfacturing integration and assembly. NEMS, nanoelectro mechanical systems.

and bottom-up approaches. Therefore, the integration of nanoparticle and nanomaterial syntheses with subsequent manufacturing steps and the consolidation and forming of nanostructures into macroscopic objects is needed.

As the technology moves forward, better control over 3D assembly, interconnection of nanostructured devices (e.g., with microcircuits), manipulation and rapid setup of material components in multistep fabrication are important goals. Reliability of nanostructures relies on the control of surface and interface composition and the structure to minimize defects and enable subsequent processing (e.g., nanoscale planarization, polishing), and on the ability to remove and repair defects in nanofabricated structures.

7.7.3 Measurements

To study features and phenomena at the nanoscale requires instruments capable of resolutions[92] at the nano-, subnano-, and even picolevels. New instruments are being developed for measuring thermal and mechanical properties of nanostructures. The challenge in nanomanufacturing remains to develop new experimental and analytical tools with a broader range of capabilities at the nanoscale for chemical analysis, surface and subsurface defects, subsurface properties, charge transport, and spectroscopy. These tools should also be capable of working *in situ*, in real time, and under the variable conditions of processing with respect to temperature, pressure, and electrical and magnetic fields. As new instruments are developed, new methods of calibration and standardization as well as calibration standards must be developed to ensure the accurate interpretation of results. Reproducible positioning and repositioning with nanometer accuracy are needed for calibration, measurement, and assembly.

7.7.4 Theory, Modeling, and Simulations

As nanotechnology builds upon the unique properties that matter exhibits in the form of small particles or structural clusters, research in nanomanufacturing is based on individual atomic/molecular (e.g., quantum) theories and bulk continuum theories. Many of the existing models and assumptions are not valid in nanoscale. There is a need for the development of new theories, models, and simulations. Fundamentally, there are basic structure–property–processing–performance relationships that must be established for nanomaterials and nanostructures.[93] Of particular interest to nanomanufacturing is process modeling in restricted spatial domains where boundary effects become pervasive. These models include coupling of various spatial scales and timescales, as well as coupling of physical/chemical phenomena, including nonequilibrium phenomena with respect to transport and growth). Other primary needs include developing potentials that are appropriate for nanostructure interactions, utilizing approaches to benchmark their accuracy, and exploring statistical mechanics approaches to thermodynamic properties and phase diagrams. As knowledge in this area evolves, new design paradigms will have to be created for nanomanufacturing.

References

1. Jiang, L.Y., Chung, T.S., Cao, C., Huang, Z., and Kulprathipanja, S. 2005. Fundamental understanding of nano-sized zeolite distribution in the formation of the mixed matrix single and dual-layer asymmetric hollow fiber membranes. *J. Membr. Sci.* 252: 89–100.
2. Jiang, L.Y., Chung, T.S., and Kulprathipanja, S. 2006. An investigation to revitalize the separation performance of hollow fibers with a thin mixed matrix composite skin for gas separation. *J. Membr. Sci.* 276: 113–125.
3. Li, Y., Chung, T.S., Huang, Z., and Kulprathipanja, S. 2006. Dual-layer polyethersulfone (PES)/BTDA-TDI/MDI co-polyimide (P84) hollow fiber membranes with a submicron PES–zeolite beta mixed matrix dense-selective layer for gas separation. *J. Membr. Sci.* 277: 28–37.
4. Pechar, T.W., Kim, S., Vaughan, B., Marand, E., Tsapatsis, M., Jeong, H.K., and Cornelius, C.J. 2006. Fabrication and characterization of polyimide–zeolite L mixed matrix membranes for gas separations. *J. Membr. Sci.* 277: 195–202.
5. Xiao, Y., Wang, K.Y., Chung, T.S., and Tan, J. 2006. Evolution of nano-particle distribution during the fabrication of mixed matrix TiO_2-polyimide hollow fiber membranes. *Chem. Eng. Sci.* 61: 6228–6233.
6. Jiang, L.Y. and Chung, T.S. 2006. Fabrication of mixed matrix hollow fibers with intimate polymer–zeolite interface for gas separation. *AIChE J.* 52: 2898–2908.
7. Kusworo, T.D., Ismail, A.F., Mustafa, A., and Matsuura, T. 2009. Dependence of membrane morphology and performance on preparation conditions: The shear rate effect in membrane casting. *Sep. Purif. Technol.* 61: 249–257.
8. Cong. H., Radosz, M., Towler, B.F., and Shen, Y. 2007. Polymer–inorganic nanocomposite membranes for gas separation. *Sep. Purif. Technol.* 55: 281–291.
9. Kim, S., Pechar, T.W., and Marand, E. 2006. Poly(imide siloxane) and carbon nanotube mixed matrix membranes for gas separation. *Desalination* 192: 330–339.
10. Zhang, Y., Li, H., Lin, J., Li, R., and Liang, X. 2006. Preparation and characterization of zirconium oxide particles filled acrylonitrile–methyl acrylate–sodium sulfonated acrylate copolymer hybrid membranes. *Desalination* 192: 198–206.
11. Genne, I., Kuypers, S., and Leysen, R. 1996. Effect of the addition of ZrO_2 to polysulfone based UF membranes. *J. Membr. Sci.* 113: 343–350.
12. Wara, N.M., Francis, L.F., and Velamakanni, B.V. 1995. Addition of alumina to cellulose acetate membranes. *J. Membr. Sci.* 104: 43–49.
13. Kim, S., Chen, L., Johnson, J.K., and Marand, E. 2007. Polysulfone and functionalized carbon nanotube mixed matrix membranes for gas separation: Theory and experiment. *J. Membr. Sci.* 294: 147–158.
14. Ahn, J., Chung, W.-J., Pinnau, I., and Guiver, M.D. 2008. Polysulfone/silica nanoparticle mixed-matrix membranes for gas separation. *J. Membr. Sci.* 314: 123–133.
15. Husain, S. and Koros, W.J. 2007. Mixed matrix hollow fiber membranes made with modified HSSZ-13 zeolite in polyetherimide polymer matrix for gas separation. *J. Membr. Sci.* 288: 195–207.
16. Rafizah, W.A.W. and Ismail, A.F. 2008. Effect of carbon molecular sieve sizing with poly (vinylpyrrolidone) K-15 on carbon molecular sieve–polysulfone mixed matrix membrane. *J. Membr. Sci.* 307: 53–61.

17. Chung, T.S., Ying Jiang, L., Lia, Y., and Kulprathipanja, S. 2007. Mixed matrix membranes (MMMs) comprising organic polymers with dispersed inorganic fillers for gas separation. *Prog. Polym. Sci.* 32: 483–507.
18. Duval, J.M. 1995. Adsorbent filled polymeric membranes, PhD thesis, University of Twente, Enschede, the Netherlands.
19. Mahajan, R. and Koros, W.J. 2000. Factors controlling successful formation of mixed matrix gas separation materials. *Ind. Eng. Chem. Res.* 39: 2692–2696.
20. Mahajan, R. and Koros, W.J. 2002. Mixed matrix membrane materials with glassy polymers. Part 1. *Polym. Eng. Sci.* 42: 1420–1431.
21. Mahajan, R. and Koros, W.J. 2002. Mixed matrix membrane materials with glassy polymers. Part 2. *Polym. Eng. Sci.* 42: 1432–1441.
22. Paul, D.R. and Kemp, D.R. 1973. Diffusion time lag in polymer membranes containing adsorptive fillers. *J. Polym. Sci. Part C Polym. Symp.* 41: 79–93.
23. Jia, M., Peinemann, K.V., and Behling, R.D. 1991. Molecular sieving effect of zeolite filled silicone rubber membranes. *J. Membr. Sci.* 57: 289–296.
24. Anson, M., Marchese, J., Garis, E., Ochoa, N., and Pagliero, C. 2004. ABS copolymer-activated carbon mixed matrix membranes for CO_2/CH_4 separation. *J. Membr. Sci.* 243: 19–28.
25. Moaddeb, M. and Koros, W.J. 1997. Gas transport properties of thin polymeric membranes in the presence of silicon dioxide particles. *J. Membr. Sci.* 125: 143–163.
26. Hu, Q., Marand, E., Dhingra, S., Fritsch, D., Wen, J., and Wilkes, G. 1997. Poly (amideimide)/TiO_2 nano-composite gas separation membranes: Fabrication and characterization. *J. Membr. Sci.* 135: 65–79.
27. Moore, T.T. and Koros, W.J. 2005. Non-ideal effects in organic–inorganic materials for gas separation membranes. *J. Mol. Struct.* 739: 87–98.
28. Aroona, M.A. and Ismail, A.F. 2010. Performance studies of mixed matrix membranes for gas separation: A review. *Sep. Purif. Technol.* 75: 229–242.
29. Vankelecom, I.F.J., Mercks, E., Luts, M., and Uytterhoeven, J.B. 1995. Incorporation of zeolite in polyimide membranes. *J. Phys. Chem.* 99: 13187–13192.
30. Li, Y., Chung, T.S., Cao, C., and Kulprathipanja, S. 2005. The effects of polymer chain rigidification, zeolite pore size and pore blockage on polyethersulfone (PES)–zeolite A mixed matrix membranes. *J. Membr. Sci.* 260: 45–55.
31. Qin, J. 1999. Explorations of forming and characterizing mixed matrix gas separation membranes, PhD dissertation, University of Texas, Austin, TX.
32. Ersolmaz, S.B.T., Senorkyan, L., Kalaonra, N., Tatlier, M., and Senatalar, A.E. 2001. n-Pentane/i-pentane separation by using zeolite PDMS mixed matrix membranes. *J. Membr. Sci.* 189: 59–67.
33. Clarizia, G., Algieri, C., and Drioli, E. 2004. Filler-polymer combination: A route to modify gas transport properties of a polymeric membrane. *Polymer* 45: 5671–5681.
34. Merkel, T.C., Freeman, B.D., Spontak, R.J., He, Z., Pinnau, I., Meakin, P., and Hill, A.J. 2003. Sorption, transport and structural evidence for enhanced free volume in poly(4-methyl-2-pentyne)/fumed silica nanocomposite membranes. *Chem. Mater.* 15: 109–123.
35. Hillock, A.M.W., Miller, A.M.W., and Koros, W.J. 2008. Cross-linked mixed matrix membranes for the purification of natural gas: Effects of sieve surface modification. *J. Membr. Sci.* 314: 193–199.

36. Li, Y., Guan, H.M., Chung, T.S., and Kulprathipanja, T.S. 2006. Effects of novel silane modification of zeolite surface on polymer chain rigidification and partial pore blockage in polyethersulfone (PES)–zeolite A mixed matrix membranes. *J. Membr. Sci.* 275: 17–28.
37. Manson, J.A. and Sperling, L.H. 1976. *Polymer Blends and Composites*, Plenum Press, New York.
38. Suer, M.G., Bac, N., and Yilmaz, L.1994. Gas permeation characteristics of polymer–zeolite mixed matrix membranes. *J. Membr. Sci.* 91: 77–86.
39. Duval, J., Kemperman, A., Folkers, B., Mulder, M., Desgrandchamps, G., and Smolders, C. 1994. Preparation of zeolite filled glassy polymer membranes. *J. Appl. Polym. Sci.* 54: 409–418.
40. Mahajan, R., Burns, R., Schaeffer, M., and Koros, W.J. 2002. Challenges in forming successful mixed matrix membranes with rigid polymeric materials. *J. Appl. Polym. Sci.* 86: 881–890.
41. Hacarlioglu, P., Toppare, L., and Yilmaz, L. 2003. Polycarbonate–polypyrrole mixed matrix gas separation membranes. *J. Membr. Sci.* 225: 51–62.
42. Pechar, T.W., Tsapatsis, M., Marand, E., and Davis, R. 2002. Preparation and characterization of glassy fluorinated polyimide zeolite-mixed matrix membrane. *Desalination* 146: 3–9.
43. Boom, J.P. 1994. Transport through zeolite filled polymeric membranes, PhD thesis, University of Twente, Enschede, the Netherlands, 1994.
44. Jia, M., Peinemann, K.V., and Behling, R.D. 1992. Preparation and characterization of thin film zeolite–PDMS composite membranes. *J. Membr. Sci.* 73: 119–128.
45. Strathmann, H. 1986. Synthetic membranes and their preparation, in *Synthetic Membranes: Science, Engineering and Applications*, P.M. Bungay, H.K. Lonsdale, and M.N. De Pinho (eds.), D. Riedel Publishers, Dordrecht, the Netherlands, 1–38.
46. Wang, I.F. and Minhas, B.S. 1991. Asymmetric polyimide membranes, US Patent 5,067,970.
47. Ismail, A.F. and Yean, L.P. 2003. Review on the development of defect-free and ultrathin skinned asymmetric membranes for gas separation through manipulation of phase inversion and rheological factors. *J. Appl. Polym. Sci.* 88: 442–451.
48. He, Z. Pinnau, I., and Morisato, A. 2002. Nanostructured poly (4-methyl-2-pentyne)/silica hybrid membranes for gas separation. *Desalination* 146: 11–15.
49. Tantekin-Ersolmaz, S.B. 2000. Effect of zeolite particle size on the performance of polymer–zeolite mixed matrix membrane. *J. Membr. Sci.* 175: 285–288.
50. Thostenson, E.T., Ren, Z., and Chou, T.W. 2001. Advances in the science and technology of carbon nanotubes and their composites: A review. *Compos. Sci. Technol.* 61: 1899–1912.
51. Shaffer, M.S.P. and Windle, A.H. 1999. Analogies between polymer solutions and carbon nanotube dispersions. *Macromolecules* 32: 6864–6866.
52. Gong, X., Liu, J., and Baskaran, S. 2000. Surfactant assisted processing of carbon nanotube/polymer composites. *Chem. Mater.* 12: 1049–1052.
53. Jin, L. and Bower, C. 1998. Alignment of carbon nanotubes in a polymer matrix by mechanical stretching. *Appl. Phys. Lett.* 73: 1197–1199.
54. Ajayan, P.M., Stephan, O., Colliex, C., and Trauth, D. 1994. Aligned carbon nanotube arrays formed by cutting a polymer resin nanotube composite. *Science* 265: 1212–1214.

55. Chen, J., Hamon, M.A., Hu, H., Chen, Y., Rao, A.M., Eklund, P.C., and Haddon, R.C. 1998. Solution properties of single-walled carbon nanotubes. *Science* 282: 95–98.
56. Mitchell, C.A., Bahr, J.L., Arepalli, S., Tour, J.M., and Krishnamoorti, R. 2002. Dispersion of functionalized carbon nanotubes in polystyrene. *Macromolecules* 35: 8825–8830.
57. Bubert, H., Haiber, S., Brandl, W., Marginean, G., Heintze, M., and Brüser, V. 2003. Characterization of the uppermost layer of plasma-treated carbon nanotubes. *Diamond Relat. Mater.* 12: 811–815.
58. Eitan, A., Jiang, K., Dukes, D., Andrews, R., and Schadler, L.S. 2003. Surface modification of multiwalled carbon nanotubes: Toward the tailoring of the interface in polymer composites. *Chem. Mater.* 15: 3198–3201.
59. Star, A., Stoddart, J.F., Steuerman, D., Diehl, M., Boukai, A., Wong, E.W., Yang, X., Chung, S.-W., Choi, H., and Heath, J.R. 2001. Preparation and properties of polymer-wrapped single-walled carbon nanotubes. *Angew. Chem. Int. Ed.* 40: 1721–1725.
60. Jia, Z., Wang, Z., Xu, C., Liang, J., We, B., Wu, D., and Zhu, S. 1999. Study on poly (methyl methacrylate)/carbon nanotube composites. *Mater. Sci. Eng.* 271: 395–400.
61. Deng, J., Ding, X., Zhang, W., Peng, Y., Wang, J., Long, X., Li, P., and Chan, A.S.C. 2002. Carbon nanotube-polyaniline hybrid materials. *Eur. Polym. J.* 38: 2497–2501.
62. Qian, D., Dickey, E.C., Andrews, R., and Rantell, T. 2000. Load transfer and deformation mechanisms in carbon nanotube-polystyrene composites. *Appl. Phys. Lett.* 76: 2868–2870.
63. Bhattacharyya, A.R., Sreekumar, T.V., Liu, T., Kumar, S., Ericson, L.M., Hauge, R.H., and Smalley, R.E. 2003. Crystallization and orientation studies in polypropylene/single wall carbon nanotube composite. *Polymer* 44: 2373–2377.
64. Potschke, P., Fornes, T.D., and Paul, D.R. 2002. Rheological behaviors of multiwalled carbon nanotube/polycarbonate composites. *Polymer* 43: 3247–3255.
65. Siochi, E.J., Working, D.C., Park, C., Lillehei, P.T., Rouse, J.H., Topping, C.C., Bhattacharyya, A.R., and Kumar, S. 2004. Melt processing of SWCNT-polyimide nanocomposite fibers. *Compos. Part B Eng.* 35: 439–446.
66. Tang, W.Z., Santare, M.H., and Advani, S.G. 2003. Melt processing and mechanical property characterization of multi-walled carbon nanotube/high density polyethylene (MWNT/HDPE) composite films. *Carbon* 41: 2779–2785.
67. Gong, X., Liu, J., Baskaran, S., Voise, R.D., and Young, S. 2000. Surfactant-assisted processing of carbon nanotube/polymer composites. *Chem. Mater.* 12: 1049–1052.
68. Shaffer, M.S.P., Fan, X., and Windle, A.H. 1998. Load transfer in carbon nanotube epoxy composites. *Carbon* 36: 1603–1612.
69. Dror, Y., Salalha, W., Khalfin, R.L., Cohen, Y., Yarin, A.L., and Zussman, E. 2003. Carbon nanotubes embedded in oriented polymer nanofibers by electrospinning. *Langmuir* 19: 7012–7020.
70. Chen, G.Z., Shaffer, M.S.P., Coleby, D., Dixon, G., Zhou, W., Fray, D.J. and Windle, A.H. 2000. Carbon nanotube and polypyrrole composites: Coating and doping. *Adv. Mater.* 12: 522–526.
71. Bin, Y.Z., Kitanaka, M., Zhu, D., and Matsuo, M. 2003. Development of highly oriented polyethylene filled with aligned carbon nanotubes by gelation/crystallization from solutions. *Macromolecules* 36: 6213–6219.

72. Jang, J., Bae, J., and Yoon, S.H. 2003. A study on the effect of surface treatment of carbon nanotubes for liquid crystalline epoxide-carbon nanotube composites. *J. Mater. Chem.* 13: 676–681.
73. Kimura, T. and Ago, H. 2002. Polymer composites of carbon nanotubes aligned by a magnetic field. *Adv. Mater.* 14: 1380–1383.
74. Wilbrink, M.W.L., Argon, A.S., Cohen, R.E., and Weinberg, M. 2001. Toughen ability of nylon-6 with $CaCO_3$ filler particles: New findings and general principles. *Polymer* 42: 10155–10180.
75. Choi, E.S., Brooks, J.S., Eaton, D.L., Al-Haik, M.S., Hussaini, M.Y., Garmestani, H., Li, D., and Dahmen, K. 2003. Enhancement of thermal and electrical properties of carbon nanotube polymer composites by magnetic field processing. *J. Appl. Phys.* 94: 6034–6039.
76. Liang, Z., Shankar, K.R., Barefield, K., Zhang, C., Wang, B., and Kramer, L. 2003. Investigation of magnetically aligned carbon nanotube bucky paper/epoxy composites, *Conference Proceedings at SAMPE*, Long Beach, CA.
77. Wang, Z., Liang, Z., Wang, B., Zhang, C., and Kramer, L. 2004. Processing and property investigation of single-walled carbon nanotube (SWNT) buckypaper/epoxy resin matrix nanocomposites. *Compos Part A* 35: 1225–1232.
78. Liang, Z., Gonnet, P., Choi, E.S., Shankar, R., Zhang, C., Brooks, J.S., Wang, B., and Kramer, L. 2005. Investigation of thermal conductivity of carbon nanotube bucky papers and nanocomposites, *Conference Proceedings at SAMPE*, Long Beach, CA.
79. Walters, D.A., Casavant, M.J., Qiin, X.C., Huffman, C.B., Boul, P.J., Erickson, L.M., and Smith, K. 2001. In plane aligned membranes of carbon nanotube. *Chem. Phys. Lett.* 338: 14–20.
80. Hibshman, C., Cornelius, C.J., and Marand, E. 2003. The gas separation effects of annealing polyimide-organosilicate hybrid membranes. *J. Membr. Sci.* 211: 25–40.
81. Vankelecom, I.F.J., Broeck, S.V., Merck., E., Geerts, H., Grobet, P., and Uytterhoeven, J.B. 1996. Silylation to improve incorporation of zeolites in polyimide films. *J. Phys. Chem.* 100: 3753–3758.
82. Hillock, A.M.W. 2005. Crosslinkable polyimide mixed matrix membranes for natural gas purification, PhD thesis, Georgia Institute of Technology, Atlanta, GA.
83. Guer, T. 1994. Permselectivity of zeolite filled polysulfone gas separation membranes. *J. Membr. Sci.* 93: 283–289.
84. Vu, D.Q., Koros, W.J., and Miller, S.J. 2003. Mixed matrix membranes using carbon molecular sieves. I. Preparation and experimental results. *J. Membr. Sci.* 211: 311–334.
85. Yong, H.H., Park, H.C., Kang, Y.S., Won, J., and Kim, W.N. 2001. Zeolite-filled polyimide membrane containing 2,4,6-triaminopyrimidine. *J. Membr. Sci.* 188: 151–163.
86. NSF-EC Workshop on Nanomanufacturing and Processing, 2002. http://www.nsf.gov/mps/dmr/nsfec_workshop_report.pdf.
87. Biscarini, F., Taliani, C., Chen, J., and Komanduri, R. 2002. Nanomanufacturing and processing—Research, education, infrastructure security, resource. http://chm.pse.umass.edu/NMSworkshop/presentations/Komanduri%20NSF%20Report.pdf.
88. National Nanotechnology Initiative. Manufacturing at the nanoscale. http://www.nano.gov/nanotech-101/what/manufacturing.

89. The Royal Society, 2004. Nanomanufacturing and the industrial application of nanotechnologies, *Nanoscience and Nanotechnologies: Opportunities and Uncertainties*, The Royal Society, London. http://www.nanotec.org.uk/report/chapter4.pdf.
90. US Department of Energy, 2014. *Nanomanufacturing Portfolio: Manufacturing Processes and Applications to Accelerate Commercial Use of Nanomaterial*, Industrial Technologies Program. https://www1.eere.energy.gov/manufacturing/industries_technologies/ nanomanufacturing/pdfs/nanomanufacturing_portfolio.pdf.
91. Huang, Q. 2012. Integrated nanomanufacturing and nanoinformatics for quality improvement http://www-bcf.usc.edu/~qianghua/papers/INN.pdf.
92. Morse, J.D. 2008. Research challenges for integrated systems nanomanufacturing. Report from the National Science Foundation Workshop February 10–11, 2008. http://www.coe.neu.edu/Research/nanophm/materialForDistribution/NMSWorkshopReport.pdf.
93. Watkins, J. and Morse, J.D. 2009. Critical National Need Idea. National nanomanufacturing testbed program, Integrated hierarchical nanomanufacturing. Center for Hierarchical Manufacturing, University of Massachusetts Amherst, MA. http://www.nist.gov/tip/wp/pswp/upload/200_national_nanomanufacturing.pdf.

8

Nanocomposite Membranes: Health, Environment, Safety, and Societal Implications

8.1 Introduction

Nanocomposite membrane has unprecedented potential of providing technological solutions to many environmental problems, including climate change, pollution abatement, and clean drinking water. It is claimed that nanotechnology enables economic growth through more efficient and durable products and new markets. However, the applicability of such a system has to be perceived after due consideration of the process and product in totality with serious attention toward the probable health and environmental risks.

It is perceived that nanocomposite membrane opts for cleaner production through green chemistry, synthesis and processing of nanoscale materials that will reduce consumption of raw materials and natural resources such as water and energy, and improved chemical reactions and catalysis. However, there is need for a proper life cycle analysis of the nanomaterials through validated nano-specific risk assessment methodologies.

Although there are now only a limited number of products in the market that contain engineered nanomaterials, the pace of nanotechnology development indicates that the market soon is going to be flooded with nano-based products. Hence, there is need to look into associated health and environment risks due to the following:

- Size of particles
- Increased reactivity and conductivity
- Routes of exposure

Due to very small size, nanoparticles can be inhaled or indigested and are capable of crossing the blood–brain barrier that protects the brain against contamination. Nanoparticles are more reactive and conductive than the conventional particles of the same material. Nanomaterials present risk management issues that are not easily characterized because of the breadth of categories of such substances An understanding of the toxicity of nanomaterials, dose metrics, probable exposure pathways, and environmental fate is needed for sound scientific information of the risk management process.

8.2 Risk Assessment

The general approach to assessing and controlling risk involves the identification of hazard potential of the substance and a structured approach for determining the probability of exposure and the associated consequences. Risk is usually controlled in practice by reducing the probability of exposure, although the first principle of risk management is to substitute more hazardous substance with less hazardous substance wherever possible. An evaluation of hazard with respect to toxicity is required for determining to what extent exposure should be controlled. Risk is controlled by limiting release of the material to air or water, and/or by interrupting the pathways by which the substance reaches the receptor where it could cause harm (e.g., an organ in the body), making an understanding of exposure pathways and likely quantities essential to risk management. As mentioned, in any new technology, foresight of possible risks depends on a consideration of the life cycle of the material being produced. This involves understanding the processes and materials used in manufacturing, the likely interactions between the product and individuals or the environment during its manufacture and useful life, and the methods used in its eventual disposal. It is important to note that nanomaterials have large surface areas per unit of volume and unique properties relative to conventional chemicals. Some of the special properties that make nanomaterials useful are also properties that may cause some nanomaterials to pose hazards to humans and the environment under specific conditions. It will be necessary to consider these unique properties and their potential impacts on fate, exposure, and toxicity in developing risk assessments for nanomaterials.[1–6]

8.3 Environmental Issues of Nanomaterials

The identification and characterization of chemical substances and materials is an important first step in assessing their risk. Understanding the physical and chemical properties in particular is necessary in the evaluation of toxicological and ecological hazards and exposure. The diversity and complexity of nanomaterials make chemical identification and characterization not only more important but also more difficult. A broader spectrum of properties is needed to sufficiently characterize a given nanomaterial for the purpose of evaluating hazard and assessing risk. Chemical properties such as molecular weight, melting point, boiling point, vapor pressure, water solubility, reactivity, and stability are important for some nanomaterials, but other properties such as particle size and distribution, surface/volume ratio, magnetic properties, coatings, and conductivity are expected to be more important for the majority of nanoparticles.

A given nanomaterial can be produced in many cases by several different processes yielding several derivatives of the same material. For example, single-walled carbon nanotubes can be produced by four different processes that can generate products with different physicochemical properties (e.g., size, shape, composition) and potentially different ecological and toxicological properties.

Potential nanomaterials release sources include (1) direct as well as indirect releases to the environment from the manufacturing and processing of nanomaterials, (2) releases from refining processes, (3) chemical and material manufacturing processes, (4) chemical cleanup activities including the remediation of soil-contaminated sites, (5) releases of nanomaterials incorporated into materials used to fabricate products for consumer use including pharmaceutical products, and (6) releases resulting from disposal of consumer products containing nanoscale materials. Figure 8.1 shows pathways for nanomaterials to move in the environment.[7]

Several processes influence the fate of airborne nanomaterials in addition to their initial dimensional and chemical characteristics: the length of time the particles remain airborne, the nature of their interaction with other airborne particles or molecules, and the distance that they may travel prior to

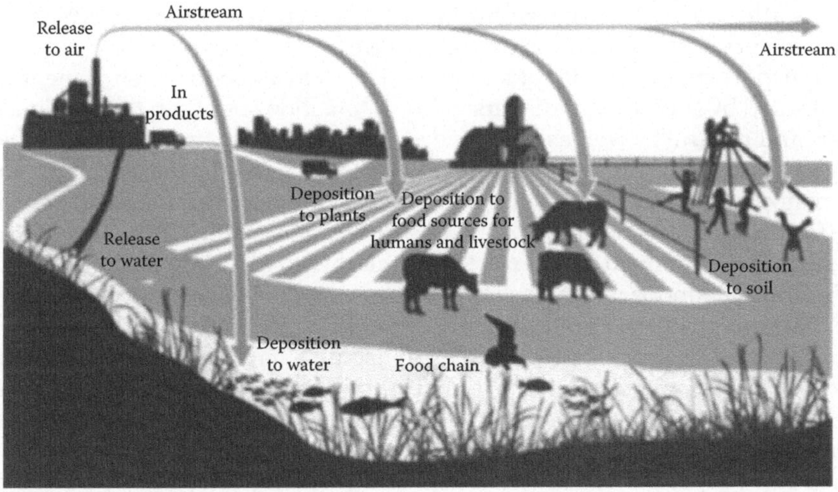

FIGURE 8.1
Potential exposure pathways for nanomaterials. (Data from Shatkin, J.A., *Nanotechnology Health and Environmental Risks*, CRC Press, Boca Raton, FL, 2008.)

deposition. The transport of nanosized particles in atmosphere takes place by diffusion, agglomeration, and gravitational settling. With respect to the length of time particles remain airborne, particles with aerodynamic diameters in the nanoscale range (<100 nm) follow the laws of gaseous diffusion when released to air. The rate of diffusion is inversely proportional to the particle diameter, whereas the rate of gravitational settling is proportional to the particle diameter.[8] Airborne particles can be classified by their size and behavior into three general groups: (1) Small particles (diameters < 80 nm) are described as being in the agglomeration mode and are short-lived because they rapidly agglomerate to form larger particles; (2) large particles (>2000 nm) are described as being in the coarse mode and are subject to gravitational settling; and (3) intermediate-sized particles (>80 and <2000 nm) are described as being in the accumulation mode and can remain suspended in air for the longest time, days to weeks, and can be removed from air via dry or wet deposition.[9–13]

The transport behavior of nanomaterials released to soil is likely to vary depending upon its physical and chemical characteristics. Nanomaterials released to soil can be strongly sorbed to soil due to their high surface areas and therefore be immobile. However, they are small enough to fit into smaller spaces between soil particles, and might therefore travel farther than larger particles before getting trapped in the soil matrix. The sorption characteristics of any intentionally produced nanoparticle to soil will depend on its size, chemistry, surface treatment, and the conditions under which it is applied. Studies have demonstrated the differences in mobility of a variety of insoluble nanosized materials in a porous medium.[14,15] Additionally, the properties of the soil and environment can affect nanomaterial mobility. For example, the mobility of mineral colloids in soils and sediments is significantly affected by their charge. Surface photoreactions provide a pathway for nanomaterial transformation on soil surfaces.

Transport characteristics of nanomaterials in an aqueous environment are controlled by solubility or dispersability, interactions between the nanomaterials and natural as well as anthropogenic chemicals in the system, and biological and abiotic processes. There are limited data on the fate and transport of nanoparticles, but the existing data show that their behavior can be very different from much larger particles of the same materials. Nanoparticles generally are retained in the water column due to diffusion and dispersion. Waterborne nanoparticles generally settle more slowly than larger particles of the same materials but can be removed from water by agglomeration or sorption and sedimentation. Dispersed insoluble nanoparticles can be stabilized in water by interactions with naturally occurring humic substances or other species. Biodegradation or association with biological materials may remove nanomaterials. Photocatalyzed reactions may alter the physical and chemical properties of nanomaterials and therefore alter their behavior in water. Processes that control transport and removal of nanoparticles in water and wastewater are under investigation. Nanoparticle photochemistry has good possibility of potential application in water treatment.

Due to their high surface area-to-mass ratios, nanosized particles have the potential to get adsorbed to suspended soil and sediment particles.[16] Certain organic and metallic nanomaterials may possibly be transformed under anaerobic conditions, such as in aquatic sediments. Particles in the upper layers of aquatic environments, on soil surfaces, and in water droplets in the atmosphere are exposed to sunlight. Light-induced photoreactions often are important in determining environmental fate of chemical substances.

The fate of nanosized particles in wastewater treatment plant is not well characterized. Wastewater may be subjected to many different types of treatment, including physical, chemical, and biological processes, depending on the characteristics of the wastewater, whether the plant is a publicly owned treatment work or on-site industrial facility.

Biodegradation of nanoparticles may result in their breakdown as typically seen in biodegradation of organic molecules, or may result in changes in the physical structure or surface characteristics of the material. The potential and mechanism of biodegradation of nanosized particles have just begun to be investigated. As is the case for other processes, the potential for biodegradation will depend strongly on the chemical and physical nature of the particle. Many of the nanomaterials in current use are composed of inherently nonbiodegradable inorganic chemicals, such as ceramics, metals, and metal oxides, and are not expected to biodegrade. However, a recent preliminary study found that C60 and C70 fullerenes were taken up by wood decay fungi after 12 weeks, suggesting that the fullerene carbon had been metabolized.[17] For other nanomaterials, biodegradability may be integral to the material's design and function. This is the case for some biodegradable polymers being investigated for use in drug transport,[18] for which biodegradability is mostly a function of chemical structure and not particle size.

Bacteria and living cells can take up nanosized particles, providing the basis for potential bioaccumulation in the food chain.[19] Aquatic and marine filter feeders near the base of the food chain feed on nanosized and larger particles. The bioavailability of specific nanomaterials in the environment will depend in part on the particle. Environmental fate processes may be too slow for effective removal of persistent nanomaterials before they can be taken up by an organism.

Certain nanomaterials are being designed for release as reactants in the environment, and therefore are expected to undergo chemical transformation. One example of this is iron (Fe^0) nanoparticles employed as reactants for the dechlorination of organic pollutants.[15] As the reaction progresses, the iron is oxidized to iron oxide. Other metal particles are also converted to oxides in the presence of air and water. Whether the oxides are more or less toxic than the free metals depends on the metal.

Mathematical models are often used to generate estimated data, which can provide a basis for making regulatory decisions. It is advantageous if such models can be applied to provide estimated properties for nanomaterials, because there are very little experimental data available for these materials. Depending on the relevance of the chemical property or transformation process, new models may have to be developed to provide estimations for these materials. However, mathematical models require experimental validation. Before the environmental fate, transport, and multimedia partitioning of nanomaterials can be effectively modeled, reliable experimental data must be acquired for a variety of intentionally produced nanomaterials.

8.4 Health Risk

Some of the possible exposure routes for nanoparticles based on current and potential future applications are illustrated in Figure 8.2.[20] The material may be inhaled directly if released in the air particularly by those involved in manufacturing nanoparticles in the work place. The material may also be inhaled by all those who are exposed to nanoparticles from sources such as combustion. In addition to inhalation, exposure to nanoparticles can occur from surface contact (such as in cosmetics) or from ingestion (such as food or drink). In the future, medicinal applications may result in particles being injected into the body. Other organisms such as bacteria and

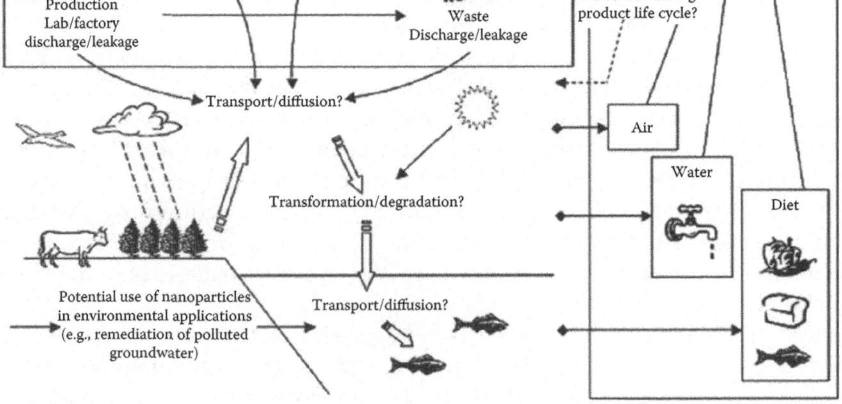

FIGURE 8.2
Possible exposure routes for nanoparticles based on current and potential future applications.

protozoa may take in nanoparticles through their cell membranes, and thus allow the particles to enter a biological food chain.

The toxicity of nanoparticles depends on the following:

- The total surface area of nanoparticle to the target organ
- The chemical reactivity
- The ability to take part in reactions that release free radicals
- The physical dimensions of the particle that allow it to penetrate into the organ or into cells

8.4.1 Toxicity of Nanomaterials

The nanoparticles have ability to induce the lung injuries because of their small size, large surface area, and ability to generate reactive oxygen species (ROS).[21] The short-term pulmonary toxicity studies in rats with ultrafine and fine carbon black, nickel, and TiO_2 particles have established enhanced lung inflammatory strength of the ultrafine particles in comparison with fine-sized particulates of similar composition.[22–24] Low-toxicity nanoparticles such as carbon black and polystyrene stimulate the macrophages via ROS and calcium signaling to make proinflammatory cytokines such as tumor necrosis factor alpha.[25] The cationic nanoparticles, including gold and polystyrene, have shown to cause hemolysis and blood clotting, whereas anionic particles are usually quite nontoxic. High exposures to diesel exhaust particles (DEPs) by inhalation caused altered heart rate in hypertensive rats interpreted as a direct effect of DEP on the pacemaker activity of the heart.[26] Exposure to single-walled carbon nanotubes results in cardiovascular effects.[27] The nanoparticles inhaled can gain access to the brain by means of two different mechanisms: trans-synaptic transport after inhalation through the olfactory epithelium and uptake through the blood–brain barrier.[28,29] *In vitro* studies have shown that multiwalled carbon nanotubes are capable of localizing within and initiating an irritation response in human epidermal keratinocytes, which are a primary route of occupational exposure.[30,31]

The change in the structural and physicochemical properties of nanoparticles with a decrease in size can be responsible for numerous material interactions that could lead to toxicological effects;[32] for example, shrinkage in size may create discontinuous crystal planes that increase the number of structural defects as well as disrupt the electronic configuration of the material and give rise to altered electronic properties. These changes could establish specific surface groups that could function as reactive sites. Chemical composition of the materials is particularly responsible for these changes and their importance. The surface groups can make nanoparticles hydrophilic or hydrophobic, lipophilic or lipophobic, or catalytically active or passive. These surface properties can lead to toxicity by the interaction of electron donor or acceptor active sites (chemically or physically activated) with molecular oxygen (O_2) and electron capture can lead to the formation of the superoxide radical, which generates additional ROS through Fenton chemistry. Various studies have been carried out to investigate the adverse effects of nanoparticle on the biological systems.[33–37] It has been demonstrated that carbon black nanoparticles produce its increased inflammatory effects via mechanisms other than the leaching of soluble components from the particle surface. Transition metals are an important source of free radicals, which are important in PM10-stimulated lung inflammation. Therefore, it is clear that nanoparticles may exert their increased pro-inflammatory effects, at least in part, by modulating intracellular calcium.

Nanomaterials themselves constitute a new generation of toxic chemicals. As particle size decreases, in many nanomaterials the production of free radicals increases, as does toxicity. Studies have shown that nanomaterials now in commercial use can damage human DNA, negatively affect cellular function, and even cause cell death. There is a small but growing body of scientific studies (termed as nanotoxicology), showing that some nanomaterials are toxic to commonly used environmental indicators such as algae, invertebrate, and fish species.[38–41] There is also evidence that some nanomaterials can impair the function or reproductive cycles of earthworms, which play a key role in nutrient cycling that underpins ecosystem function.[42]

When introduced into the lungs of rodents, certain carbon nanotubes cause inflammation, granuloma development, fibrosis, artery *plaque* responsible for heart attacks, and DNA damage.[43–45] Two independent studies—(1) carbon nanotubes introduced into the abdominal cavity of mice[46] and (2) induction of mesothelioma in p53+/– mouse by intraperitoneal application of multiwall carbon nanotube[47]—have shown that some carbon nanotubes can also cause the onset of the mesothelioma cancer previously thought to be only associated with asbestos exposure.[46,47]

Very little is known about the safety risks presented by engineered nanomaterials. Given their unique properties, particularly their high reactivity and electrical conductivity, safety concerns are focusing on whether nanomaterials can cause fires or explosions. Nanoparticles behave differently compared to conventional particles. Small amount of release of nanoparticles can pollute the water supply or damage crops and get into the air, soil, or water.

Laboratory and production staff may also be exposed to health and safety risks during the manufacturing of nanomaterials. People in these occupations should be aware of the potential risk and hazards of using nanomaterials and take appropriate measures to mitigate the risks.

Owing to the interdisciplinary nature of nanotechnology and nanocomposite membranes, it is viewed as an enabling technology for the existing technologies in the field of water purification, textile, aerospace, health care, and electronics. Keeping in view the unique behavior of nanomaterials in the environment, nanotechnology may pose challenges to the existing waste management systems. Knowledge on the mobility, persistence, and bioaccumulation potential in the environment is scarcely available. Hence, risk assessment on the possible impact of nanowastes is critical and needs to be pursued.

8.5 Societal and Ethical Issues

Nanocomposite membrane is made of nanomaterials embedded in membrane material. Nanotechnology plays an important role in the development of nanocomposite membrane. As yet the potential impact of nanomaterial exposure on humans or the environment is poorly understood. Data on their possible impact are needed for expanded development and use of nanotechnology. The environmental fate and toxicity of a nanomaterial are critical issues with respect to material selection and design for several applications. Systematic investigations of the oxidative, photochemical, and biological stability of nanomaterials (dendrimers, carbonaceous nanoparticles, metal oxides, etc.) in natural and engineered environmental systems need to be carried out. Assessing the risk of using nanomaterials presents unique challenges due to scarcity of published data. As with any new technology that offers significant benefits to humankind, there are some risks of adverse and unintended consequences with nanotechnology. Interdisciplinary aspects of ethical and social dimensions of nanotechnology must be explored in detail.

New technologies come into being through a complex interplay of technical and social factors. The process of innovation that will produce nanotechnology and diffuse its benefits into society is complex and only partially understood. The greatest difficulty in predicting the societal impacts of nanocomposite membrane technology is the fact that once the technical and commercial feasibility[48] of the innovation is demonstrated, subsequent development is much in the hands of users as in those of the innovators. The diffusion and impact of technological innovations often depend on the development of complementary technologies and the user network.

To assess societal issues with respect to the development of nanocomposite membrane technology, the entire system through its life cycle should be examined.[48] Initially, the impact of nanocomposite membrane is likely to be limited to a few specific products and services. Nanocomposite membrane may be introduced in an accelerated manner to those markets where consumers are willing to pay a premium for new or improved performance. Nanocomposite membrane technology will coexist with conventional membrane technologies rather than suddenly displacing them.

An important aim of a societal impact investigation of nanotechnology is to identify harms, conflicts over justice and fairness, and issues concerning respect for persons. For example, changes in workforce needs and human resources are likely to bring benefits to some and harm to others. Other examples of potential issues include safeguards for workers engaged with hazardous production processes, equity disputes raised by intellectual property protection, and questions about relationships between government, industry, and universities. In addition, attention needs to be given to the individual responsibility of engineers, scientists, and others involved in the processes of generating powerful new nanotechnologies. Professional societies have a role to play in providing opportunities for discussing and devising guidelines that incorporate relevant ethical principles into emerging issues. Perhaps most importantly, ethics must be incorporated effectively into the curriculum for training new nanoscientists, nanotechnologists, and nanofabrication technicians.

It is important for the social scientists[49] to study the processes by which nanocomposite membranes are developed. The knowledge gained will help them as well as policy-makers and the public to understand how nanoscience and nanotechnology are advancing for making the membranes, how these advances are being diffused to the ground level, and how to make necessary course corrections. Insight into the innovation process will also grow. A challenging, but important, area for social science research is the social acceptance, resistance, or rejection of nanotechnology. Representative sample surveys, supplemented by focus groups and open-ended interviews, can measure affective, cognitive, and psychosocial parameters. As more and more new nanotechnologies are developed and actually appear in the market, the degree of social acceptance will become ever more important. Indicators to measure social acceptance of nanotechnology will be needed in the following areas: economic, political, religious, and cultural.

Investment in nanotechnology along with carefully addressing the issues related to adverse impact on human health and the environment is both necessary. Nanotechnology will present opportunities to integrate science and technology with social science and humanities. Education must provide mechanism for updating scientists and engineers on new technologies as well as help organize intelligent debates about societal effects of nanotechnology.[50]

8.6 Public Awareness and Participation

Although there may not be any moral or ethical duty, requirement, or obligation for government and industry to engage the public before adopting and marketing a technology, there are many pragmatic reasons to do so. Research and development is expensive and time consuming with multiple opportunity trade-off costs. Unless public resources are shifted to purchase, install, and maintain new technologies such as water treatment, they can be cost prohibitive for many markets. In addition, public money often tracks public sentiment. Public acts as consumers and shareholders of industries in the business of providing potable water. Public support for high-cost facilities involves public contracts and bonds. Neutral attitudes, if not palpable opposition, should be serious concerns to public service providers.[51]

It would be easy to argue that the assessment and control of the impacts of nanotechnologies including nanocomposite membrane technology—as a highly technical and complex subject—should be an expert-led process, restricted primarily to the peer community of scientists and engineers within academia, industry, and government. However, some of the social and ethical concerns, that certain applications of nanotechnologies are likely to raise, stretch well beyond the basic science or engineering of the matter. In this respect, it is recommended[52] that the government communicate with, and involve as far as possible, the public in the decision-making process in the area of nanotechnologies. This view is also in line with that of the European Commission,[53] set out in its communication "Towards a European Strategy for Nanotechnology," in which coherent action "to integrate societal considerations into the R&D process at an early stage" is endorsed.

Possible approaches to dialog are as follows:

- Participatory and constructive technology assessment with stakeholders[54]
- Scenario analysis with stakeholders to identify significant uncertainties[55]
- Direct public engagement such as citizen juries or panels for identifying at an early stage broad *desired futures* for nanotechnologies, significant ethical concerns, or the acceptability of key applications and options
- Formal approaches for framing problems, as well as for identifying preferred options and their attributes[56]
- Multistage methods, which combine different approaches to framing, option appraisal, and final choice in a sequence of linked activities, with different groups of stakeholders and the public at various stages[57]
- Research into public attitudes, both qualitative and quantitative, to generate good-quality *social intelligence*[58] about nanotechnologies and public concerns

For nanotechnologies, decisions will need to be sensitive to public opinion where significant ethical issues arise. For example, the concerns about the future trajectory of the technology and some of the issues associated with the convergence of nanotechnologies with other technologies, and in particular developments in bio-nanotechnologies, are likely to raise novel ethical questions in the future requiring wide public debate. There is need for periodic reflection on possible emerging ethical questions and initiating appropriate forms of dialog with stakeholders or the public as appropriate, as the technology matures and its tangible applications become clearer.

Awareness or educating people is a particularly important task highlighting the positive aspects of the nanochnologies. At a broad societal level, there is a need for a mature debate addressing different aspects, including issues associated with the convergence of nanotechnologies.

Thus, information exchange should aim at more than just educating the public.[59]

Unlike with some of other mature technologies, nanotechnologies have so far not generated conflict among stakeholders. As more and more applications emerge, the situation might change. Government agencies need to play an important role in supporting public dialog about nanotechnologies. Public debate and technology assessment need to be carried out as widely as possible. It is recommended that the public dialogue be initiated on the development and associated issues of nanotechnologies.

8.7 Regulatory Issues

As nanotechnology and nanocomposite materials continue to be developed for commercial domain and as financial aspects become more important, legal environmental aspects and issues are likely to play an important role. Particularly as nanotechnology may be increasingly used in commercial products, financial aspects will become more important. Government agencies may also increasingly become involved with integrating nanotechnology into a broader legal framework, which allows for a variety of different concerns to be addressed. It is certain that as nanotechnology and nanocomposite materials get more and more included in commercial products, government agencies may bring out legislation targeted to issues specific to nanotechnology.

The bright prospects of nanotechnology has potential to encourage scientists and technologists to use their intellectual abilities for developing the science and engineering used in producing nanotechnology products such as nanocomposite membranes for different applications. Stringent laws may come into play to deal with environmental aspects. A number of issues are relevant for assigning liability for the harm in such circumstances, with criminal laws focusing on punishment by the state and civil laws mostly focusing on providing private compensation to injured parties. Nanoscience and nanotechnology are pursued in different research institutions and industrial sectors for various applications and end use. Thus it is likely that regulators may need to consider the impact that nanotechnologies may have on each of their areas of coverage.[60]

Regulation requires assessment of hazard and assessment of the likelihood or duration of exposure. These factors combine to produce the risk to any exposed biological or human population. The overall aim is to determine the risk management measures needed to eliminate risks or reduce them to acceptable levels.

Several regulatory options are available such as the following:

- Workplace controls
- Classification and labeling measures

- Control of emissions to air, water, and land
- Waste disposal restrictions
- Marketing and use restrictions
- Prohibition

Regulatory measures are not static; the regulator collaborates with research institutions and industries in seeking to identify further measures that are reasonably practicable to reduce risks. The most likely place of exposure to nanoparticles and nanotubes is the workplace, including academic research laboratories. There is need to formulate the responsibilities for health and safety that employers have toward employees and members of the public, and employees have to themselves and to each other. The chemical industry is likely to be the major producer of nanomaterials, currently in the form of bulk nanoparticles such as titanium dioxide and eventually more advanced functional materials as research and development progresses. Although nanomaterials currently account for only a tiny fraction of the total quantity of chemicals manufactured, production is expected to increase over the coming years.

Manufacturers of consumer products may utilize the advantages derived from nanomaterials in their products to give improved performance and additional functionality. Here the nanoparticles will essentially be free rather than fixed, although their reactivity (and thus toxicity) may be influenced by coatings.

Research is being pursued to explore the possibility of using nanomaterials in medical diagnosis and treatment. Although such materials would be subject to the stringent regulatory regime that governs all new interventions in medicine, some of the special properties of nanoparticles may lead to the possibility of unforeseen toxicity if introduced into the body in large numbers.

There is knowledge gap that must be addressed for evolving appropriate regulation, with respect to hazard exposure and measurement. Possible toxic hazards associated with nanoparticles and nanotubes may be viewed in light of two important facts: (1) such materials are currently being produced in very low volumes and their use involve as yet little or no exposure to population outside the workplace, and (2) the well-publicized adverse effects of particulate air pollution are related to exposures of very high concentrations of particles, usually in susceptible individuals. Thus, any assessment of risk needs to take into account not just the toxic potential but also likely exposures.

Even when, as in the present, the magnitude and mechanism of risks associated with the production, use and disposal of nanoparticles, nanotubes, and nanocomposites remain uncertain, it should nevertheless be possible to manage the overall level of risk through careful control of exposure. Indeed, the history of the regulatory process shows that delays have in the past occurred from a desire to understand detailed mechanisms of toxicity before firm action to reduce exposures is taken. Steps can be taken by regulators to control possible risks from new manufactured nanoparticles without the need for a curtailment of development activity. Such steps may be taken along with the efforts for further understanding of the possible mechanism of toxicity.

Because of the small size of manufactured nanoparticles and nanotubes, there are several technical challenges surrounding the measurement of their physical and chemical properties. These challenges become particularly important when measurement is required in *real-world* situations, compared to carefully controllable laboratory conditions (as might be used for quality control or toxicity experiments). Such measurement problems arise in the field as fluctuating environmental conditions (e.g., wind speed, temperature, humidity) can modify readings, and background nanoparticles already present in the environment (e.g., from pollution) may mask the manufactured nanoparticles of interest. In addition to the development of measurement techniques for regulatory purposes, there is a growing need for international measurement standards for nanoscalar metrics of dimension, chemical composition, force, and electrical quanta. Monitoring of nanoparticles will also require a high level of traceability to ensure that any future agreed exposure levels are accurately adhered to.

References

1. Maynard, A.D., Baron, P.A., Foley, M., Shvedova, A.A., Kisin, E.R., and Castranova, V. 2004. Exposure to carbon nanotube material: Aerosol release during handling of unrefined single walled carbon nanotube material. *J. Toxicol. Environ. Health Part A* 67: 87–107.
2. Dreher, K.L. 2004. Health and environmental impact of nanotechnology: Toxicological assessment of manufactured nanoparticles. *Toxicol. Sci.* 77: 3–5.
3. Hett, A. et al. 2004. Nanotechnology-small matter, many unknowns, Swiss Reinsurance Company, Zurich, Switzerland. http://www.denix.osd.mil/cmrmd/upload/SWISSREPUBL04_NANO_EN.PDF.
4. United Kingdom Royal Society. 2004. The Royal Society and the Royal Academy of Engineering. Nanoscience and nanotechnologies: Opportunities and uncertainties. http://www.nanotec.org.uk/finalreport.htm.
5. European Commission. 2004. European Commission, Community Health and Consumer Protection. Nanotechnologies: A preliminary risk analysis. Workshop Organized in Brussels on March 1–2, 2004 by the Health and Consumer Protection Directorate General of the European Commission. http://europa.eu.int/comm/health/ph_risk/events_risk_en.htm.
6. US Environmental Protection Agency (USEPA), 2007. Nanotechnology white paper. http://es.epa.gov/ncer/nano/publications/whitepaper 12022005.pdf.
7. Shatkin, J.A. 2008. *Nanotechnology Health and Environmental Risks*, CRC Press, Boca Raton, FL.
8. Aitken, R.J., Creely, K.S., and Tran, C.L. 2004. Nanoparticles: An occupational hygiene review. Research Report 274. Prepared by the Institute of Occupational Medicine for the Health and Safety Executive, North Riccarton, Edinburgh.

9. Atkinson, R. 2000. Atmospheric oxidation, *Handbook of Property Estimation Methods for Chemicals, Environmental and Health Sciences*, R.S. Boethling and D. Mackay (eds.), CRC Press, Boca Raton, FL, 335–354.
10. Bidleman, T.F. 1988. Atmospheric processes, wet and dry deposition of organic compounds are controlled by their vapor-particle partitioning. *Environ. Sci. Technol.* 22(4): 361–367.
11. Dennekamp, M., Mehenni, O.H., Cherrie, J., and Seaton, A. 2002. Exposure to ultrafine particles and PM2.5 in different micro-environments. *Ann. Occup. Hyg.* 46(suppl. 1): 412–414.
12. Preining, O. 1998. The physical nature of very, very small particles and its impact on their behaviour. *J. Aerosol Sci.* 29(5/6): 481–495.
13. Spurny, K.R. 1998. On the physics, chemistry and toxicology of ultrafine anthropogenic, atmospheric aerosols (UAAA). *New Adv. Toxicol. Lett.* 96–97: 253–261.
14. Lecoanet, H.F. and Wiesner, M.R. 2004. Velocity effects on fullerene and oxide nanoparticle deposition in porous media. *Environ. Sci. Technol.* 38: 4377–4382.
15. Zhang, W. 2003. Nanoscale iron particles for environmental remediation: An overview. *J. Nanopart. Res.* 5: 323–332.
16. Oberdorster, G., Oberdorster, E., and Oberdorster, J. 2005. Nanotoxicology: An emerging discipline evolving from studies of ultrafine particles. *Environ. Health Perspect.* 113(7): 823–839.
17. Filley, T.R., Ahn, M., Held, B.W., and Blanchette, R.A. 2005. Investigations of fungal mediated (C60-C70) fullerene decomposition. *Preprints of Extended Abstracts Presented at the ACS National Meeting*, Vol. 45, American Chemical Society, Division of Environmental Chemistry, San Diego, CA, 446–450.
18. Brzoska, M., Langer, K., Coester, C., Loitsch, S., Wagner, T.O., and Mallinckrodt, C. 2004. Incorporation of biodegradable nanoparticles into human airway epithelium cells—In vitro study of the suitability as a vehicle for drug or gene delivery in pulmonary diseases. *Biochem. Biophys. Res. Commun.* 318(2): 562–570.
19. Biswas, P. and Wu, C.-Y. 2005. Nanoparticles and the environment. *J. Air Waste Manage. Assoc.* 55: 708–746.
20. National Institute for Resources and Environment (Japan). 2004. Possible exposure routes for nanoparticles/materials based on current and future potential applications. http://www.nire.go.jp/eco_tec_e/hyouka_e.htm.
21. Wani, M.Y., Hashim, M.A., Nabi, F., and Malik, M.A. 2011. Nanotoxicity: Dimensional and morphological concerns. *Adv. Phys. Chem.* 2011: 1–15.
22. Warheit, D.B., Laurence, B.R., Reed, K.L., Roach, D.H., Reynolds, G.A.M., and Webb, T.R. 2004. Comparative pulmonary toxicity assessment of single-wall carbon nanotubes in rats. *Toxicol. Sci.* 77: 117–125.
23. Pettibone, J.M., Adamcakova-Dodd, A., Thorne, P.S., O'Shaughnessy, P.T., Weydert, J.A., and Grassian, V.H. 2008. Inflammatory response of mice following inhalation exposure to iron and copper nanoparticles. *Nanotoxicology* 2(4): 189–204.
24. Grassian, V.H., O'Shaughnessy, P.T., Adamcakova-Dodd, A., Pettibone, J.M., and Thorne, P.S. 2007. Inhalation exposure study of titanium dioxide nanoparticles with a primary particle size of 2 to 5 nm. *Environ. Health Perspect.*, 115: 397–402.
25. Brown, D.M., Donaldson, K., Borm, P.J., Schins, R.P., Dehnhardt, M., Gilour, P., Jimenez, L.A., and Stone, V. 2004. Calcium and ROS-mediated activation of transcription factors and TNF-α cytokine gene expression in macrophages exposed to ultrafine particles. *Am. J. Physiol.* 286: L344–L353.

26. Hansen, C.S., Sheykhzade, M., Moller, P., Folkmann, J.K., Amtorp, O., Jonassen, T., and Loft, S. 2007. Diesel exhaust particles induce endothelial dysfunction in apoE−/− mice. *Toxicol. Appl. Pharmacol.* 219: 24–32.
27. Li, Z., Hulderman, T., Salmen, R., Chapman, R., Leonard, S.S., Young, S.H., Shvedova, A., Luster, M.I., and Simeonova, P.P. 2007. Cardiovascular effects of pulmonary exposure to single-wall carbon nanotubes. *Environ. Health Perspect.* 115(3): 377–382.
28. Lockman, P.R., Koziara, J.M., Mumper, R.J., and Allen, D. 2004. Nanoparticle surface charges alter blood-brain barrier integrity and permeability. *J. Drug. Target.* 12: 635–641.
29. Jallouli, Y., Paillard, A., Chang, J., Sevin, E., and Betbeder, D. 2007. Influence of surface charge and inner composition of porous nanoparticles to cross blood-brain barrier in vitro. *Int. J. Pharm.* 344: 103–109.
30. Zvyagin, A.V., Zhao, X., Gierden, A., Sanchez, W., Ross, J.A., and Roberts, M.S. 2008. Imaging of zinc oxide nanoparticle penetration in human skin in vitro and in vivo. *J. Biomed. Opt.*, 13, Article ID 064031.
31. Baroli, B., Ennas, M.G., Loffredo, F., Isola, M., Pinna, R., and López-Quintela, M.A. 2007. Penetration of metallic nanoparticles in human full-thickness skin. *J. Invest. Dermatol.* 127: 1701–1712.
32. Nel, A., Xia, T., Madler, L., and Li, N. 2006. Toxic potential of materials at the nanolevel. *Science* 311: 622–627.
33. Li, N., Sioutas, C., Cho, A., Schmitz, D., Misra, C., Sempf, J., Wang, M., Oberley, T., Froines, J., and Andre, N. 2003. Ultrafine particulate pollutants induce oxidative stress and mitochondrial damage. *Environ. Health Perspect.* 111: 455–460.
34. Baker, G.L., Gupta, A., Clark, M.L., Valenzuela, B.R., Staska, L.M., Harbo, S.J., Pierce, J.T., and Dill, J.A. 2008. Inhalation toxicity and lung toxicokinetics of C60 fullerene nanoparticles and microparticles. *Toxicol. Sci.* 101(1): 122–131.
35. Lyon, D.Y., Adams, L.K., Falkner, J.C., and Alvarez, P.J.J. 2006. Antibacterial activity of fullerene water suspensions: Effects of preparation method and particle size. *Environ. Sci. Technol.* 40: 4360–4366.
36. Peters, A., Veronesi, B., Calderon-Garciduenas, L., Gehr, P., Chen, L.C., Geiser, M., Reed, W., Rothen-Rutishauser, B., Schurch, S., and Schultz, H. 2006. Translocation and potential neurological effects of fine and ultrafine particles a critical update. *Part. Fiber. Toxicol.* 3: 13.
37. Hoshino, A., Fujioka, K., Oku, T., Nakamura, S., Suga, M., Yamaguchi, Y., Suzuki, K., Yasuhara, M., and Yamamoto, K. 2004. Quantum dots targeted to the assigned organelle in living cells. *Microbiol. Immunol.*, 48: 985–994.
38. Hund-Rinke, K. and Simon, M. 2006. Ecotoxic effect of photocatalytic active nanoparticles (TiO_2) on algae and daphnids. *Environ. Sci. Poll. Res.* 13(4): 225–232.
39. Lovern, B. and Klaper, R. 2006. Daphnia magna mortality when exposed to titanium dioxide and fullerene (C60) nanoparticles. *Environ. Toxicol. Chem.* 25(4): 1132–1137.
40. Templeton, R., Ferguson, P., Washburn, K., Scrivens, W., and Chandler, G. 2006. Life-cycle effects of single-walled carbon nanotubes (SWNTs) on an estuarine meiobenthic copepod. *Environ. Sci. Technol.* 40: 7387–7393.
41. Federici, G., Shaw, B., and Handy, R. 2007. Toxicity of titanium dioxide nanoparticles to rainbow trout (*Oncorhynchus mykiss*): Gill injury, oxidative stress, and other physiological effects. *Aquat. Toxicol.* 84: 415–430.

42. Fordsmand, J.C., Krogh, P., Schaefer, M., and Johansen, A. 2008. The toxicity testing of double-walled nanotubes contaminated food to *Eisenia veneta* earthworms. *Ecotoxicol. Environ. Safety* 71: 616–619.
43. Donaldson, K., Aitken, R., Tran, L., Stone, V., Duffin, R., Forrest, G., and Alexander, A. 2006. Carbon nanotubes: A review of their properties in relation to pulmonary toxicology and workplace safety. *Toxicol. Sci.* 92(1): 5–22.
44. Lam, C.W., James, J.T., McCluskey, R., and Hunter, R.L. 2004. Pulmonary toxicity of single-wall carbon nanotubes in mice 7 and 90 days after intractracheal instillation. *Toxicol. Sci.* 77: 126–134.
45. Muller, J., Huaux, F., and Lison, D. 2006. Respiratory toxicity of carbon nanotubes: How worried should we be. *Carbon* 44: 1048–1056.
46. Poland, C., Duffin, R., Kinloch, I., Maynard, A., Wallace, Q., Seaton, A., Stone, V., Brown, S., MacNee, S.W., and Donaldson, K. 2008. Carbon nanotubes introduced into the abdominal cavity of mice show asbestos like pathogenicity in a pilot study. *Nat. Nanotechnol.* 3: 423–428.
47. Takagi, A., Hirose, A., Nishimura, T., Fukumori, N., Ogata, A., Ohashi, N., Kitajima, S., and Kanno, J. 2008. Induction of mesothelioma in p53+/- mouse by intraperitoneal application of multi-wall carbon nanotube. *J. Toxicol. Sci.* 33(1): 105–116.
48. Roco, M.C. and Bainbridge, W.S. (eds.). March 2001. *Societal Implications of Nanoscience and Nanotechnology*, National Science Foundation, Arlington, VA.
49. Allhoff, F. and Lin, P. (eds.). 2008. *Nanotechnology & Society Current and Emerging Ethical Issues*, Springer, Dordrecht, the Netherlands.
50. Doyle, M.E. 2006. *Nanotechnology: A Brief Literature Review.* http://fri.wisc.edu/docs/pdf/FRIBrief_Nanotech_Lit_Rev.pdf.
51. Savage, N. and Diallo, M. (eds.). 2009. *Nanotechnology Applications for Clean Water*, William Andrew.
52. Better Regulation Task Force. 2003. *Scientific Research: Innovation with Controls. Better Regulation Task Force Report*, Cabinet Office Publications and Publicity Team, London.
53. European Commission. 2004. *Communication of the European Commission: Towards a European Strategy for Nanotechnology*, Brussels, Belgium. COM(2004) 338.
54. Rip, A., Misa, T.J., and Schot, J. 1995. *Managing Technology in Society: The Approach of Constructive Technology Assessment*, Pinter Press, London.
55. Cabinet Office. 2003. *Field Work: Weighing up the Costs and Benefits of GM Crops*, Cabinet Office Strategy Unit, London.
56. Arvai, J., Gregory, R., and McDaniels, T. 2001. Testing a structured decision-aiding approach: Value-focused thinking for deliberative risk communication. *Risk Anal.* 21: 1065–1076.
57. Renn, O. 1999. A model for an analytic-deliberative process in risk management. *Environ. Sci. Technol.* 33: 3049–3055.
58. Grove-White, R., Mcnaghten, P., and Wynne, B. 2000. *Wising up: The Public and New Technologies*, Centre for the Study of Environmental Change, University of Lancaster, Lancaster.
59. Nanotechnology Work Group, EPA's Science Policy Council, Nanotechnology white paper, U.S. Environmental Protection Agency, December 2, 2005 External Review Draft.
60. Boucher, P.M. 2008. *Nanotechnology Legal Aspects*, CRC Press, Boca Raton, FL.

Index

Note: Locator followed by *'f'* and *'t'* denotes figure and table in the text

D

E

O

P